The Stem Cell

Edited by Paul F. Kisak

Contents

Chapter 1

Stem cell

This article is about the cell type. For the medical therapy, see Stem cell therapy.

Stem cells are undifferentiated biological cells that can differentiate into specialized cells and can divide (through mitosis) to produce more stem cells. They are found in multicellular organisms. In mammals, there are two broad types of stem cells: embryonic stem cells, which are isolated from the inner cell mass of blastocysts, and adult stem cells, which are found in various tissues. In adult organisms, stem cells and progenitor cells act as a repair system for the body, replenishing adult tissues. In a developing embryo, stem cells can differentiate into all the specialized cells—ectoderm, endoderm and mesoderm (see induced pluripotent stem cells)—but also maintain the normal turnover of regenerative organs, such as blood, skin, or intestinal tissues.

There are three known accessible sources of autologous adult stem cells in humans:

1. Bone marrow, which requires extraction by *harvesting*, that is, drilling into bone (typically the femur or iliac crest).

2. Adipose tissue (lipid cells), which requires extraction by liposuction.

3. Blood, which requires extraction through apheresis, wherein blood is drawn from the donor (similar to a blood donation), and passed through a machine that extracts the stem cells and returns other portions of the blood to the donor.

Stem cells can also be taken from umbilical cord blood just after birth. Of all stem cell types, autologous harvesting involves the least risk. By definition, autologous cells are obtained from one's own body, just as one may bank his or her own blood for elective surgical procedures.

Adult stem cells are frequently used in medical therapies, for example in bone marrow transplantation. Stem cells can now be artificially grown and transformed (differentiated) into specialized cell types with characteristics consistent with cells of various tissues such as muscles or nerves. Embryonic cell lines and autologous embryonic stem cells generated through somatic cell nuclear transfer or dedifferentiation have also been proposed as promising candidates for future therapies.[1] Research into stem cells grew out of findings by Ernest A. McCulloch and James E. Till at the University of Toronto in the 1960s.[2][3]

1.1 Properties

The classical definition of a stem cell requires that it possess two properties:

- *Self-renewal*: the ability to go through numerous cycles of cell division while maintaining the undifferentiated state.

- *Potency*: the capacity to differentiate into specialized cell types. In the strictest sense, this requires stem cells to be either totipotent or pluripotent—to be able to give rise to any mature cell type, although multipotent or unipotent progenitor cells are sometimes referred to as stem cells. Apart from this it is said that stem cell function is regulated in a feed back mechanism.

1.1.1 Self-renewal

Two mechanisms exist to ensure that a stem cell population is maintained:

1. Obligatory asymmetric replication: a stem cell divides into one mother cell that is identical to the original stem cell, and another daughter cell that is differentiated.

2. Stochastic differentiation: when one stem cell develops into two differentiated daughter cells, another

stem cell undergoes mitosis and produces two stem cells identical to the original.

1.1.2 Potency definition

Main article: Cell potency
Potency specifies the differentiation potential (the potential

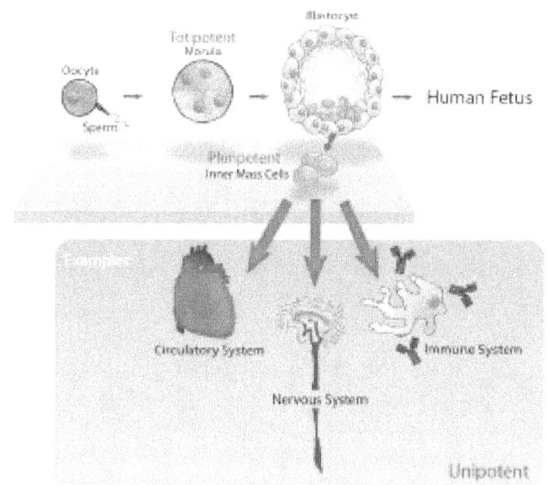

Pluripotent, embryonic stem cells originate as inner cell mass (ICM) cells within a blastocyst. These stem cells can become any tissue in the body, excluding a placenta. Only cells from an earlier stage of the embryo, known as the morula, are totipotent, able to become all tissues in the body and the extraembryonic placenta.

to differentiate into different cell types) of the stem cell.[4]

- Totipotent (a.k.a. omnipotent) stem cells can differentiate into embryonic and extraembryonic cell types. Such cells can construct a complete, viable organism.[4] These cells are produced from the fusion of an egg and sperm cell. Cells produced by the first few divisions of the fertilized egg are also totipotent.[5]

- Pluripotent stem cells are the descendants of totipotent cells and can differentiate into nearly all cells,[4] i.e. cells derived from any of the three germ layers.[6]

- Multipotent stem cells can differentiate into a number of cell types, but only those of a closely related family of cells.[4]

- Oligopotent stem cells can differentiate into only a few cell types, such as lymphoid or myeloid stem cells.[4]

- Unipotent cells can produce only one cell type, their own,[4] but have the property of self-renewal, which distinguishes them from non-stem cells (e.g. progenitor cells, muscle stem cells).

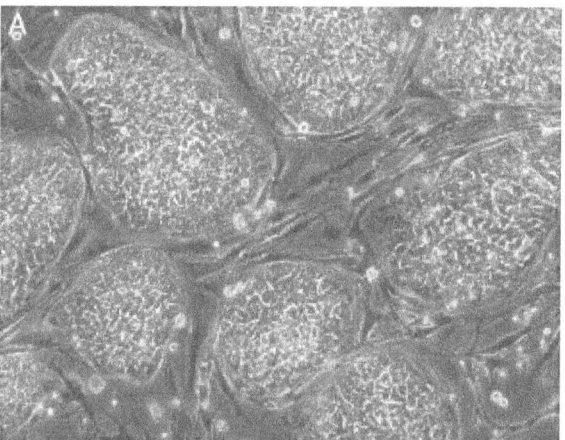

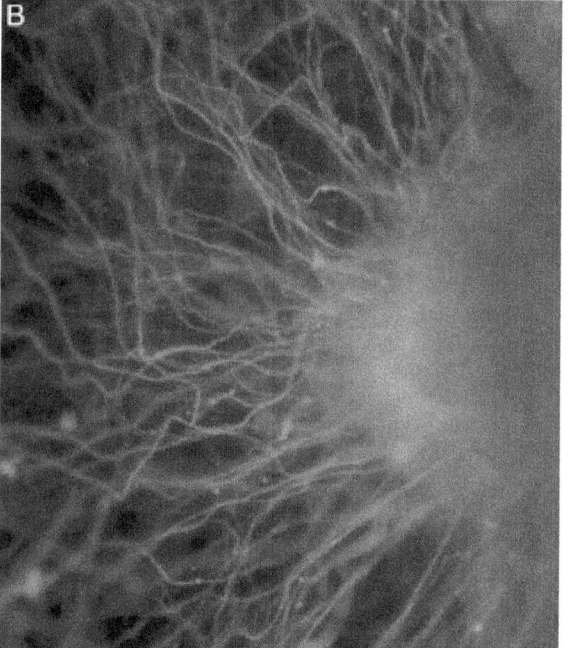

Human embryonic stem cells
A: Stem cell colonies that are not yet differentiated.
B: Nerve cells, an example of a cell type after differentiation.

1.1.3 Identification

In practice, stem cells are identified by whether they can regenerate tissue. For example, the defining test for bone marrow or hematopoietic stem cells (HSCs) is the ability to transplant the cells and save an individual without HSCs. This demonstrates that the cells can produce new blood cells over a long term. It should also be possible to isolate stem cells from the transplanted individual, which can themselves be transplanted into another individual without HSCs, demonstrating that the stem cell was able to self-renew.

Properties of stem cells can be illustrated *in vitro*, using methods such as clonogenic assays, in which single cells are assessed for their ability to differentiate and self-renew.[7][8] Stem cells can also be isolated by their possession of a distinctive set of cell surface markers. However, *in vitro* culture conditions can alter the behavior of cells, making it unclear whether the cells will behave in a similar manner *in vivo*. There is considerable debate as to whether some proposed adult cell populations are truly stem cells.

1.2 Embryonic

Main article: Embryonic stem cell

Embryonic stem (ES) cells are stem cells derived from the inner cell mass of a blastocyst, an early-stage embryo.[9] Human embryos reach the blastocyst stage 4–5 days post fertilization, at which time they consist of 50–150 cells. ES cells are pluripotent and give rise during development to all derivatives of the three primary germ layers: ectoderm, endoderm and mesoderm. In other words, they can develop into each of the more than 200 cell types of the adult body when given sufficient and necessary stimulation for a specific cell type. They do not contribute to the extra-embryonic membranes or the placenta.

Nearly all research to date has made use of mouse embryonic stem cells (mES) or human embryonic stem cells (hES). Both have the essential stem cell characteristics, yet they require very different environments in order to maintain an undifferentiated state. Mouse ES cells are grown on a layer of gelatin as an extracellular matrix (for support) and require the presence of leukemia inhibitory factor (LIF). Human ES cells are grown on a feeder layer of mouse embryonic fibroblasts (MEFs) and require the presence of basic fibroblast growth factor (bFGF or FGF-2).[10] Without optimal culture conditions or genetic manipulation,[11] embryonic stem cells will rapidly differentiate.

A human embryonic stem cell is also defined by the expression of several transcription factors and cell surface proteins. The transcription factors Oct-4, Nanog, and Sox2 form the core regulatory network that ensures the suppression of genes that lead to differentiation and the maintenance of pluripotency.[12] The cell surface antigens most commonly used to identify hES cells are the glycolipids stage specific embryonic antigen 3 and 4 and the keratan sulfate antigens Tra-1-60 and Tra-1-81. The molecular definition of a stem cell includes many more proteins and continues to be a topic of research.[13]

There are currently no approved treatments using embryonic stem cells. The first human trial was approved by the US Food and Drug Administration in January 2009.[14] However, the human trial was not initiated until October 13, 2010 in Atlanta for Spinal cord injury research. On November 14, 2011 the company conducting the trial announced that it will discontinue further development of its stem cell programs.[15] ES cells, being pluripotent cells, require specific signals for correct differentiation—if injected directly into another body, ES cells will differentiate into many different types of cells, causing a teratoma. Differentiating ES cells into usable cells while avoiding transplant rejection are just a few of the hurdles that embryonic stem cell researchers still face.[16] Many nations currently have moratoria on either ES cell research or the production of new ES cell lines. Because of their combined abilities of unlimited expansion and pluripotency, embryonic stem cells remain a theoretically potential source for regenerative medicine and tissue replacement after injury or disease.

- Mouse embryonic stem cells with fluorescent marker
- Human embryonic stem cell colony on mouse embryonic fibroblast feeder layer

1.3 Fetal

The primitive stem cells located in the organs of fetuses are referred to as fetal stem cells.[17] There are two types of fetal stem cells:

1. Fetal proper stem cells come from the tissue of the fetus proper, and are generally obtained after an abortion. These stem cells are not immortal but have a high level of division and are multipotent.

2. Extraembryonic fetal stem cells come from extraembryonic membranes, and are generally not distinguished from adult stem cells. These stem cells are acquired after birth, they are not immortal but have a high level of cell division, and are pluripotent.[18]

1.4 Adult

Main article: Adult stem cell

Adult stem cells, also called somatic (from Greek σωματικός, "of the body") stem cells, are stem cells which maintain and repair the tissue in which they are found.[19] They can be found in children, as well as adults.[20]

Pluripotent adult stem cells are rare and generally small in number, but they can be found in umbilical cord blood and other tissues.[21] Bone marrow is a rich source of adult stem cells,[22] which have been used in treating several conditions including spinal cord injury,[23] liver cirrhosis,[24] chronic limb ischemia [25] and endstage heart failure.[26] The quantity of bone marrow stem cells declines with age and is

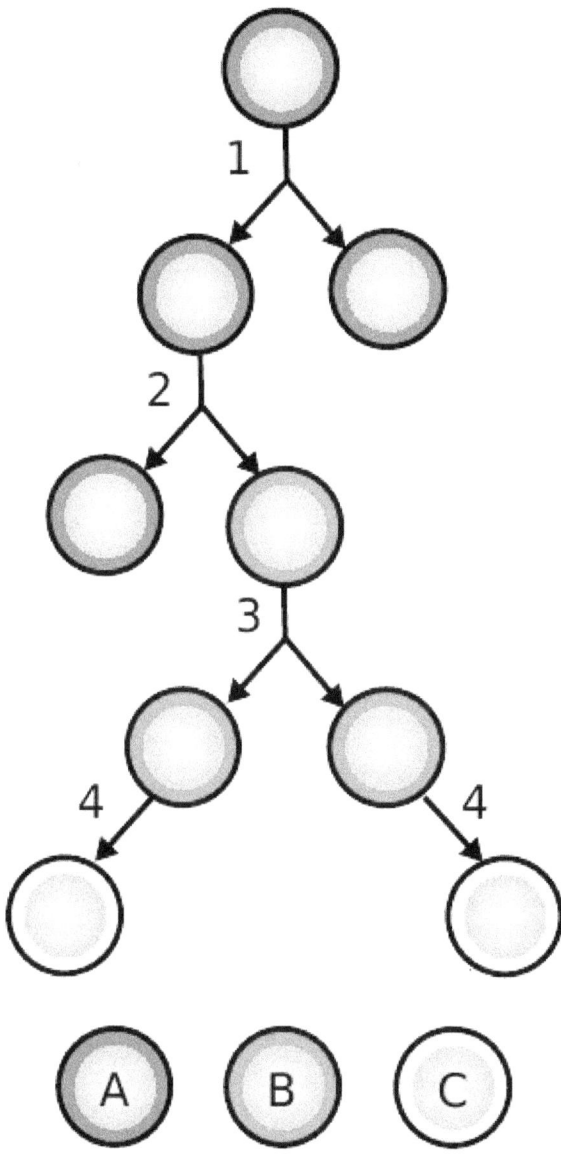

Stem cell division and differentiation. A: stem cell; B: progenitor cell; C: differentiated cell; 1: symmetric stem cell division; 2: asymmetric stem cell division; 3: progenitor division; 4: terminal differentiation

greater in males than females during reproductive years.[27] Much adult stem cell research to date has aimed to characterize their potency and self-renewal capabilities.[28] DNA damage accumulates with age in both stem cells and the cells that comprise the stem cell environment. This accumulation is considered to be responsible, at least in part, for increasing stem cell dysfunction with aging (see DNA damage theory of aging).[29]

Most adult stem cells are lineage-restricted (multipotent) and are generally referred to by their tissue origin (mesenchymal stem cell, adipose-derived stem cell,

endothelial stem cell, dental pulp stem cell, etc.).[30][31]

Adult stem cell treatments have been successfully used for many years to treat leukemia and related bone/blood cancers through bone marrow transplants.[32] Adult stem cells are also used in veterinary medicine to treat tendon and ligament injuries in horses.[33]

The use of adult stem cells in research and therapy is not as controversial as the use of embryonic stem cells, because the production of adult stem cells does not require the destruction of an embryo. Additionally, in instances where adult stem cells are obtained from the intended recipient (an autograft), the risk of rejection is essentially non-existent. Consequently, more US government funding is being provided for adult stem cell research.[34]

1.5   Amniotic

Multipotent stem cells are also found in amniotic fluid. These stem cells are very active, expand extensively without feeders and are not tumorigenic. Amniotic stem cells are multipotent and can differentiate in cells of adipogenic, osteogenic, myogenic, endothelial, hepatic and also neuronal lines.[35] Amniotic stem cells are a topic of active research.

Use of stem cells from amniotic fluid overcomes the ethical objections to using human embryos as a source of cells. Roman Catholic teaching forbids the use of embryonic stem cells in experimentation; accordingly, the Vatican newspaper "Osservatore Romano" called amniotic stem cells "the future of medicine".[36]

It is possible to collect amniotic stem cells for donors or for autoluguous use: the first US amniotic stem cells bank [37][38] was opened in 2009 in Medford, MA, by Biocell Center Corporation[39][40][41] and collaborates with various hospitals and universities all over the world.[42]

1.6   Induced pluripotent

Main article: Induced pluripotent stem cell

These are not adult stem cells, but rather adult cells (e.g. epithelial cells) reprogrammed to give rise to pluripotent capabilities. Using genetic reprogramming with protein transcription factors, pluripotent stem cells equivalent to embryonic stem cells have been derived from human adult skin tissue.[43][44][45] Shinya Yamanaka and his colleagues at Kyoto University used the transcription factors Oct3/4, Sox2, c-Myc, and Klf4[43] in their experiments on cells from human faces. Junying Yu, James Thomson, and their colleagues at the University of Wisconsin–Madison used a different set of factors, Oct4, Sox2, Nanog and Lin28,[43] and carried out their experiments using cells from human foreskin.

As a result of the success of these experiments, Ian Wilmut, who helped create the first cloned animal Dolly the Sheep, has announced that he will abandon somatic cell nuclear transfer as an avenue of research.[46]

Frozen blood samples can be used as a source of induced pluripotent stem cells, opening a new avenue for obtaining the valued cells.[47]

1.7 Lineage

Main article: Stem cell line

To ensure self-renewal, stem cells undergo two types of cell division (see *Stem cell division and differentiation* diagram). Symmetric division gives rise to two identical daughter cells both endowed with stem cell properties. Asymmetric division, on the other hand, produces only one stem cell and a progenitor cell with limited self-renewal potential. Progenitors can go through several rounds of cell division before terminally differentiating into a mature cell. It is possible that the molecular distinction between symmetric and asymmetric divisions lies in differential segregation of cell membrane proteins (such as receptors) between the daughter cells.[48]

An alternative theory is that stem cells remain undifferentiated due to environmental cues in their particular niche. Stem cells differentiate when they leave that niche or no longer receive those signals. Studies in *Drosophila* germarium have identified the signals decapentaplegic and adherens junctions that prevent germarium stem cells from differentiating.[49][50]

1.8 Treatments

Main article: Stem cell therapy
 Diseases and conditions where stem cell treatment is being investigated include:

- Diabetes[51]

- Rheumatoid arthritis[51]

- Parkinson's disease[51]

- Alzheimer's disease[51]

- Osteoarthritis[51]

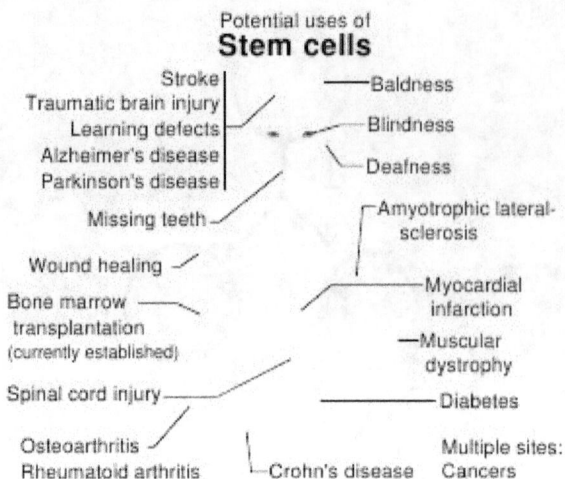

Diseases and conditions where stem cell treatment is being investigated.

- Stroke and traumatic brain injury repair[52]

- Learning disability due to congenital disorder [53]

- Spinal cord injury repair [54]

- Heart infarction [55]

- Anti-cancer treatments [56]

- Baldness reversal[57]

- Replace missing teeth [58]

- Repair hearing [59]

- Restore vision [60] and repair damage to the cornea[61]

- Amyotrophic lateral sclerosis [62]

- Crohn's disease [63]

- Wound healing [64]

Stem cell therapy is the use of stem cells to treat or prevent a disease or condition. Bone marrow transplant is a crude form of stem cell therapy that has been used clinically for many years without controversy. No stem cell therapies other than bone marrow transplant are widely used.[65][66]

Research is underway to develop various sources for stem cells, and to apply stem cell treatments for neurodegenerative diseases and conditions, diabetes, heart disease, and other conditions.[67]

In more recent years, with the ability of scientists to isolate and culture embryonic stem cells, and with scientists' growing ability to create stem cells using somatic cell nuclear transfer and techniques to created induced pluripotent stem cells, controversy has crept in, both related to abortion politics and to human cloning.

1.8.1 Disadvantages

Stem cell treatments may require immunosuppression because of a requirement for radiation before the transplant to remove the patient's previous cells, or because the patient's immune system may target the stem cells. One approach to avoid the second possibility is to use stem cells from the same patient who is being treated.

Pluripotency in certain stem cells could also make it difficult to obtain a specific cell type. It is also difficult to obtain the exact cell type needed, because not all cells in a population differentiate uniformly. Undifferentiated cells can create tissues other than desired types.[68]

Some stem cells form tumors after transplantation;[69] pluripotency is linked to tumor formation especially in embryonic stem cells, fetal proper stem cells, induced pluripotent stem cells. Fetal proper stem cells form tumors despite multipotency.

Hepatotoxicity and drug-induced liver injury account for a substantial number of failures of new drugs in development and market withdrawal, highlighting the need for screening assays such as stem cell-derived hepatocyte-like cells, that are capable of detecting toxicity early in the drug development process.[70]

1.9 Research patents

Further information: Consumer Watchdog vs. Wisconsin Alumni Research Foundation

Some of the fundamental patents covering human embryonic stem cells are owned by the Wisconsin Alumni Research Foundation (WARF) - they are patents 5,843,780, 6,200,806, and 7,029,913 invented by James A. Thomson. WARF does not enforce these patents against academic scientists, but does enforce them against companies.[71]

In 2006, a request for the US Patent and Trademark Office (USPTO) to re-examine the three patents was filed by the Public Patent Foundation on behalf of its client, the non-profit patent-watchdog group Consumer Watchdog (formerly the Foundation for Taxpayer and Consumer Rights).[71] In the re-examination process, which involves several rounds of discussion between the USTPO and the parties, the USPTO initially agreed with Consumer Watchdog and rejected all the claims in all three patents,[72] however in response, WARF amended the claims of all three patents to make them more narrow, and in 2008 the USPTO found the amended claims in all three patents to be patentable. The decision on one of the patents (7,029,913) was appealable, while the decisions on the other two were

not.[73][74] Consumer Watchdog appealed the granting of the '913 patent to the USTPO's Board of Patent Appeals and Interferences (BPAI) which granted the appeal, and in 2010 the BPAI decided that the amended claims of the '913 patent were not patentable.[75] However, WARF was able to re-open prosecution of the case and did so, amending the claims of the '913 patent again to make them more narrow, and in January 2013 the amended claims were allowed.[76]

In July 2013, Consumer Watchdog announced that it would appeal the decision to allow the claims of the '913 patent to the US Court of Appeals for the Federal Circuit (CAFC), the federal appeals court that hears patent cases.[77] At a hearing in December 2013, the CAFC raised the question of whether Consumer Watchdog had legal standing to appeal; the case could not proceed until that issue was resolved.[78]

1.10 See also

- Cell bank

- Human genome

- Meristem

- Partial cloning

- Plant stem cell

- Stem cell controversy

- Stem cell marker

- Shinya Yamanaka

1.11 References

[1] Tuch BE (2006). "Stem cells—a clinical update". *Australian Family Physician* **35** (9): 719–21. PMID 16969445.

[2] Becker AJ, McCulloch EA, Till JE (1963). "Cytological demonstration of the clonal nature of spleen colonies derived from transplanted mouse marrow cells". *Nature* **197** (4866): 452–4. Bibcode:1963Natur.197..452B. doi:10.1038/197452a0. PMID 13970094.

[3] Siminovitch L, Mcculloch EA, Till JE (1963). "The distribution of colony-forming cells among spleen colonies". *Journal of Cellular and Comparative Physiology* **62** (3): 327–36. doi:10.1002/jcp.1030620313. PMID 14086156.

[4] Schöler, Hans R. (2007). "The Potential of Stem Cells: An Inventory". In Nikolaus Knoepffler, Dagmar Schipanski, and Stefan Lorenz Sorgner. *Humanbiotechnology as Social Challenge*. Ashgate Publishing. p. 28. ISBN 978-0-7546-5755-2.

[5] Mitalipov S, Wolf D (2009). "Totipotency, pluripotency and nuclear reprogramming". *Adv. Biochem. Eng. Biotechnol.* Advances in Biochemical Engineering/Biotechnology **114**: 185–99. Bibcode:2009esc..book..185M. doi:10.1007/10_2008_45. ISBN 978-3-540-88805-5. PMC 2752493. PMID 19343304.

[6] Ulloa-Montoya F, Verfaillie CM, Hu WS (2005). "Culture systems for pluripotent stem cells". *J Biosci Bioeng.* **100** (1): 12–27. doi:10.1263/jbb.100.12. PMID 16233846.

[7] Friedenstein AJ, Deriglasova UF, Kulagina NN, Panasuk AF, Rudakowa SF, Luriá EA, Ruadkow IA (1974). "Precursors for fibroblasts in different populations of hematopoietic cells as detected by the *in vitro* colony assay method". *Experimental Hematology* **2** (2): 83–92. ISSN 0301-472X. PMID 4455512.

[8] Friedenstein AJ, Gorskaja JF, Kulagina NN (1976). "Fibroblast precursors in normal and irradiated mouse hematopoietic organs". *Experimental Hematology* **4** (5): 267–74. PMID 976387.

[9] Thomson JA, Itskovitz-Eldor J, Shapiro SS, Waknitz MA, Swiergiel JJ, Marshall VS, Jones JM (1998). "Blastocysts Embryonic Stem Cell Lines Derived from Human". *Science* **282** (5391): 1145–1147. Bibcode:1998Sci...282.1145T. doi:10.1126/science.282.5391.1145. PMID 9804556.

[10] "Culture of Human Embryonic Stem Cells (hESC)". National Institutes of Health. Archived from the original on 2010-01-06. Retrieved 2010-03-07.

[11] Chambers I, Colby D, Robertson M, Nichols J, Lee S, Tweedie S, Smith A (2003). "Functional expression cloning of Nanog, a pluripotency sustaining factor in embryonic stem cells". *Cell* **113** (5): 643–55. doi:10.1016/S0092-8674(03)00392-1. PMID 12787505.

[12] Boyer LA, Lee TI, Cole MF, Johnstone SE, Levine SS, Zucker JP, Guenther MG, Kumar RM, Murray HL, Jenner RG, Gifford DK, Melton DA, Jaenisch R, Young RA (2005). "Core transcriptional regulatory circuitry in human embryonic stem cells". *Cell* **122** (6): 947–56. doi:10.1016/j.cell.2005.08.020. PMC 3006442. PMID 16153702.

[13] Adewumi O, Aflatoonian B, Ahrlund-Richter L, Amit M, Andrews PW, Beighton G, Bello PA, Benvenisty N, Berry LS, Bevan S, Blum B, Brooking J, Chen KG, Choo AB, Churchill GA, Corbel M, Damjanov I, Draper JS, Dvorak P, Emanuelsson K, Fleck RA, Ford A, Gertow K, Gertsenstein M, Gokhale PJ, Hamilton RS, Hampl A, Healy LE, Hovatta O, Hyllner J, Imreh MP, Itskovitz-Eldor J, Jackson J, Johnson JL, Jones M, Kee K, King BL, Knowles BB, Lako M, Lebrin F, Mallon BS, Manning D, Mayshar Y, McKay RD, Michalska AE, Mikkola M, Mileikovsky M, Minger SL, Moore HD, Mummery CL, Nagy A, Nakatsuji N, O'Brien CM, Oh SK, Olsson C, Otonkoski T, Park KY, Passier R, Patel H, Patel M, Pedersen R, Pera MF, Piekarczyk MS, Pera RA, Reubinoff BE, Robins AJ, Rossant J, Rugg-Gunn P, Schulz TC, Semb H, Sherrer ES, Siemen H, Stacey GN, Stojkovic M, Suemori H, Szatkiewicz J, Turetsky T, Tuuri T, van den Brink S, Vintersten K, Vuoristo S, Ward D, Weaver TA, Young LA, Zhang W (2007). "Characterization of human embryonic stem cell lines by the International Stem Cell Initiative". *Nat. Biotechnol.* **25** (7): 803–16. doi:10.1038/nbt1318. PMID 17572666.

[14] Ron Winslow (2009). "First Embryonic Stem-Cell Trial Gets Approval from the FDA". *The Wall Street Journal.* 23. January 2009.

[15] "Embryonic Stem Cell Therapy At Risk? Geron Ends Clinical Trial". ScienceDebate.com. Retrieved 2011-12-11.

[16] Wu DC, Boyd AS, Wood KJ (2007). "Embryonic stem cell transplantation: potential applicability in cell replacement therapy and regenerative medicine". *Front Biosci* **12** (8–12): 4525–35. doi:10.2741/2407. PMID 17485394.

[17] Ariff Bongso: Eng Hin Lee, ed. (2005). "Stem cells: their definition, classification and sources". *Stem Cells: From Benchtop to Bedside.* World Scientific. p. 5. ISBN 981-256-126-9. OCLC 443407924.

[18] Moore, K.L., T.V.N. Persaud, and A.G. Torchia. Before We Are Born: Essentials of Embryology and Birth Defects. Philadelphia, PA: Saunders, Elsevier. 2013. Print

[19] "Stem Cells" Mayo Clinic. Mayo foundation for medical education and research n.d Web. March 23, 2013

[20] Jiang Y, Jahagirdar BN, Reinhardt RL, Schwartz RE, Keene CD, Ortiz-Gonzalez XR, Reyes M, Lenvik T, Lund T, Blackstad M, Du J, Aldrich S, Lisberg A, Low WC, Largaespada DA, Verfaillie CM (2002). "Pluripotency of mesenchymal stem cells derived from adult marrow". *Nature* **418** (6893): 41–9. doi:10.1038/nature00870. PMID 12077603.

[21] Ratajczak MZ, Machalinski B, Wojakowski W, Ratajczak J, Kucia M (2007). "A hypothesis for an embryonic origin of pluripotent Oct-4(+) stem cells in adult bone marrow and other tissues". *Leukemia* **21** (5): 860–7. doi:10.1038/sj.leu.2404630. PMID 17344915.

[22] Narasipura SD, Wojciechowski JC, Charles N, Liesveld JL, King MR (2008). "P-Selectin coated microtube for enrichment of CD34+ hematopoietic stem and progenitor cells from human bone marrow". *Clin Chem* **54** (1): 77–85. doi:10.1373/clinchem.2007.089896. PMID 18024531.

[23] William JB, Prabakaran R , Ayyappan S, Puskhinraj H, Rao D, Manjunath SR, Thamaraikannan P, Dedeepiya VD, Kuroda S, Yoshioka H, Mori Y, Preethy SK, Abraham SJK (2011). "Functional Recovery of Spinal Cord Injury Following Application of Intralesional Bone Marrow Mononuclear Cells Embedded in Polymer Scaffold – Two Year Follow-up in a Canine". *Journal of Stem Cell Research & Therapy* **1** (3). doi:10.4172/2157-7633.1000110.

[24] Terai S, Ishikawa T, Omori K, Aoyama K, Marumoto Y, Urata Y, Yokoyama Y, Uchida K, Yamasaki T, Fujii Y, Okita K, Sakaida I (2006). "Improved liver function in patients with liver cirrhosis after autologous bone marrow cell infusion therapy". Stem Cells 24 (10): 2292–8. doi:10.1634/stemcells.2005-0542. PMID 16778155.

[25] Subrammaniyan R, Amalorpavanathan J, Shankar R, et al. (September 2011). "Application of autologous bone marrow mononuclear cells in six patients with advanced chronic critical limb ischemia as a result of diabetes: our experience". Cytotherapy 13 (8): 993–9. doi:10.3109/14653249.2011.579961. PMID 21671823.

[26] Madhusankar N. "Use of Bone Marrow derived Stem Cells in Patients with Cardiovascular Disorders". Journal of Stem Cells and Regenerative Medicine.

[27] Dedeepiya VD, Rao YY, Jayakrishnan GA, Parthiban JK, Baskar S, Manjunath SR, Senthilkumar R, Abraham SJ (2012). "Index of CD34+ Cells and Mononuclear Cells in the Bone Marrow of Spinal Cord Injury Patients of Different Age Groups: A Comparative Analysis". Bone Marrow Res 2012: 787414. doi:10.1155/2012/787414. PMC 3398573. PMID 22830032.

[28] Gardner RL (2002). "Stem cells: potency, plasticity and public perception". Journal of Anatomy 200 (3): 277–82. doi:10.1046/j.1469-7580.2002.00029.x. PMC 1570679. PMID 12033732.

[29] Behrens A, van Deursen JM, Rudolph KL, Schumacher B (2014). "Impact of genomic damage and ageing on stem cell function". Nat. Cell Biol. 16 (3): 201–7. doi:10.1038/ncb2928. PMC 4214082. PMID 24576896.

[30] Barrilleaux B, Phinney DG, Prockop DJ, O'Connor KC (2006). "Review: ex vivo engineering of living tissues with adult stem cells". Tissue Eng 12 (11): 3007–19. doi:10.1089/ten.2006.12.3007. PMID 17518617.

[31] Gimble JM, Katz AJ, Bunnell BA (2007). "Adipose-derived stem cells for regenerative medicine". Circ Res 100 (9): 1249–60. doi:10.1161/01.RES.0000265074.83288.09. PMID 17495232.

[32] "Bone Marrow Transplant".

[33] Kane, Ed (2008-05-01). "Stem-cell therapy shows promise for horse soft-tissue injury, disease". DVM Newsmagazine. Retrieved 2008-06-12.

[34] "Stem Cell FAQ". US Department of Health and Human Services. 2004-07-14. Archived from the original on 2009-01-09.

[35] De Coppi P, Bartsch G, Siddiqui MM, Xu T, Santos CC, Perin L, Mostoslavsky G, Serre AC, Snyder EY, Yoo JJ, Furth ME, Soker S, Atala A (2007). "Isolation of amniotic stem cell lines with potential for therapy". Nature Biotechnology 25 (5): 100–106. doi:10.1038/nbt1274. PMID 17206138.

[36] "Vatican newspaper calls new stem cell source 'future of medicine' :: Catholic News Agency (CNA)". Catholic News Agency. 2010-02-03. Retrieved 2010-03-14.

[37] "European Biotech Company Biocell Center Opens First U.S. Facility for Preservation of Amniotic Stem Cells in Medford, Massachusetts". Reuters. 2009-10-22. Retrieved 2010-03-14.

[38] "Europe's Biocell Center opens Medford office – Daily Business Update". The Boston Globe. 2009-10-22. Retrieved 2010-03-14.

[39] "The Ticker". BostonHerald.com. 2009-10-22. Retrieved 2010-03-14.

[40] "Biocell Center opens amniotic stem cell bank in Medford". Mass High Tech Business News. 2009-10-23. Retrieved 2012-08-26.

[41] "News » World's First Amniotic Stem Cell Bank Opens In Medford". wbur.org. Retrieved 2010-03-14.

[42] "Biocell Center Corporation Partners with New England's Largest Community-Based Hospital Network to Offer a Unique... – MEDFORD, Mass., March 8 /PRNewswire/". Massachusetts: Prnewswire.com. Retrieved 2010-03-14.

[43] "Making human embryonic stem cells". The Economist. 2007-11-22.

[44] Brand, Madeleine; Palca, Joe and Cohen, Alex (2007-11-20). "Skin Cells Can Become Embryonic Stem Cells". National Public Radio.

[45] "Breakthrough Set to Radically Change Stem Cell Debate". News Hour with Jim Lehrer. 2007-11-20.

[46] "His inspiration comes from the research by Prof Shinya Yamanaka at Kyoto University, which suggests a way to create human embryo stem cells without the need for human eggs, which are in extremely short supply, and without the need to create and destroy human cloned embryos, which is bitterly opposed by the pro life movement."Highfield, Roger (2007-11-16). "Dolly creator Prof Ian Wilmut shuns cloning". London: The Telegraph.

[47] Frozen blood a source of stem cells, study finds. newsdaily.com (2010-07-01)

[48] Beckmann J, Scheitza S, Wernet P, Fischer JC, Giebel B (2007). "Asymmetric cell division within the human hematopoietic stem and progenitor cell compartment: identification of asymmetrically segregating proteins". Blood 109 (12): 5494–501. doi:10.1182/blood-2006-11-055921. PMID 17332245.

[49] Xie T, Spradling AC (1998). "decapentaplegic is essential for the maintenance and division of germline stem cells in the Drosophila ovary". Cell 94 (2): 251–60. doi:10.1016/S0092-8674(00)81424-5. PMID 9695953.

[50] Song X, Zhu CH, Doan C, Xie T (2002). "Germline stem cells anchored by adherens junctions in the Drosophila ovary niches". *Science* **296** (5574): 1855–7. Bibcode:2002Sci...296.1855S. doi:10.1126/science.1069871. PMID 12052957.

[51] Stem Cell Basics: What are the potential uses of human stem cells and the obstacles that must be overcome before these potential uses will be realized?. In Stem Cell Information World Wide Web site. Bethesda, MD: National Institutes of Health, U.S. Department of Health and Human Services, 2009. cited Sunday, April 26, 2009

[52] Steinberg, Douglas (November 2000) Stem Cells Tapped to Replenish Organs thescientist.com

[53] ISRAEL21c: Israeli scientists reverse brain birth defects using stem cells December 25, 2008. (Researchers from the Hebrew University of Jerusalem-Hadassah Medical led by Prof. Joseph Yanai)

[54] Kang KS, Kim SW, Oh YH, Yu JW, Kim KY, Park HK, Song CH, Han H (2005). "A 37-year-old spinal cord-injured female patient, transplanted of multipotent stem cells from human UC blood, with improved sensory perception and mobility, both functionally and morphologically: a case study". *Cytotherapy* **7** (4): 368–73. doi:10.1080/14653240500238160. PMID 16162459.

[55] Strauer BE, Schannwell CM, Brehm M (2009). "Therapeutic potentials of stem cells in cardiac diseases". *Minerva Cardioangiol* **57** (2): 249–67. PMID 19274033.

[56] Stem Cells Tapped to Replenish Organs thescientist.com. Nov 2000. By Douglas Steinberg

[57] *Hair Cloning Nears Reality as Baldness Cure* WebMD November 2004

[58] Yen AH, Sharpe PT (2008). "Stem cells and tooth tissue engineering". *Cell Tissue Res.* **331** (1): 359–72. doi:10.1007/s00441-007-0467-6. PMID 17938970.

[59] "Gene therapy is first deafness 'cure'". New Scientist. February 14, 2005.

[60] "BBC NEWS - UK - England - Southern Counties - Stem cells used to restore vision".

[61] Hanson, Charles; Hardarson, Thorir; Ellerström, Catharina; Nordberg, Markus; Caisander, Gunilla; Rao, Mahendra; Hyllner, Johan; Stenevi, Ulf (2013-03-01). "Transplantation of human embryonic stem cells onto a partially wounded human cornea in vitro". *Acta Ophthalmologica* **91** (2): 127–130. doi:10.1111/j.1755-3768.2011.02358.x. ISSN 1755-3768. PMC 3660785. PMID 22280565.

[62] Vastag B (2001). "Stem Cells Step Closer to the Clinic: Paralysis Partially Reversed in Rats with ALS-like Disease". *JAMA: the Journal of the American Medical Association* **285** (13): 1691–1693. doi:10.1001/jama.285.13.1691. PMID 11277806.

[63] Anderson, Querida (2008-06-15). "Osiris Trumpets Its Adult Stem Cell Product". *Genetic Engineering & Biotechnology News* (Mary Ann Liebert, Inc.). p. 13. Retrieved 2008-07-06. (subtitle) Procymal is being developed in many indications, GvHD being the most advanced

[64] Gurtner GC, Callaghan MJ, Longaker MT (2007). "Progress and potential for regenerative medicine". *Annu. Rev. Med.* **58**: 299–312. doi:10.1146/annurev.med.58.082405.095329. PMID 17076602. Bone marrow transplantation is, as of 2009, the only established use of stem cells.

[65] Ian Murnaghan for Explore Stem Cells. Updated: 16 December 2013 Why Perform a Stem Cell Transplant?

[66] Bone Marrow Transplantation and Peripheral Blood Stem Cell Transplantation In National Cancer Institute Fact Sheet web site. Bethesda, MD: National Institutes of Health, U.S. Department of Health and Human Services, 2010. Cited August 24, 2010

[67] Bubela T, Li MD, Hafez M, Bieber M, Atkins H (2012). "Is belief larger than fact: Expectations, optimism and reality for translational stem cell research". *BMC Med* **10**: 133. doi:10.1186/1741-7015-10-133. PMC 3520764. PMID 23131007.

[68] Moore, Keith L., T.V.N. Persaud, and Mark G. Torchia. Before We Are Born: Essentials of Embryology and Birth Defects. Philadelphia, PA: Saunders, Elsevier, 2013 Print.

[69] Bernadine Healy, M.D.. "Why Embryonic Stem Cells are obsolete" *US News and world report*. Retrieved on Aug 17, 2015.

[70] Greenhough S, Hay DC. (2012). "Stem Cell-Based Toxicity Screening: Recent Advances in Hepatocyte Generation". *Pharm Med* **26** (2): 85–89. doi:10.1007/BF03256896.

[71] Regalado, Antonio, David P. Hamilton (July 2006). "How a University's Patents May Limit Stem-Cell Researcher." *The Wall Street Journal*. Retrieved on July 24, 2006.

[72] Stephen Jenei for Patent Baristas, April 3, 2007 WARF Stem Cell Patents Knocked Down in Round One

[73] Stephen Jenei for Patent Baristas, March 3, 2008 Ding! WARF Wins Round 2 As Stem Cell Patent Upheld

[74] Constance Holden for Science Now. March 12, 2008 WARF Goes 3 for 3 on Patents

[75] Stephen G. Kunin for Patents Post Grant. May 10, 2010 BPAI Rejects WARF Stem Cell Patent Claims in Inter Partes Reexamination Appeal

[76] United States Patent And Trademark Office. Board Of Patent Appeals and Interferences. The Foundation For Taxpayer & Consumer Rights, Requester And Appellant V. Patent Of Wisconsin Alumni Research Foundation, Patent Owner And Respondent. Appeal 2012-011693, Reexamination Control 95/000,154. Patent 7,029,913 Decision on Appeal

[77] GenomeWeb staff, July 03, 2013 Consumer Watchdog, PPF Seek Invalidation of WARF's Stem Cell Patent

[78] Antoinette Konski for Personalized Medicine Bulletin, February 3, 2014 U.S. Government and USPTO Urges Federal Circuit to Dismiss Stem Cell Appeal

1.12   External links

General

- Stem Cell Basics Courtesy of the National Institutes of Health

- Nature Reports Stem Cells: Introductory material, research advances and debates concerning stem cell research.

- Understanding Stem Cells: A View of the Science and Issues from the National Academies

- Scientific American Magazine (June 2004 Issue) The Stem Cell Challenge

- Scientific American Magazine (July 2006 Issue) Stem Cells: The Real Culprits in Cancer?

- Ethics of Stem Cell Research entry by Andrew Siegel in the *Stanford Encyclopedia of Philosophy*

- Stem Cell Research at Johns Hopkins

- StemBook

Chapter 2

Stem-cell therapy

This article is about the medical therapy. For the cell type, see Stem cell.

Stem-cell therapy is the use of stem cells to treat or prevent a disease or condition.

Bone marrow transplant is the most widely used stem-cell therapy, but some therapies derived from umbilical cord blood are also in use. Research is underway to develop various sources for stem cells, and to apply stem-cell treatments for neurodegenerative diseases and conditions, diabetes, heart disease, and other conditions.

With the ability of scientists to isolate and culture embryonic stem cells, and with scientists' growing ability to create stem cells using somatic cell nuclear transfer and techniques to create induced pluripotent stem cells, controversy has crept in, both related to abortion politics and to human cloning. Additionally, efforts to market treatments based on transplant of stored umbilical cord blood have been controversial.

2.1 Medical uses

Further information: Hematopoietic stem cell transplantation

For over 30 years, bone marrow has been used to treat cancer patients with conditions such as leukaemia and lymphoma; this is the only form of stem-cell therapy that is widely practiced.[1][2][3] During chemotherapy, most growing cells are killed by the cytotoxic agents. These agents, however, cannot discriminate between the leukaemia or neoplastic cells, and the hematopoietic stem cells within the bone marrow. It is this side effect of conventional chemotherapy strategies that the stem-cell transplant attempts to reverse; a donor's healthy bone marrow reintroduces functional stem cells to replace the cells lost in the host's body during treatment. The transplanted cells also

generate an immune response that helps to kill off the cancer cells; this process can go too far, however, leading to graft vs host disease, the most serious side effect of this treatment.[4]

Another stem-cell therapy called Prochymal, was conditionally approved in Canada in 2012 for the management of acute graft-vs-host disease in children who are unresponsive to steroids.[5] It is an allogenic stem therapy based on mesenchymal stem cells (MSCs) derived from the bone marrow of adult donors. MSCs are purified from the marrow, cultured and packaged, with up to 10,000 doses derived from a single donor. The doses are stored frozen until needed.[6]

The FDA has approved five hematopoietic stem-cell products derived from umbilical cord blood, for the treatment of blood and immunological diseases.[7]

In 2014, the European Medicines Agency recommended approval of Holoclar, a treatment involving stem cells, for use in the European Union. Holoclar is used for people with severe limbal stem cell deficiency due to burns in the eye.[8]

2.2 Research

2.2.1 Neurodegeneration

Research has been conducted to learn whether stem cells may be used to treat brain degeneration, such as in Parkinson's, Amyotrophic lateral sclerosis, and Alzheimer's disease.[9][10][11] There have been preliminary studies related to multiple sclerosis.[12]

Healthy adult brains contain neural stem cells which divide to maintain general stem-cell numbers, or become progenitor cells. In healthy adult animals, progenitor cells migrate within the brain and function primarily to maintain neuron populations for olfaction (the sense of smell). Pharmacological activation of endogenous neural stem cells has been reported to induce neuroprotection

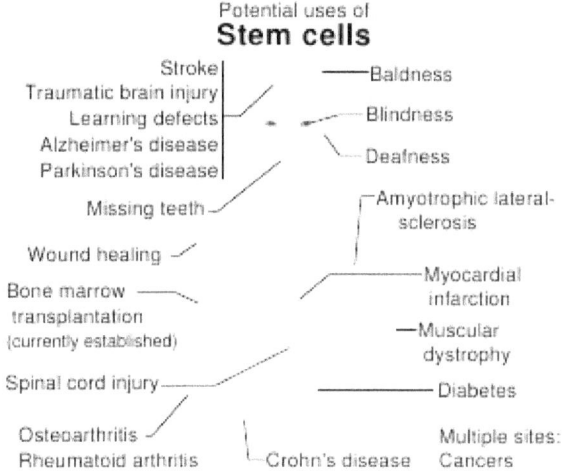

Potential uses of
Stem cells

Stroke
Traumatic brain injury ——— Baldness
Learning defects
Alzheimer's disease ——— Blindness
Parkinson's disease
                        Deafness
Missing teeth ——— Amyotrophic lateral-
                        sclerosis
Wound healing
Bone marrow ——— Myocardial
transplantation   infarction
(currently established) ——— Muscular
                        dystrophy
Spinal cord injury ——— Diabetes

Osteoarthritis   Multiple sites:
Rheumatoid arthritis ——— Crohn's disease   Cancers

Diseases and conditions where stem cell treatment is promising or emerging.

and behavioral recovery in adult rat models of neurological disorder.[13][14][15]

2.2.2   Brain and spinal cord injury

Stroke and traumatic brain injury lead to cell death, characterized by a loss of neurons and oligodendrocytes within the brain. A small clinical trial was underway in Scotland in 2013, in which stem cells were injected into the brains of stroke patients.[16]

Clinical and animal studies have been conducted into the use of stem cells in cases of spinal cord injury.[17][18][19]

2.2.3   Heart

The pioneering work[20] by Bodo-Eckehard Strauer has now been discredited by the identification of hundreds of factual contradictions.[21] Among several clinical trials that have reported that adult stem-cell therapy is safe and effective, powerful effects have been reported from only a few laboratories, but this has covered old[22] and recent[23] infarcts as well as heart failure not arising from myocardial infarction.[24] While initial animal studies demonstrated remarkable therapeutic effects,[25][26] later clinical trials achieved only modest, though statistically significant, improvements.[27][28] Possible reasons for this discrepancy are patient age,[29] timing of treatment[30] and the recent occurrence of a myocardial infarction.[31] It appears that these obstacles may be overcome by additional treatments which increase the effectiveness of the treatment[32] or by optimizing the methodology although these too can be controversial. Current studies vary greatly in cell-procuring techniques, cell types, cell-administration timing and pro-

cedures, and studied parameters, making it very difficult to make comparisons. Comparative studies are therefore currently needed.

Stem-cell therapy for treatment of myocardial infarction usually makes use of autologous bone-marrow stem cells (a specific type or all), however other types of adult stem cells may be used, such as adipose-derived stem cells.[33] Adult stem cell therapy for treating heart disease was commercially available in at least five continents as of 2007.

Possible mechanisms of recovery include:[9]

- Generation of heart muscle cells

- Stimulation of growth of new blood vessels to repopulate damaged heart tissue

- Secretion of growth factors

- Assistance via some other mechanism

It may be possible to have adult bone-marrow cells differentiate into heart muscle cells.[9]

The first successful integration of human embryonic stem cell derived cardiomyocytes in guinea pigs (mouse hearts beat too fast) was reported in August 2012. The contraction strength was measured four weeks after the guinea pigs underwent simulated heart attacks and cell treatment. The cells contracted synchronously with the existing cells, but it is unknown if the positive results were produced mainly from paracrine as opposed to direct electromechanical effects from the human cells. Future work will focus on how to get the cells to engraft more strongly around the scar tissue. Whether treatments from embryonic or adult bone marrow stem cells will prove more effective remains to be seen.[34]

In 2013 the pioneering reports of powerful beneficial effects of autologous bone marrow stem cells on ventricular function were found to contain "hundreds" of discrepancies.[35] Critics report that of 48 reports there seemed to be just five underlying trials, and that in many cases whether they were randomized or merely observational accepter-versus-rejecter, was contradictory between reports of the same trial. One pair of reports of identical baseline characteristics and final results, was presented in two publications as, respectively, a 578 patient randomized trial and as a 391 patient observational study. Other reports required (impossible) negative standard deviations in subsets of patients, or contained fractional patients, negative NYHA classes. Overall there were many more patients published as having receiving stem cells in trials, than the number of stem cells processed in the hospital's laboratory during that time. A university investigation, closed in 2012 without reporting, was reopened in July 2013.[36]

One of the most promising benefits of stem cell therapy is the potential for cardiac tissue regeneration to reverse the tissue loss underlying the development of heart failure after cardiac injury.[37]

Initially, the observed improvements were attributed to a transdifferentiation of BM-MSCs into cardiomyocyte-like cells.[25] Given the apparent inadequacy of unmodified stem cells for heart tissue regeneration, a more promising modern technique involves treating these cells to create cardiac progenitor cells before implantation to the injured area.[38]

2.2.4   Blood-cell formation

The specificity of the human immune-cell repertoire is what allows the human body to defend itself from rapidly adapting antigens. However, the immune system is vulnerable to degradation upon the pathogenesis of disease, and because of the critical role that it plays in overall defense, its degradation is often fatal to the organism as a whole. Diseases of hematopoietic cells are diagnosed and classified via a subspecialty of pathology known as hematopathology. The specificity of the immune cells is what allows recognition of foreign antigens, causing further challenges in the treatment of immune disease. Identical matches between donor and recipient must be made for successful transplantation treatments, but matches are uncommon, even between first-degree relatives. Research using both hematopoietic adult stem cells and embryonic stem cells has provided insight into the possible mechanisms and methods of treatment for many of these ailments.

Fully mature human red blood cells may be generated *ex vivo* by hematopoietic stem cells (HSCs), which are precursors of red blood cells. In this process, HSCs are grown together with stromal cells, creating an environment that mimics the conditions of bone marrow, the natural site of red-blood-cell growth. Erythropoietin, a growth factor, is added, coaxing the stem cells to complete terminal differentiation into red blood cells.[39] Further research into this technique should have potential benefits to gene therapy, blood transfusion, and topical medicine.

2.2.5   Baldness

Hair follicles also contain stem cells, and some researchers predict these follicle stem cells may lead to successes in treating baldness through activation of progenitor stem cells. This treatment is expected to work by activating already existing stem cells on the scalp. Later treatments may be able to simply signal follicle stem cells to give off chemical signals to nearby follicle cells which have shrunk during the aging process, which in turn respond to these signals by regenerating and once again making healthy hair.

2.2.6   Missing teeth

In 2004, scientists at King's College London discovered a way to cultivate a complete tooth in mice[40] and were able to grow bioengineered teeth stand-alone in the laboratory. Researchers are confident that the tooth regeneration technology can be used to grow live teeth in human patients.

In theory, stem cells taken from the patient could be coaxed in the lab turning into a tooth bud which, when implanted in the gums, will give rise to a new tooth, and would be expected to be grown in a time over three weeks.[41] It will fuse with the jawbone and release chemicals that encourage nerves and blood vessels to connect with it. The process is similar to what happens when humans grow their original adult teeth. Many challenges remain, however, before stem cells could be a choice for the replacement of missing teeth in the future.[42][43]

Research is ongoing in different fields, alligators which are polyphyodonts grow up to 50 times a successional tooth (a small replacement tooth) under each mature functional tooth for replacement once a year.[44]

2.2.7   Deafness

Heller has reported success in re-growing cochlea hair cells with the use of embryonic stem cells.[45]

2.2.8   Blindness and vision impairment

Since 2003, researchers have successfully transplanted corneal stem cells into damaged eyes to restore vision. "Sheets of retinal cells used by the team are harvested from aborted fetuses, which some people find objectionable." When these sheets are transplanted over the damaged cornea, the stem cells stimulate renewed repair, eventually restore vision.[46] The latest such development was in June 2005, when researchers at the Queen Victoria Hospital of Sussex, England were able to restore the sight of forty patients using the same technique. The group, led by Sheraz Daya, was able to successfully use adult stem cells obtained from the patient, a relative, or even a cadaver. Further rounds of trials are ongoing.[47]

In April 2005, doctors in the UK transplanted corneal stem cells from an organ donor to the cornea of Deborah Catlyn, a woman who was blinded in one eye when acid was thrown in her eye at a nightclub. The cornea, which is the transparent window of the eye, is a particularly suitable site for transplants. In fact, the first successful human transplant was a cornea transplant. The absence of blood vessels within the cornea makes this area a relatively easy target for transplantation. The majority of corneal transplants carried out today are due to a degenerative disease called keratoconus.

The University Hospital of New Jersey reports that the success rate for growth of new cells from transplanted stem cells varies from 25 percent to 70 percent.[48]

In 2014, researchers demonstrated that stem cells collected as biopsies from donor human corneas can prevent scar formation without provoking a rejection response in mice with corneal damage.[49]

In January 2012, The Lancet published a paper by Steven Schwartz, at UCLA's Jules Stein Eye Institute, reporting two women who had gone legally blind from macular degeneration had dramatic improvements in their vision after retinal injections of human embryonic stem cells.[50]

In June 2015, the Stem Cell Ophthalmology Treatment Study (SCOTS), the largest adult stem cell study in ophthalmology (www.clinicaltrials.gov NCT # 01920867) published initial results on a patient with optic nerve disease who improved from 20/2000 to 20/40 following treatment with bone marrow derived stem cells.[51]

2.2.9   Diabetes

Diabetes patients lose the function of insulin-producing beta cells within the pancreas.[52] In recent experiments, scientists have been able to coax embryonic stem cell to turn into beta cells in the lab. In theory if the beta cell is transplanted successfully, they will be able to replace malfunctioning ones in a diabetic patient.[53]

Transplantation

Human embryonic stem cells may be grown in cell culture and stimulated to form insulin-producing cells that can be transplanted into the patient.

However, clinical success is highly dependent on the development of the following procedures:[9]

- Transplanted cells should proliferate

- Transplanted cells should differentiate in a site-specific manner

- Transplanted cells should survive in the recipient (prevention of transplant rejection)

- Transplanted cells should integrate within the targeted tissue

- Transplanted cells should integrate into the host circuitry and restore function

2.2.10   Orthopaedics

Clinical case reports in the treatment orthopaedic conditions have been reported. To date, the focus in the literature for musculoskeletal care appears to be on mesenchymal stem cells. Centeno et al. have published MRI evidence of increased cartilage and meniscus volume in individual human subjects.[54][55] The results of trials that include a large number of subjects, are yet to be published. However, a published safety study conducted in a group of 227 patients over a 3-4-year period shows adequate safety and minimal complications associated with mesenchymal cell transplantation.[56]

Wakitani has also published a small case series of nine defects in five knees involving surgical transplantation of mesenchymal stem cells with coverage of the treated chondral defects.[57]

2.2.11   Wound healing

Stem cells can also be used to stimulate the growth of human tissues. In an adult, wounded tissue is most often replaced by scar tissue, which is characterized in the skin by disorganized collagen structure, loss of hair follicles and irregular vascular structure. In the case of wounded fetal tissue, however, wounded tissue is replaced with normal tissue through the activity of stem cells.[58] A possible method for tissue regeneration in adults is to place adult stem cell "seeds" inside a tissue bed "soil" in a wound bed and allow the stem cells to stimulate differentiation in the tissue bed cells. This method elicits a regenerative response more similar to fetal wound-healing than adult scar tissue formation.[58] Researchers are still investigating different aspects of the "soil" tissue that are conducive to regeneration.[58]

2.2.12   Infertility

Culture of human embryonic stem cells in mitotically inactivated porcine ovarian fibroblasts (POF) causes differentiation into germ cells (precursor cells of oocytes and spermatozoa), as evidenced by gene expression analysis.[59]

Human embryonic stem cells have been stimulated to form Spermatozoon-like cells, yet still slightly damaged or malformed.[60] It could potentially treat azoospermia.

In 2012, oogonial stem cells were isolated from adult mouse and human ovaries and demonstrated to be capable of forming mature oocytes.[61] These cells have the potential to treat infertility.

2.2.13   HIV/AIDS

Destruction of the immune system by the HIV is driven by the loss of CD4+ T cells in the peripheral blood and lymphoid tissues. Viral entry into CD4+ cells is mediated by the interaction with a cellular chemokine receptor, the most common of which are CCR5 and CXCR4.1 Because subsequent viral replication requires cellular gene expression processes, activated CD4+ cells are the primary targets of productive HIV infection.[62] Recently scientists have been investigating an alternative approach to treating HIV-1/AIDS, based on the creation of a disease-resistant immune system through transplantation of autologous, gene-modified (HIV-1-resistant) hematopoietic stem and progenitor cells (GM-HSPC).[63]

2.2.14   Clinical trials

Further information: Human embryonic stem cells clinical trials

GRNOPC1

On 23 January 2009, the US Food and Drug Administration gave clearance to Geron Corporation for the initiation of the first clinical trial of an embryonic stem-cell-based therapy on humans. The trial aimed evaluate the drug GRNOPC1, embryonic stem cell-derived oligodendrocyte progenitor cells, on patients with acute spinal cord injury. The trial was discontinued in November 2011 so that the company could focus on therapies in the "current environment of capital scarcity and uncertain economic conditions".[64] In 2013 biotechnology and regenerative medicine company BioTime (NYSE MKT: BTX) acquired Geron's stem cell assets in a stock transaction, with the aim of restarting the clinical trial.[65]

Cryopreserved mesenchymal stromal cells (MSCs)

Scientists have reported that MSCs when transfused immediately within few hours post thawing may show reduced function or show decreased efficacy in treating diseases as compared to those MSCs which are in log phase of cell growth(fresh), so cryopreserved MSCs should be brought back into log phase of cell growth in invitro culture before these are administered for clinical trials or experimental therapies, re-culturing of MSCs will help in recovering from the shock the cells get during freezing and thawing. Various clinical trials on MSCs have failed which used cryopreserved product immediately post thaw as compared to those clinical trials which used fresh MSCs.[66]

2.3   Other animals

2.3.1   Veterinary applications

Potential contributions to veterinary medicine

Research currently conducted on horses, dogs, and cats can benefit the development of stem cell treatments in veterinary medicine and can target a wide range of injuries and diseases such as myocardial infarction, stroke, tendon and ligament damage, osteoarthritis, osteochondrosis and muscular dystrophy both in large animals, as well as humans.[67][68][69][70] While investigation of cell-based therapeutics generally reflects human medical needs, the high degree of frequency and severity of certain injuries in racehorses has put veterinary medicine at the forefront of this novel regenerative approach.[71] Companion animals can serve as clinically relevant models that closely mimic human disease.[72][73]

Development of regenerative treatment models

Stem cells are thought to mediate repair via five primary mechanisms: 1) providing an anti-inflammatory effect, 2) homing to damaged tissues and recruiting other cells, such as endothelial progenitor cells, that are necessary for tissue growth, 3) supporting tissue remodeling over scar formation, 4) inhibiting apoptosis, and 5) differentiating into bone, cartilage, tendon, and ligament tissue.[74][75]

To further enrich blood supply to the damaged areas, and consequently promote tissue regeneration, platelet-rich plasma could be used in conjunction with stem cell transplantation.[76][77] The efficacy of some stem cell populations may also be affected by the method of delivery; for instance, to regenerate bone, stem cells are often introduced in a scaffold where they produce the minerals necessary for generation of functional bone.[76][77][78][79]

Stem cells have also been shown to have a low immunogenicity due to the relatively low number of MHC molecules found on their surface. In addition, they have been found to secrete chemokines that alter the immune response and promote tolerance of the new tissue. This allows for allogeneic treatments to be performed without a high rejection risk.[80]

Sources of stem cells

Veterinary applications of stem cell therapy as a means of tissue regeneration have been largely shaped by research that began with the use of adult-derived mesenchymal stem cells to treat animals with injuries or defects affecting bone, cartilage, ligaments and/or tendons.[81][82][83] There are two main categories of stem cells used for treatments: allogeneic stem cells derived from a genetically different donor within the same species[79][84] and autologous mesenchymal stem cells, derived from the patient prior to use in various treatments.[76] A third category, xenogenic stem cells, or stem cells derived from different species, are used primarily for research purposes, especially for human treatments.[85]

Most stem cells intended for regenerative therapy are generally isolated either from the patient's bone marrow or from adipose tissue.[77][79] Mesenchymal stem cells can differentiate into the cells that make up bone, cartilage, tendons, and ligaments, as well as muscle, neural and other progenitor tissues, they have been the main type of stem cells studied in the treatment of diseases affecting these tissues.[82][86] The number of stem cells transplanted into damaged tissue may alter efficacy of treatment. Accordingly, stem cells derived from bone marrow aspirates, for instance, are cultured in specialized laboratories for expansion to millions of cells.[77][79] Although adipose-derived tissue also requires processing prior to use, the culturing methodology for adipose-derived stem cells is not as extensive as that for bone marrow-derived cells.[87][88] While it is thought that bone-marrow derived stem cells are preferred for bone, cartilage, ligament, and tendon repair, others believe that the less challenging collection techniques and the multi-cellular microenvironment already present in adipose-derived stem cell fractions make the latter the preferred source for autologous transplantation.[76]

New sources of mesenchymal stem cells are being researched, including stem cells present in the skin and dermis which are of interest because of the ease at which they can be harvested with minimal risk to the animal.[89] Hematopoetic stem cells have also been discovered to be travelling in the blood stream and possess equal differentiating ability as other mesenchymal stem cells, again with a very non-invasive harvesting technique.[90]

There has been more recent interest in the use of extra embryonic mesenchymal stem cells. Research is underway to examine the differentiating capabilities of stem cells found in the umbilical cord, yolk sac and placenta of different animals. These stem cells are thought to have more differentiating ability than their adult counterparts, including the ability to more readily form tissues of endodermal and ectodermal origin.[80]

Stem cells and hard-tissue repair

Because of the general positive healing capabilities of stem cells, they have gained interest for the treatment of cutaneous wounds. This is important interest for those with reduced healing capabilities, like diabetics and those undergoing chemotherapy. In one trial, stem cells were isolated from the Wharton's jelly of the umbilical cord. These cells were injected directly into the wounds. Within a week, full re-epithelialization of the wounds had occurred, compared to minor re-epithelialization in the control wounds. This showed the capabilities of mesenchymal stem cells in the repair of epidermal tissues.[91]

Soft-palate defects in horses are caused by a failure of the embryo to fully close at the midline during embryogenesis. These are often not found until after they have become worse because of the difficulty in visualizing the entire soft palate. This lack of visualization is thought to also contribute to the low success rate in surgical intervention to repair the defect. As a result, the horse often has to be euthanized. Recently, the use of mesenchymal stem cells has been added to the conventional treatments. After the surgeon has sutured the palate closed, autologous mesenchymal cells are injected into the soft palate. The stem cells were found to be integrated into the healing tissue especially along the border with the old tissue. There was also a large reduction in the number of inflammatory cells present, which is thought to aid in the healing process.[92]

Stem cells and orthopedic repairs

Autologous stem cell-based treatments for ligament injury, tendon injury, osteoarthritis, osteochondrosis, and sub-chondral bone cysts have been commercially available to practicing veterinarians to treat horses since 2003 in the United States and since 2006 in the United

Kingdom. Autologous stem cell based treatments for tendon injury, ligament injury, and osteoarthritis in dogs have been available to veterinarians in the United States since 2005. Over 3000 privately owned horses and dogs have been treated with autologous adipose-derived stem cells. The efficacy of these treatments has been shown in double-blind clinical trials for dogs with osteoarthritis of the hip and elbow and horses with tendon damage.[93][94]

Tendon repair

Race horses are especially prone to injuries of the tendon and ligaments. Conventional therapies are very unsuccessful in returning the horse to full functioning potential. Natural healing, guided by the conventional treatments, leads to the formation of fibrous scar tissue that reduces flexibility and full joint movement. Traditional treatments prevented a large number of horses from returning to full activity and also have a high incidence of re-injury due to the stiff nature of the scarred tendon. Introduction of both bone marrow and adipose derived stem cells, along with natural mechanical stimulus promoted the regeneration of tendon tissue. The natural movement promoted the alignment of the new fibers and tendocytes with the natural alignment found in uninjured tendons. Stem cell treatment not only allowed more horses to return to full duty and also greatly reduced the re-injury rate over a three-year period.[80]

The use of embryonic stem cells has also been applied to tendon repair. The embryonic stem cells were shown to have a better survival rate in the tendon as well as better migrating capabilities to reach all areas of damaged tendon. The overall repair quality was also higher, with better tendon architecture and collagen formed. There was also no tumor formation seen during the three month experimental period. Long-term studies need to be carried out to examine the long-term efficacy and risks associated with the use of embryonic stem cells.[80] Similar results have been found in small animals.[80]

Joint repair

Osteoarthritis is the main cause of joint pain both in animals and humans. Horses and dogs are most frequently affected arthritis. Natural cartilage regeneration is very limited and no current drug therapies are curative, but rather look to reduce the symptoms associated with the degeneration. Different types of mesenchymal stem cells and other additives are still being researched to find the best type of cell and method for long-term treatment.[80]

Adipose-derived mesenchymal cells are currently the most often used because of the non-invasive harvesting. There has been a lot of success recently injecting mesenchymal stem cells directly into the joint. This is a recently developed, non-invasive technique developed for easier clinical use. Dogs receiving this treatment showed greater flexibility in their joints and less pain.[95]

Bone defect repair

While further studies are necessary to fully characterize the use of cell-based therapeutics for treatment of bone fractures, s

Bone has a unique and well documented natural healing process that normally is sufficient to repair fractures and other common injuries. Misaligned breaks due to severe trauma, as well as things like tumor resections of bone cancer, are prone to improper healing if left to the natural process alone. Scaffolds composed of natural and artificial components are seeded with mesenchymal stem cells and placed in the defect. Within four weeks of placing the scaffold, newly formed bone begins to integrate with the old bone and within 32 weeks, full union is achieved.[96] Further studies are necessary to fully characterize the use of cell-based therapeutics for treatment of bone fractures.

Stem cells have been used to treat degenerative bone diseases. The normally recommended treatment for dogs that have Legg–Calve–Perthes disease is to remove the head of the femur after the degeneration has progressed. Recently, mesenchymal stem cells have been injected directly in to the head of the femur, with success not only in bone regeneration, but also in pain reduction.[96]

Stem cells and muscle repairs

Stem cells have successfully been used to ameliorate healing in the heart after myocardial infarction in dogs. Adipose and bone marrow derived stem cells were removed and induced to a cardiac cell fate before being injected into the heart. The heart was found to have improved contractility and a reduction in the damaged area four weeks after the stem cells were applied.[97]

A different trial is underway for a patch made of a porous substance on to which the stem cells are "seeded" in order to induce tissue regeneration in heart defects. Tissue was regenerated and the patch was well incorporated into the heart tissue. This is thought to be due, in part, to improved angiogenesis and reduction of inflammation. Although cardiomyocytes were produced from the mesenchymal stem cells, they did not appear to be contractile. Other treatments that induced a cardiac fate in the cells before transplanting had greater success at creating contractile heart tissue.[98]

Stem cells and nervous system repairs

Spinal cord injuries are one of the most common traumas brought into veterinary hospitals.[96] Spinal injuries occur in two ways after the trauma: the primary mechanical damage, and in secondary processes, like inflammation and scar formation, in the days following the trauma. These cells involved in the secondary damage response secrete factors that promote scar formation and inhibit cellular regeneration. Mesenchymal stem cells that are induced to a neural cell fate are loaded on to a porous scaffold and are then implanted at the site of injury. The cells and scaffold secrete factors that counteract those secreted by scar forming cells and promote neural regeneration. Eight weeks later, dogs treated with stem cells showed immense improvement over those treated with conventional therapies. Dogs treated with stem cells were able to occasionally support their own weight, which has not been seen in dogs undergoing conventional therapies.[99][100][101]

Treatments are also in clinical trials to repair and regenerate peripheral nerves. Peripheral nerves are more likely to be damaged, but the effects of the damage are not as widespread as seen in injuries to the spinal cord. Treatments are currently in clinical trials to repair severed nerves, with early success. Stem cells induced to a neural fate injected in to a severed nerve. Within four weeks, regeneration of previously damaged

stem cells and completely formed nerve bundles were observed.[89]

Stem cells are also in clinical phases for treatment in ophthalmology. Hematopoietic stem cells have been used to treat corneal ulcers of different origin of several horses. These ulcers were resistant to conventional treatments available, but quickly responded positively to the stem cell treatment. Stem cells were also able to restore sight in one eye of a horse with retinal detachment, allowing the horse to return to daily activities.[90]

Keratoconjunctivitis Sicca (KCS)

Pre-clinical models of Sjögrens syndrome[102][103] have culminated in allogeneic MSCs implanted around the lacrimal glands in KSC dogs that were refractory to current therapy. Significantly improved scores in ocular discharge, conjunctival hyperaemia, corneal changes and Schirmer tear tests (STT) were seen.[104]

Current areas of research

Stems cells in the lab

The ability to grow up functional adult tissues indefinitely in culture through Directed differentiation creates new opportunities for drug research. Researchers are able to grow up differentiated cell lines and then test new drugs on each cell type to examine possible interactions *in vitro* before performing *in vivo* studies. This is critical in the development of drugs for use in veterinary research because of the possibilities of species specific interactions. The hope is that having these cell lines available for research use will reduce the need for research animals used because effects on human tissue *in vitro* will provide insight not normally known before the animal testing phase.[85]

With the advent of induced pluripotent stem cells (iPSC), treatments being explored and created for the used in endangered low production animals possible. Rather than needing to harvest embryos or eggs, which are limited, the researchers can remove mesenchymal stem cells with greater ease and greatly reducing the danger to the animal due to noninvasive techniques. This allows the limited eggs to be put to use for reproductive purposes only.[85]

Stem cells and conservation

Stem cells are being explored for use in conservation efforts. Spermatogonial stem cells have been harvested from a rat and placed into a mouse host and fully mature sperm were produced with the ability to produce viable offspring. Currently research is underway to find suitable hosts for the introduction of donor spermatogonial stem cells. If this becomes a viable option for conservationists, sperm can be produced from high genetic quality individuals who die before reaching sexual maturity, preserving a line that would otherwise be lost.[105]

Future clinical uses

The use of stem cells for the treatment of liver disease in both humans and animals has been the focus of considerable interest. The liver has some natural regenerative properties, but is often insufficient to deal with the extent of some liver diseases. Hepatocytes have been formed from some sources of MSC, but they have not been applied clinically currently.[96] There is a large effort to create stem cells differentiated along the pancreatic line as a possible cure for diabetes, but no line has been well established.[96]

Mesenchymal stem cells are currently under clinical trials as a possible treatment for graft v. host disease and graft rejection after experiments on various animals showing that allogenic stem cell treatments were not rejected and showed no difference in healing capabilities compared with autologous stem cells. This is being further researched for creating off-the-shelf allogenic stem cell treatments for various aspects in regenerative veterinary medicine.[80] Clinical trials are underway to explore the low immunogenic properties of stem cells and their possible use for treatment of problems with an overactive immune system seen with allergies and autoimmune disorders.[85]

In recent years, US-based stem-cell clinics have emerged that treat patients with their own bone marrow or adipose derived adult stem cells as part of clinical trials or FDA authorized same day outpatient IRB programs, most notably for athletes to recover from osteoskeletal (bone, joint and connective tissue) related injuries. This emergence of US based human adult stem cell therapy is discussed by Rudderham in his 2012 article Adult Stem Cell US Therapy.[106]

The long-term impact of these treatments will need to be examined outside of their contribution to medicine.[80] Vast improvements in veterinary medicine has allowed for companion and farm animals to live longer lives. This, however, has contributed to the rise in injury and chronic illness in companion animals.[85] Stem cell treatments, especially for the treatment of orthopedic issues in horses, allows for working animals to return to a normal state of activity at a faster rate with a reduction in the re-injury rate.[80]

2.4 Embryonic stem-cell controversy

Main article: Stem-cell controversy

There is widespread controversy over the use of human embryonic stem cells. This controversy primarily targets the techniques used to derive new embryonic stem cell lines, which often requires the destruction of the blastocyst. Opposition to the use of human embryonic stem cells in research is often based on philosophical, moral, or religious objections.[107] There is other stem cell research that does not involve the destruction of a human embryo, and such research involves adult stem cells, amniotic stem cells, and induced pluripotent stem cells.

2.5 Around the world

2.5.1 China

Stem-cell research and treatment was practiced in the People's Republic of China. The Ministry of Health of the People's Republic of China has permitted the use of stem-cell therapy for conditions beyond those approved of in Western countries. The Western World has scrutinized China for its failed attempts to meet international documentation standards of these trials and procedures.[108]

State-funded companies based in the Shenzhen Hi-Tech Industrial Zone treat the symptoms of numerous disorders with adult stem-cell therapy. Development companies are currently focused on the treatment of neurodegenerative and cardiovascular disorders. The most radical successes of Chinese adult stem cell therapy have been in treating the brain. These therapies administer stem cells directly to the brain of patients with cerebral palsy, Alzheimer's, and brain injuries.

2.5.2 Middle East

Since 2008 many universities, centers and doctors tried a diversity of methods; in Lebanon proliferation for stem cell therapy, in-vivo and in-vitro techniques were used. Thus this country is considered the launching place of the **Regentime**[109] procedure. http://www.researchgate.net/publication/281712114_ Treatment_of_Long_Standing_Multiple_Sclerosis_with_ Regentime_Stem_Cell_Technique   The regenerative medicine also took place in Jordan and Egypt.

2.5.3 Mexico

Stem-cell treatment is currently being practiced at a clinical level in Mexico. An International Health Department Permit (COFEPRIS) is required. Authorized centers are found in Tijuana, Guadalajara and Cancun. Currently undergoing the approval process is Los Cabos. This permit allows the use of stem cell.

2.5.4 South Korea

In 2005, South Korean scientists claimed to have generated stem cells that were tailored to match the recipient. Each of the 11 new stem cell lines was developed using somatic cell nuclear transfer (SCNT) technology. The resultant cells were thought to match the genetic material of the recipient, thus suggesting minimal to no cell rejection.[110]

2.5.5 Thailand

As of 2013, Thailand still considers Hematopoietic stem cell transplants as experimental. Kampon Sriwatanakul began with a clinical trial in October 2013 with 20 patients. 10 are going to receive stem-cell therapy for Type-2 diabetes and the other 10 will receive stem-cell therapy for emphysema. Chotinantakul's research is on Hematopoietic cells and their role for the hematopoietic system function in homeostasis and immune response.[111]

2.5.6 Ukraine

Today, Ukraine is permitted to perform clinical trials of stem-cell treatments (Order of the MH of Ukraine № 630 "About carrying out clinical trials of stem cells", 2008) for the treatment of these pathologies: pancreatic necrosis, cirrhosis, hepatitis, burn disease, diabetes, multiple sclerosis, critical lower limb ischemia. The first medical institution granted the right to conduct clinical trials became the "Institute of Cell Therapy"(Kiev).

2.5.7 Other countries

Other countries where doctors did stem cells research, trials, manipulation, storage, therapy: Brazil, Cyprus, Germany, Italy, Israel, Japan, Pakistan, Philippines, Russia, Switzerland, Turkey, United Kingdom, India, and many others.

2.6 See also

- Autologous stem-cell transplantation
- Cardiovascular Cell Therapy Research Network (CC-TRN)
- Fetal tissue implant
- Human Stem Cells Institute
- Induced pluripotent stem cell
- Induced stem cells

2.7 References

[1] Ian Murnaghan for Explore Stem Cells. Updated: 16 December 2013 Why Perform a Stem Cell Transplant?

[2] Bone Marrow Transplantation and Peripheral Blood Stem Cell Transplantation In National Cancer Institute Fact Sheet web site. Bethesda, MD: National Institutes of Health, U.S. Department of Health and Human Services, 2010. Cited 24 August 2010

[3] Karanes C, Nelson GO, Chitphakdithai P, Agura E, Ballen KK, Bolan CD, Porter DL, Uberti JP, King RJ, Confer DL; Nelson; Chitphakdithai; Agura; Ballen; Bolan; Porter; Uberti; King; Confer (2008). "Twenty years of unrelated donor hematopoietic cell transplantation for adult recipients facilitated by the National Marrow Donor Program". *Biology of Blood and Marrow Transplantation* **14** (9 Suppl): 8–15. doi:10.1016/j.bbmt.2008.06.006. PMID 18721775.

[4] Malard F, Mohty M (2014). "New Insight for the Diagnosis of Gastrointestinal Acute Graft-versus-Host Disease". *Mediators Inflamm* **2014**: 701013. doi:10.1155/2014/701013. PMID 24733964.

[5] "Prochymal – First Stem Cell Drug Approved". 22 May 2012.

[6] "A Stem-Cell-Based Drug Gets Approval in Canada". 17 May 2012.

[7] Rosemann A (Dec 2014). "Why regenerative stem cell medicine progresses slower than expected". *J Cell Biochem* **115** (12): 2073–6. doi:10.1002/jcb.24894. PMID 25079695.

[8] European Medicines Agency. "First stem-cell therapy recommended for approval in EU". Retrieved 12 December 2014.

[9] Cell Basics: What are the potential uses of human stem cells and the obstacles that must be overcome before these potential uses will be realized?. In Stem Cell Information World Wide Web site. Bethesda, MD: National Institutes of Health, U.S. Department of Health and Human Services, 2009. cited Sunday, 26 April 2009

[10] Neural Stem Cells May Rescue Memory In Advanced Alzheimer's, Mouse Study Suggests

[11] Vastag B (April 2001). "Stem cells step closer to the clinic: paralysis partially reversed in rats with ALS-like disease". *JAMA* **285** (13): 1691–3. doi:10.1001/jama.285.13.1691. PMID 11277806.

[12] Rebeiro P, Moore J. The role of autologous haemopoietic stem cell transplantation in the treatment of autoimmune disorders. Intern Med J. 2016 Jan;46(1):17-28. PMID 26524106

[13] Androutsellis-Theotokis A, Leker RR, Soldner F, et al. (August 2006). "Notch signalling regulates stem cell numbers in vitro and in vivo". *Nature* **442** (7104): 823–6. doi:10.1038/nature04940. PMID 16799564.

[14] Androutsellis-Theotokis A, Rueger MA, Park DM, et al. (August 2009). "Targeting neural precursors in the adult brain rescues injured dopamine neurons". *Proc. Natl. Acad. Sci. U.S.A.* **106** (32): 13570–5. doi:10.1073/pnas.0905125106. PMC 2714762. PMID 19628689.

[15] Androutsellis-Theotokis A, Rueger MA, Mkhikian H, Korb E, McKay RD (2008). "Signaling pathways controlling neural stem cells slow progressive brain disease". *Cold Spring Harb. Symp. Quant. Biol.* **73**: 403–10. doi:10.1101/sqb.2008.73.018. PMID 19022746.

[16] Ghosh, Pallab (27 May 2013) Stroke patients see signs of recovery in stem cell trial BBC News health. Retrieved 27 May 2013

[17] Kang KS, Kim SW, Oh YH, et al. (2005). "A 37-year-old spinal cord-injured female patient, transplanted of multipotent stem cells from human UC blood, with improved sensory perception and mobility, both functionally and morphologically: a case study". *Cytotherapy* **7** (4): 368–73. doi:10.1080/14653240500238160. PMID 16162459.

[18] Team co-headed by researchers at Chosun University, Seoul National University and the Seoul Cord Blood Bank (SCB) Umbilical cord cells 'allow paralysed woman to walk' By Roger Highfield, Science Editor. Last Updated: 1:28AM GMT 30 November 2004

[19] Cummings BJ, Uchida N, Tamaki SJ, et al. (September 2005). "Human neural stem cells differentiate and promote locomotor recovery in spinal cord-injured mice".

Proc. Natl. Acad. Sci. U.S.A. **102** (39): 14069–74. doi:10.1073/pnas.0507063102. PMC 1216836. PMID 16172374.

[20] Strauer, Bodo; Steinhoff G (September 2011). "10 years of intracoronary and intramyocardial bone marrow stem cell therapy of the heart: from the methodological origin to clinical practice". *J Am Coll Cardiol* **58** (11): 1095–1104. doi:10.1016/j.jacc.2011.06.016. PMID 21884944.

[21] Francis, DP; Mielewczik, M; Zargaran, D; Cole, GD (26 June 2013). "Autologous bone marrow-derived stem cell therapy in heart disease: Discrepancies and contradictions". *International journal of cardiology* **168** (4): 3381–403. doi:10.1016/j.ijcard.2013.04.152. PMID 23830344. Retrieved 21 July 2013.

[22] Strauer, BE; Yousef, M; Schannwell, CM (July 2010). "The acute and long-term effects of intracoronary Stem cell Transplantation in 191 patients with chronic heARt failure: the STAR-heart study". *European journal of heart failure* **12** (7): 721–9. doi:10.1093/eurjhf/hfq095. PMID 20576835.

[23] Yousef, M; Schannwell, CM; Köstering, M; Zeus, T; Brehm, M; Strauer, BE (16 June 2009). "The BALANCE Study: clinical benefit and long-term outcome after intracoronary autologous bone marrow cell transplantation in patients with acute myocardial infarction". *Journal of the American College of Cardiology* **53** (24): 2262–9. doi:10.1016/j.jacc.2009.02.051. PMID 19520249.

[24] Schannwell CM, Kostering M, Zeus T, Brehm M, Erdmann G, Fleissner T, Yosef M, Kogler G, Wernet P, Strauer BE. (2008). "Hmane autologe Stammzelltransplantation zur Myokardregeneration bei dilatativer Kardiomyopathie (NYHA Stadium II bis III)" (PDF). *Austrian Journal of Cardiology* **15** (1): 23–30.

[25] Orlic D.; et al. (2001). "Bone marrow cells regenerate infarcted myocardium". *Nature* **410** (6829): 701–705. doi:10.1038/35070587. PMID 11287958. |first2= missing |last2= in Authors list (help); |first3= missing |last3= in Authors list (help); |first4= missing |last4= in Authors list (help); |first5= missing |last5= in Authors list (help); |first6= missing |last6= in Authors list (help); |first7= missing |last7= in Authors list (help); |first8= missing |last8= in Authors list (help); |first9= missing |last9= in Authors list (help); |first10= missing |last10= in Authors list (help); |first11= missing |last11= in Authors list (help); |first12= missing |last12= in Authors list (help)

[26] Kocher A. A.; et al. "Neovascularization of ischemic myocardium by human bone-marrow-derived angioblasts prevents cardiomyocyte apoptosis, reduces remodeling and improves cardiac function". *Nature Medicine* **7**: 430–436. doi:10.1038/86498.

[27] Kuswardhani R. A., Soejitno A. (2011). "Bone marrow-derived stem cells as an adjunctive treatment for acute myocardial infarction: a systematic review and meta-analysis". *Acta medica Indonesiana* **43** (3): 168–177. PMID 21979282.

[28] Malliaras K., Kreke M., Marban E. (2011). "The stuttering progress of cell therapy for heart disease". *Clinical pharmacology and therapeutics* **90** (4): 532–541. doi:10.1038/clpt.2011.175. PMID 21900888.

[29] Ayala-Lugo A., et al. "Age-dependent availability and functionality of bone marrow stem cells in an experimental model of acute and chronic myocardial infarction". *Cell transplantation* **20**: 407–419. doi:10.3727/096368909X519283.

[30] Zhang Y. et al. "Timing of bone marrow cell therapy is more important than repeated injections after myocardial infarction". *Cardiovascular pathology : the official journal of the Society for Cardiovascular Pathology* **20**: 204–212. doi:10.1016/j.carpath.2010.06.007.

[31] Wang X.; et al. "Donor myocardial infarction impairs the therapeutic potential of bone marrow cells by an interleukin-1-mediated inflammatory response". *Science Translational Medicine* **3**: 100ra90. doi:10.1126/scitranslmed.3002814.

[32] Krishnamurthy P.; et al. (2011). "Interleukin-10 deficiency impairs bone marrow-derived endothelial progenitor cell survival and function in ischemic myocardium". *Circulation Research* **109** (11): 1280–1289. doi:10.1161/CIRCRESAHA.111.248369. PMC 3235675. PMID 21959218. |first2= missing |last2= in Authors list (help); |first3= missing |last3= in Authors list (help); |first4= missing |last4= in Authors list (help); |first5= missing |last5= in Authors list (help); |first6= missing |last6= in Authors list (help); |first7= missing |last7= in Authors list (help); |first8= missing |last8= in Authors list (help); |first9= missing |last9= in Authors list (help)

[33] Paul A., Srivastava S., Chen G., Shum-Tim D., Prakash S. "Functional Assessment of Adipose Stem Cells for Xenotransplantation Using Myocardial Infarction Immunocompetent Models: Comparison with Bone Marrow Stem Cells". *Cell biochemistry and biophysics* **67**: 263–273. doi:10.1007/s12013-011-9323-0.

[34] Guinea pig hearts beat with human cells : Nature News & Comment

[35] Francis, Darrel P (Oct 2013). "Autologous bone marrow-derived stem cell therapy in heart disease: Discrepancies and contradictions". *Int J Cardiol* (Elsevier) **168**: 3381–403. doi:10.1016/j.ijcard.2013.04.152. PMID 23830344. Retrieved 6 July 2013.

[36] Berndt, Christina. "A minefield of contradictions". Suddeutsche Zeitung. Retrieved 6 July 2013.

[37] Haider, HKh; Ashraf, M (June 2005). "Bone marrow stem cell transplantation for cardiac repair.". *American Journal of Physiology. Heart and Circulatory Physiology* **288** (6): H2557–67. doi:10.1152/ajpheart.01215.2004. PMID 15897328.

[38] Ptaszek LM, Mansour M, Ruskin JN, Chien KR (2012). "Towards regenerative therapy for cardiac disease". *The*

Lancet **379** (9819): 933–942. doi:10.1016/s0140-6736(12)60075-0.

[39] Giarratana MC, Kobari L, Lapillonne H, et al. (January 2005). "Ex vivo generation of fully mature human red blood cells from hematopoietic stem cells". *Nat. Biotechnol.* **23** (1): 69–74. doi:10.1038/nbt1047. PMID 15619619.

[40] Archer, Graeme. "Technology". *The Daily Telegraph* (London). Retrieved 24 May 2010.

[41] Anglin, Ian. "Scientists Grow Teeth Using Stem Cells". *SingularityHUB*. Retrieved 31 July 2014.

[42] Yen AH, Sharpe PT (January 2008). "Stem cells and tooth tissue engineering". *Cell Tissue Res.* **331** (1): 359–72. doi:10.1007/s00441-007-0467-6. PMID 17938970.

[43] "Stem cell-based biological tooth repair and regeneration". *Trends Cell Biol.* **20**: 715–22. 2010. doi:10.1016/j.tcb.2010.09.012. PMC 3000521. PMID 21035344.

[44] Wu, Ping; Wu, Xiaoshan; Jiang, Ting-Xin; Elsey, Ruth M.; Temple, Bradley L.; Divers, Stephen J.; Glenn, Travis C.; Yuan, Kuo; Chen, Min-Huey; Widelitz, Randall B.; Chuon, Cheng-Ming (2013). "Specialized stem cell niche enables repetitive renewal of alligator teeth" (PDF). *Proceedings of the National Academy of Sciences of the United States of America* **110** (22): E2009–E2018. doi:10.1073/pnas.1213202110. PMC 3670376. PMID 23671090.

[45] Gene therapy is first deafness 'cure' – health – 14 February 2005 – New Scientist

[46] Fetal tissue restores lost sight MedicalNewsToday. Article Date: 28 October 2004 – 10:00 PDT

[47] BBC NEWS | England | Southern Counties | Stem cells used to restore vision

[48] The University Hospital of New Jersey, 2002

[49] "Human limbal biopsy–derived stromal stem cells prevent corneal scarring". *Science Translational Medicine*. 12 December 2014. Retrieved 2015-08-02.

[50] "Embryonic stem cells improve sight of legally blind women". *CNN.com*. 23 January 2012. Retrieved 2012-01-12.

[51] Weiss JN, Levy S, Malkin A. Stem Cell Ophthalmology Treatment Study (SCOTS) for retinal and optic nerve diseases: a preliminary report. Neural Regen Res [serial online] 2015 [cited 2015 Sep 21];10:982-8. Available from: http://www.nrronline.org/text.asp?2015/10/6/982/158365

[52] Saki N, Jalalifar MA, Soleimani M, Hajizamani S, Rahim F (2013). "Adverse Effect of High Glucose Concentration on Stem Cell Therap". *Int J Hematol Oncol Stem Cell Res* **7** (3): 34–40. PMC 3913149. PMID 24505533.

[53] Goldstein, Ron (2007). *Embryonic stem cell research is necessary to find a diabetes cure.* Greenhaven Press. p. 44.

[54] Centeno CJ, Busse D, Kisiday J, Keohan C, Freeman M, Karli D (December 2008). "Regeneration of meniscus cartilage in a knee treated with percutaneously implanted autologous mesenchymal stem cells". *Med. Hypotheses* **71** (6): 900–8. doi:10.1016/j.mehy.2008.06.042. PMID 18786777.

[55] Centeno CJ, Busse D, Kisiday J, Keohan C, Freeman M, Karli D (2008). "Increased knee cartilage volume in degenerative joint disease using percutaneously implanted, autologous mesenchymal stem cells". *Pain Physician* **11** (3): 343–53. PMID 18523506.

[56] Centeno CJ, Schultz JR, Cheever M, Robinson B, Freeman M, Marasco W (March 2010). "Safety and complications reporting on the re-implantation of culture-expanded mesenchymal stem cells using autologous platelet lysate technique". *Curr Stem Cell Res Ther* **5** (1): 81–93. doi:10.2174/157488810790442796. PMID 19951252.

[57] Wakitani S, Nawata M, Tensho K, Okabe T, Machida H, Ohgushi H (2007). "Repair of articular cartilage defects in the patello-femoral joint with autologous bone marrow mesenchymal cell transplantation: three case reports involving nine defects in five knees". *J Tissue Eng Regen Med* **1** (1): 74–9. doi:10.1002/term.8. PMID 18038395.

[58] Gurtner GC, Callaghan MJ, Longaker MT (2007). "Progress and potential for regenerative medicine". *Annu. Rev. Med* **58** (1): 299–312. doi:10.1146/annurev.med.58.082405.095329. PMID 17076602.

[59] Richards M, Fong CY, Bongso A (December 2008). "Comparative evaluation of different in vitro systems that stimulate germ cell differentiation in human embryonic stem cells". *Fertil. Steril.* **93** (3): 986–94. doi:10.1016/j.fertnstert.2008.10.030. PMID 19064262.

[60] Ledford H (7 July 2009). "Sperm-like cells made from human embryonic stem cells". *Nature News.* doi:10.1038/news.2009.646. Archived from the original on 9 May 2011.

[61] White, YAR; Woods DC; Takai Y; Ishihara O; Seki H; Tilly JL. (2012). "Oocyte formation by mitotically active germ cells purified from ovaries of reproductive-age women". *Nature Medicine* **18** (3): 413–421. doi:10.1038/nm.2669. PMC 3296965. PMID 22366948.

[62] Allers, Kristinia; Hütter, Gero; Hofmann, Jörg; Loddenkemper, Chrtoph; Rieger, Kathrin; Thiel, Eckhard; Schneider, Thomas (14 July 2014). "Evidence for the cure of HIV infection by CCR5Δ32/Δ32 stem cell transplantation". *Blood* **117** (10): 2791–2799. doi:10.1182/blood-2010-09-309591. PMID 21148083.

[63] DiGiusto, David; Stan, Rodica; Krishnan, Amrita; Li, Haitang; Rossi, John; Zaia, John (22 November 2013). "Development of Hematopoietic Stem Cell Based Gene Therapy for HIV-1 Infection: Considerations for Proof of Concept Studies and Translation to Standard Medical Practice". *Viruses* **2013** (5): 2898–2919. doi:10.3390/v5112898.

[64] O'Connell, Claire (27 January 2012). "Stem cells – where are we now?". *The Irish Times.*

[65] "BioTime acquires stem cell assets from Geron, raises $10 million". *San Francisco Business Times.* 7 January 2013.

[66] Francois M et al., Cytotherapy.2012;14:147–152

[67] Chen J, Li Y, Wang L, et al. (April 2001). "Therapeutic benefit of intravenous administration of bone marrow stromal cells after cerebral ischemia in rats". *Stroke* **32** (4): 1005–11. doi:10.1161/01.STR.32.4.1005. PMID 11283404.

[68] Assmus B, Schächinger V, Teupe C, et al. (December 2002). "Transplantation of Progenitor Cells and Regeneration Enhancement in Acute Myocardial Infarction (TOPCARE-AMI)". *Circulation* **106** (24): 3009–17. doi:10.1161/01.CIR.0000043246.74879.CD. PMID 12473544.

[69] Murphy JM, Fink DJ, Hunziker EB, Barry FP (December 2003). "Stem cell therapy in a caprine model of osteoarthritis". *Arthritis Rheum.* **48** (12): 3464–74. doi:10.1002/art.11365. PMID 14673997.

[70] Sampaolesi M, Blot S, D'Antona G, et al. (November 2006). "Mesoangioblast stem cells ameliorate muscle function in dystrophic dogs". *Nature* **444** (7119): 574–9. doi:10.1038/nature05282. PMID 17108972.

[71] Taylor SE, Smith RK, Clegg PD (March 2007). "Mesenchymal stem cell therapy in equine musculoskeletal disease: scientific fact or clinical fiction?". *Equine Vet. J.* **39** (2): 172–80. doi:10.2746/042516407X180868. PMID 17378447.

[72] Tecirlioglu RT, Trounson AO (2007). "Embryonic stem cells in companion animals (horses, dogs and cats): present status and future prospects". *Reprod. Fertil. Dev.* **19** (6): 740–7. doi:10.1071/RD07039. PMID 17714628.

[73] Koch TG, Betts DH (November 2007). "Stem cell therapy for joint problems using the horse as a clinically relevant animal model". *Expert Opin Biol Ther* **7** (11): 1621–6. doi:10.1517/14712598.7.11.1621. PMID 17961087.

[74] Richardson LE, Dudhia J, Clegg PD, Smith R (September 2007). "Stem cells in veterinary medicine—attempts at regenerating equine tendon after injury". *Trends Biotechnol.* **25** (9): 409–16. doi:10.1016/j.tibtech.2007.07.009. PMID 17692415.

[75] Csaki C, Matis U, Mobasheri A, Ye H, Shakibaei M (December 2007). "Chondrogenesis, osteogenesis and adipogenesis of canine mesenchymal stem cells: a biochemical,

morphological and ultrastructural study". *Histochem. Cell Biol.* **128** (6): 507–20. doi:10.1007/s00418-007-0337-z. PMID 17922135.

[76] Kane, Ed (May 2008). Stem-cell therapy shows promise for soft-tissue injury, disease. DVM Newsmagazine. 6E-10E.

[77] Yamada Y, Ueda M, Naiki T, Takahashi M, Hata K, Nagasaka T (2004). "Autogenous injectable bone for regeneration with mesenchymal stem cells and platelet-rich plasma: tissue-engineered bone regeneration". *Tissue Eng.* **10** (5–6): 955–64. doi:10.1089/1076327041348284. PMID 15265313.

[78] Singec I, Jandial R, Crain A, Nikkhah G, Snyder EY (2007). "The leading edge of stem cell therapeutics". *Annu. Rev. Med.* **58** (1): 313–28. doi:10.1146/annurev.med.58.070605.115252. PMID 17100553.

[79] Zachos TA, Smith TJ (September 2008). Use of adult stem cells in clinical orthopedics. DVM Newsmagazine. 36–39.

[80] Walter Brehm (2012). "Stem cell-based tissue engineering in veterinary orthopaedics". *Cell Tissue Research* **347** (3): 677–688. doi:10.1007/s00441-011-1316-1. PMID 22287044. |first2= missing |last2= in Authors list (help); |first3= missing |last3= in Authors list (help); |first4= missing |last4= in Authors list (help); |first5= missing |last5= in Authors list (help)

[81] Young RG, Butler DL, Weber W, Caplan AI, Gordon SL, Fink DJ (July 1998). "Use of mesenchymal stem cells in a collagen matrix for Achilles tendon repair". *J. Orthop. Res.* **16** (4): 406–13. doi:10.1002/jor.1100160403. PMID 9747780.

[82] Awad HA, Butler DL, Boivin GP, et al. (June 1999). "Autologous mesenchymal stem cell-mediated repair of tendon". *Tissue Eng.* **5** (3): 267–77. doi:10.1089/ten.1999.5.267. PMID 10434073.

[83] Bruder SP, Kraus KH, Goldberg VM, Kadiyala S (July 1998). "The effect of implants loaded with autologous mesenchymal stem cells on the healing of canine segmental bone defects". *J Bone Joint Surg Am* **80** (7): 985–96. PMID 9698003.

[84] Kraus KH, Kirker-Head C (April 2006). "Mesenchymal stem cells and bone regeneration". *Vet Surg* **35** (3): 232–42. doi:10.1111/j.1532-950X.2006.00142.x. PMID 16635002.

[85] Daniela Gattegno-Ho (2012). "Stem cells and veterinary medicine: Tools to understand diseases and enable tissue regeneration and drug discovery". *The Veterinary Journal* **191** (1): 19–27. doi:10.1016/j.tvj.2011.08.007. PMID 21958722. |first2= missing |last2= in Authors list (help); |first3= missing |last3= in Authors list (help)

[86] Nathan S, Das De S, Thambyah A, Fen C, Goh J, Lee EH (August 2003). "Cell-based therapy in the repair of osteochondral defects: a novel use for adipose tissue". *Tissue Eng.* **9** (4): 733–44. doi:10.1089/107632703768247412. PMID 13678450.

[87] Fraser JK, Wulur I, Alfonso Z, Hedrick MH (April 2006). "Fat tissue: an underappreciated source of stem cells for biotechnology". *Trends Biotechnol.* **24** (4): 150–4. doi:10.1016/j.tibtech.2006.01.010. PMID 16488036.

[88] Nakagami H, Morishita R, Maeda K, Kikuchi Y, Ogihara T, Kaneda Y (April 2006). "Adipose tissue-derived stromal cells as a novel option for regenerative cell therapy". *J. Atheroscler. Thromb.* **13** (2): 77–81. doi:10.5551/jat.13.77. PMID 16733294.

[89] Bong Wook Park (2012). "Peripheral nerve regeneration using autologous porcine skin-derived mesenchymal stem cells". *Journal of Tissue Engineering and Regenerative Medicine* **6** (2): 113–124. doi:10.1002/term.404. PMID 21337707. |first2= missing |last2= in Authors list (help); |first3= missing |last3= in Authors list (help); |first4= missing |last4= in Authors list (help); |first5= missing |last5= in Authors list (help); |first6= missing |last6= in Authors list (help); |first7= missing |last7= in Authors list (help)

[90] Gabriella Marfe (2012). "Blood derived stem cells: An ameliorative therapy in veterinary ophthalmology". *Journal of Cellular Physiology* **227** (3): 1250–1256. doi:10.1002/jcp.22953. PMID 21792938. |first2= missing |last2= in Authors list (help); |first3= missing |last3= in Authors list (help); |first4= missing |last4= in Authors list (help); |first5= missing |last5= in Authors list (help); |first6= missing |last6= in Authors list (help); |first7= missing |last7= in Authors list (help); |first8= missing |last8= in Authors list (help)

[91] Omid Azari (2011). "Effects of transplanted mesenchymal stem cells isolated from Wharton's jelly of caprine umbilical cord on cutaneous wound healing; histopathological evaluation". *Veterinary Research Communications* **35** (4): 211–222. doi:10.1007/s11259-011-9464-z. PMID 21340694. |first2= missing |last2= in Authors list (help); |first3= missing |last3= in Authors list (help); |first4= missing |last4= in Authors list (help); |first5= missing |last5= in Authors list (help); |first6= missing |last6= in Authors list (help)

[92] Carstanjen B, Desbois C, Hekmati M, Behr L (2006). "Successful engraftment of cultured autologous mesenchymal stem cells in a surgically repaired soft palate defect in an adult horse". *The Canadian Journal of Veterinary Research* **70** (2): 143–147. PMC 1410720. PMID 16639947.

[93] Black LL, Gaynor J, Adams C, et al. (2008). "Effect of intraarticular injection of autologous adipose-derived mesenchymal stem and regenerative cells on clinical signs of chronic osteoarthritis of the elbow joint in dogs". *Vet. Ther.* **9** (3): 192–200. PMID 19003780.

[94] Nixon AJ, Dahlgren LA, Haupt JL, Yeager AE, Ward DL (July 2008). "Effect of adipose-derived nucleated

cell fractions on tendon repair in horses with collagenase-induced tendinitis". *Am. J. Vet. Res.* **69** (7): 928–37. doi:10.2460/ajvr.69.7.928. PMID 18593247.

[95] Annalisa Guercio (2012). "Production of canine mesenchymal stem cells from adipose tissue and their application in dogs with chronic osteoarthritis of the humeroradial joints". *Cell Biology International* **36** (2): 189–194. doi:10.1042/CBI20110304. PMID 21936851. |first2= missing |last2= in Authors list (help); |first3= missing |last3= in Authors list (help); |first4= missing |last4= in Authors list (help); |first5= missing |last5= in Authors list (help); |first6= missing |last6= in Authors list (help); |first7= missing |last7= in Authors list (help); |first8= missing |last8= in Authors list (help)

[96] I Ribitsch (2010). "Basic Science and Clinical Application of Stem Cells in Veterinary Medicine". *Advanced Biochemical Engineering and Biotechnology* **123**: 219–263. doi:10.1007/10_2010_66. ISBN 978-3-642-16050-9. |first2= missing |last2= in Authors list (help); |first3= missing |last3= in Authors list (help); |first4= missing |last4= in Authors list (help); |first5= missing |last5= in Authors list (help); |first6= missing |last6= in Authors list (help); |first7= missing |last7= in Authors list (help)

[97] Ung Kim (2011). "Homing of adipose-derived stem cells to radiofrequency catheter ablated canine atrium and differentiation into cardiomyocyte-like cells". *International Journal of Cardiology* **146** (3): 371–378. doi:10.1016/j.ijcard.2009.07.016. PMID 19683815. |first2= missing |last2= in Authors list (help); |first3= missing |last3= in Authors list (help); |first4= missing |last4= in Authors list (help); |first5= missing |last5= in Authors list (help); |first6= missing |last6= in Authors list (help); |first7= missing |last7= in Authors list (help)

[98] Yen Chang (2007). "Tissue regeneration observed in a basic fibroblast growth factor–loaded porous acellular bovine pericardium populated with mesenchymal stem cells". *Surgery for Acquired Cardiovascular Disease* **134**: 65–73. doi:10.1016/j.jtcvs.2007.02.019. |first2= missing |last2= in Authors list (help); |first3= missing |last3= in Authors list (help); |first4= missing |last4= in Authors list (help); |first5= missing |last5= in Authors list (help); |first6= missing |last6= in Authors list (help); |first7= missing |last7= in Authors list (help); |first8= missing |last8= in Authors list (help)

[99] Sung Su Park; et al. (2012). "Functional recovery after spinal cord injury in dogs treated with a combination of Matrigel and neural-induced adipose-derived mesenchymal Stem cells". *Cytotherapy* **14** (5): 584–597. doi:10.319/14653249.2012.658913. PMID 22348702.

[100] Hak-Hyun Ryu (2009). "Functional recovery and neural differentiation after transplantation of allogenic adipose-derived stem cells in a canine model of acute spinal cord injury". *Journal of Veterinary Science* **10** (4): 273–284. doi:10.4142/jvs.2009.10.4.273. PMC 2807262. PMID 19934591. |first2= missing |last2= in Authors list (help);

|first3= missing |last3= in Authors list (help); |first4= missing |last4= in Authors list (help); |first5= missing |last5= in Authors list (help); |first6= missing |last6= in Authors list (help); |first7= missing |last7= in Authors list (help); |first8= missing |last8= in Authors list (help); |first9= missing |last9= in Authors list (help)

[101] Hidetaka Nishida; et al. (2012). *Veterinary Surgery* **00**: 1–6. doi:10.111/j.1532-950X.2011.00959.x. Missing or empty |title= (help)

[102] Xu, J., Wang, D., Liu, D., Fan, Z., Zhang, H., Liu, O., et al. (2012). Allogeneic mesenchymal stem cell treatment alleviates experimental and clinical Sjogren syndrome. Blood, 120(15), 3142–3151.

[103] Beyazyıldız, E., Pınarlı, F. A., Beyazyıldız, Ö., Hekimoğlu, E. R., Acar, U., Demir, M. N., et al. (2014). Efficacy of Topical Mesenchymal Stem Cell Therapy in the Treatment of Experimental Dry Eye Syndrome Model. Stem Cells International, 2014(3), 250230–9

[104] Villatoro, A. J., Fernández, V., Claros, S., Rico-Llanos, G. A., Becerra, J., & Andrades, J. A. (2015). Use of adipose-derived mesenchymal stem cells in keratoconjunctivitis sicca in a canine model. BioMed Research International, 2015(3), 527926–10. http://doi.org/10.1155/2015/527926

[105] I. Dobrinski and A.J. Travis (2007). "Germ cell transplantation for the propagation of companion animals, non-domestic and endangered species". *Reproduction, Fertility, and Development* **19**: 732–739. doi:10.1071/RD07036.

[106] Health, Wellness, Nutrition and Anti-Aging Integrative Medicine Experts & Information – Superior Raw Super Food Supplements – Online Wellness Community

[107] Mlsna, Lucas J. (2010). "Stem Cell Based Treatments and Novel Considerations for Conscience Clause Legislation". *Indiana Health Law Review* (United States: Indiana University Robert H. McKinney School of Law) **8** (2): 471–496. ISSN:1549–3199. LCCN:2004212209. OCLC:OCLC 54703225.

[108] Dobkin BH, Curt A, Guest J; Curt; Guest (March 2006). "Cellular transplants in China: observational study from the largest human experiment in chronic spinal cord injury". *Neurorehabil Neural Repair* **20** (1): 5–13. doi:10.1177/1545968305284675. PMC 4169140. PMID 16467274.

[109] http://www.regentime.com

[110] "Stem cells tailored to patients". *BBC News*. 20 May 2005. Retrieved 24 May 2010.

[111] Chotinantakul K and Leeanansaksiri W (June 2012). "Hematopoietic stem cell development, niches, and signaling pathways". *Bone Marrow Research* **10** (6): 12–16. doi:10.1155/2012/270425. PMC 3413998. PMID 22900188.

2.8   External links

- Fiona Murray PhD, Debora Spar PhD "Bit Player Or Powerhouse? China And Stem-Cell Research", "New England Journal of Medicine" 21 September 2006. (Accessed 30 July 2007)

- Clive Cookson "Generous Staffing And Permissive Laws Aid Asia's Largest Stem Cell Effort", "Scientific American" 27 June 2005. (Accessed 30 July 2007)

- Stem cell research & therapy: types of stem cells and their current uses

Chapter 3

Cell (biology)

This article is about the term in biology. For other uses, see Cell (disambiguation).

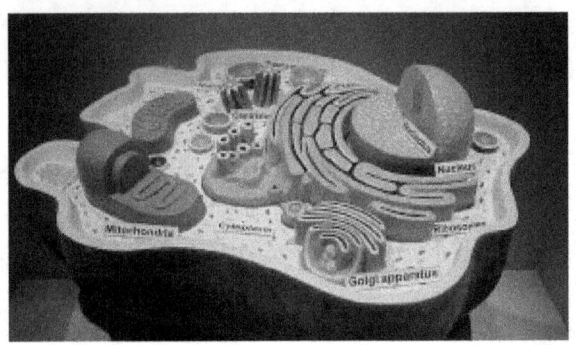

Structure of an animal cell

The **cell** (from Latin *cella*, meaning "small room"[11]) is the basic structural, functional, and biological unit of all known living organisms. A cell is the smallest unit of life that can replicate independently, and cells are often called the "building blocks of life". The study of cells is called cell biology.

Cells consist of cytoplasm enclosed within a membrane, which contains many biomolecules such as proteins and nucleic acids.[2] Organisms can be classified as unicellular (consisting of a single cell; including bacteria) or multicellular (including plants and animals). While the number of cells in plants and animals varies from species to species, humans contain more than 10 trillion (10^{13}) cells.[3] Most plant and animal cells are visible only under the microscope, with dimensions between 1 and 100 micrometres.[4]

The cell was discovered by Robert Hooke in 1665, who named the biological unit for its resemblance to cells inhabited by Christian monks in a monastery.[5][6] Cell theory, first developed in 1839 by Matthias Jakob Schleiden and Theodor Schwann, states that all organisms are composed of one or more cells, that cells are the fundamental unit of structure and function in all living organisms, that

all cells come from preexisting cells, and that all cells contain the hereditary information necessary for regulating cell functions and for transmitting information to the next generation of cells.[7] Cells emerged on Earth at least 3.5 billion years ago.[8][9][10]

3.1 Anatomy

Cells are of two types, eukaryotic, which contain a nucleus, and prokaryotic, which do not. Prokaryotes are single-celled organisms, while eukaryotes can be either single-celled or multicellular.

3.1.1 Prokaryotic cells

Main article: Prokaryote
Prokaryotic cells were the first form of life on Earth, char-

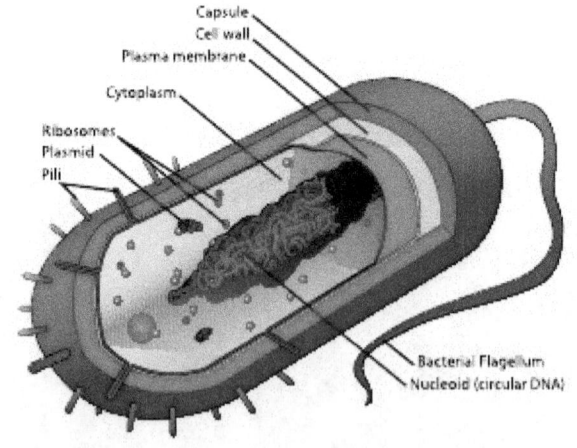

Structure of a typical prokaryotic cell

acterised by having vital biological processes including cell signaling and being self-sustaining. They are simpler and smaller than eukaryotic cells, and lack membrane-bound organelles such as the nucleus. Prokaryotes include two of the domains of life, bacteria and archaea. The DNA of a prokaryotic cell consists of a single chromosome that is in direct contact with the cytoplasm. The nuclear region in the cytoplasm is called the nucleoid. Most prokaryotes are the smallest of all organisms ranging from 0.5 to 2.0 μm in diameter.[12]

A prokaryotic cell has three architectural regions:

- Enclosing the cell is the cell envelope – generally consisting of a plasma membrane covered by a cell wall which, for some bacteria, may be further covered by a third layer called a capsule. Though most prokaryotes have both a cell membrane and a cell wall, there are exceptions such as *Mycoplasma* (bacteria) and *Thermoplasma* (archaea) which only possess the cell membrane layer. The envelope gives rigidity to the cell and separates the interior of the cell from its environment, serving as a protective filter. The cell wall consists of peptidoglycan in bacteria, and acts as an additional barrier against exterior forces. It also prevents the cell from expanding and bursting (cytolysis) from osmotic pressure due to a hypotonic environment. Some eukaryotic cells (plant cells and fungal cells) also have a cell wall.

- Inside the cell is the cytoplasmic region that contains the genome (DNA), ribosomes and various sorts of inclusions. The genetic material is freely found in the cytoplasm. Prokaryotes can carry extrachromosomal DNA elements called plasmids, which are usually circular. Linear bacterial plasmids have been identified in several species of spirochete bacteria, including members of the genus Borrelia notably *Borrelia burgdorferi*, which causes Lyme disease.[13] Though not forming a *nucleus*, the DNA is condensed in a *nucleoid*. Plasmids encode additional genes, such as antibiotic resistance genes.

- On the outside, flagella and pili project from the cell's surface. These are structures (not present in all prokaryotes) made of proteins that facilitate movement and communication between cells.

3.1.2   Eukaryotic cells

Main article: Eukaryote

Plants, animals, fungi, slime moulds, protozoa, and algae are all eukaryotic. These cells are about fifteen times wider than a typical prokaryote and can be as much as a thousand times greater in volume. The main distinguishing feature of eukaryotes as compared to prokaryotes is compartmentalization: the presence of membrane-bound

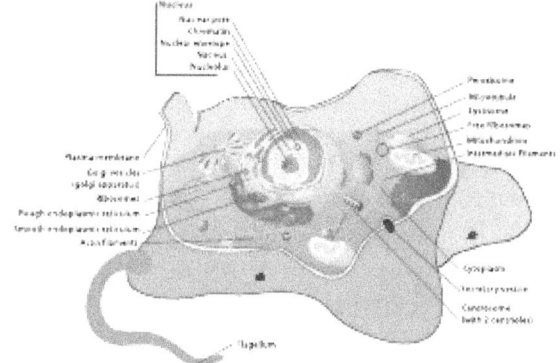

Structure of a typical animal cell

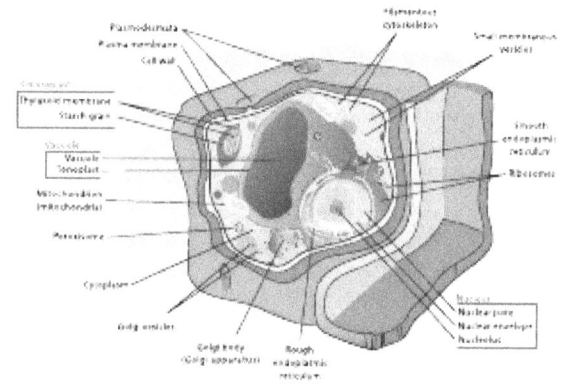

Structure of a typical plant cell

organelles (compartments) in which specific metabolic activities take place. Most important among these is a cell nucleus, an organelle that houses the cell's DNA. This nucleus gives the eukaryote its name, which means "true kernel (nucleus)". Other differences include:

- The plasma membrane resembles that of prokaryotes in function, with minor differences in the setup. Cell walls may or may not be present.

- The eukaryotic DNA is organized in one or more linear molecules, called chromosomes, which are associated with histone proteins. All chromosomal DNA is stored in the *cell nucleus*, separated from the cytoplasm by a membrane. Some eukaryotic organelles such as mitochondria also contain some DNA.

- Many eukaryotic cells are ciliated with *primary cilia*. Primary cilia play important roles in chemosensation, mechanosensation, and thermosensation. Cilia may thus be "viewed as a sensory cellular antennae that coordinates a large number of cellular signaling pathways, sometimes coupling the signaling to ciliary motility or alternatively to cell division and differentiation."[14]

- Motile cells of eukaryotes can move using *motile cilia* or *flagella*. Motile cells are absent in conifers and flowering plants.[15] Eukaryotic flagella are less complex than those of prokaryotes.

3.2 Subcellular components

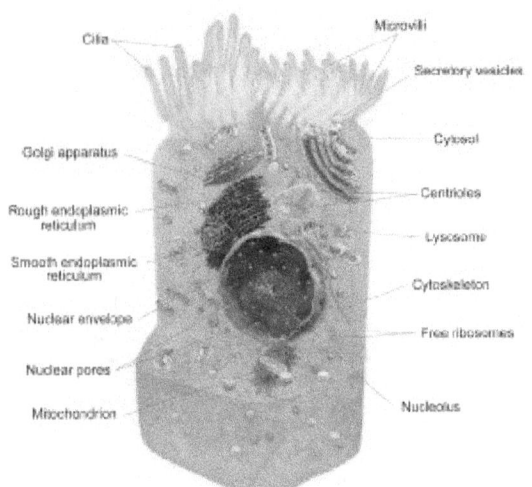

Anatomy of a Cell

Illustration depicting major structures inside a eukaryotic animal cell

All cells, whether prokaryotic or eukaryotic, have a membrane that envelops the cell, regulates what moves in and out (selectively permeable), and maintains the electric potential of the cell. Inside the membrane, the cytoplasm takes up most of the cell's volume. All cells (except red blood cells which lack a cell nucleus and most organelles to accommodate maximum space for hemoglobin) possess DNA, the hereditary material of genes, and RNA, containing the information necessary to build various proteins such as enzymes, the cell's primary machinery. There are also other kinds of biomolecules in cells. This article lists these primary components of the cell, then briefly describes their function.

3.2.1 Membrane

Main article: Cell membrane

The cell membrane, or plasma membrane, is a biological membrane that surrounds the cytoplasm of a cell. In animals, the plasma membrane is the outer boundary of the cell, while in plants and prokaryotes it is usually covered by a cell wall. This membrane serves to separate and protect a cell from its surrounding environment and is made mostly from a double layer of phospholipids, which are amphiphilic (partly hydrophobic and partly hydrophilic). Hence, the layer is called a phospholipid bilayer, or sometimes a fluid mosaic membrane. Embedded within this membrane is a variety of protein molecules that act as channels and pumps that move different molecules into and out of the cell. The membrane is said to be 'semi-permeable', in that it can either let a substance (molecule or ion) pass through freely, pass through to a limited extent or not pass through at all. Cell surface membranes also contain receptor proteins that allow cells to detect external signaling molecules such as hormones.

3.2.2 Cytoskeleton

Main article: Cytoskeleton

The cytoskeleton acts to organize and maintain the

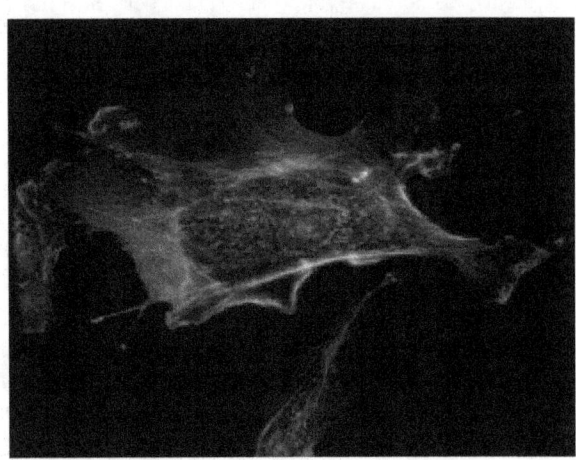

A fluorescent image of an endothelial cell. Nuclei are stained blue, mitochondria are stained red, and microfilaments are stained green.

cell's shape; anchors organelles in place; helps during endocytosis, the uptake of external materials by a cell, and cytokinesis, the separation of daughter cells after cell division; and moves parts of the cell in processes of growth and mobility. The eukaryotic cytoskeleton is composed of microfilaments, intermediate filaments and microtubules. There are a great number of proteins associated with them, each controlling a cell's structure by directing, bundling, and aligning filaments. The prokaryotic cytoskeleton is less well-studied but is involved in the maintenance of cell shape, polarity and cytokinesis.[16] The subunit protein of microfilaments is a small, monomeric protein called actin.

The subunit of microtubules is a dimeric molecule called tubulin. Intermediate filaments are heteropolymers whose subunits vary among the cell types in different tissues. But some of the subunit protein of intermediate filaments include vimentin, desmin, lamin (lamins A, B and C), keratin (multiple acidic and basic keratins), neurofilament proteins (NF - L, NF - M).

3.2.3   Genetic material

Two different kinds of genetic material exist: deoxyribonucleic acid (DNA) and ribonucleic acid (RNA). Cells use DNA for their long-term information storage. The biological information contained in an organism is encoded in its DNA sequence. RNA is used for information transport (e.g., mRNA) and enzymatic functions (e.g., ribosomal RNA). Transfer RNA (tRNA) molecules are used to add amino acids during protein translation.

Prokaryotic genetic material is organized in a simple circular DNA molecule (the bacterial chromosome) in the nucleoid region of the cytoplasm. Eukaryotic genetic material is divided into different, linear molecules called chromosomes inside a discrete nucleus, usually with additional genetic material in some organelles like mitochondria and chloroplasts (see endosymbiotic theory).

A human cell has genetic material contained in the cell nucleus (the nuclear genome) and in the mitochondria (the mitochondrial genome). In humans the nuclear genome is divided into 46 linear DNA molecules called chromosomes, including 22 homologous chromosome pairs and a pair of sex chromosomes. The mitochondrial genome is a circular DNA molecule distinct from the nuclear DNA. Although the mitochondrial DNA is very small compared to nuclear chromosomes, it codes for 13 proteins involved in mitochondrial energy production and specific tRNAs.

Foreign genetic material (most commonly DNA) can also be artificially introduced into the cell by a process called transfection. This can be transient, if the DNA is not inserted into the cell's genome, or stable, if it is. Certain viruses also insert their genetic material into the genome.

3.2.4   Organelles

Main article: Organelle

Organelles are parts of the cell which are adapted and/or specialized for carrying out one or more vital functions, analogous to the organs of the human body (such as the heart, lung, and kidney, with each organ performing a different function). Both eukaryotic and prokaryotic cells have organelles, but prokaryotic organelles are generally simpler and are not membrane-bound.

There are several types of organelles in a cell. Some (such as the nucleus and golgi apparatus) are typically solitary, while others (such as mitochondria, chloroplasts, peroxisomes and lysosomes) can be numerous (hundreds to thousands). The cytosol is the gelatinous fluid that fills the cell and surrounds the organelles.

Eukaryotic

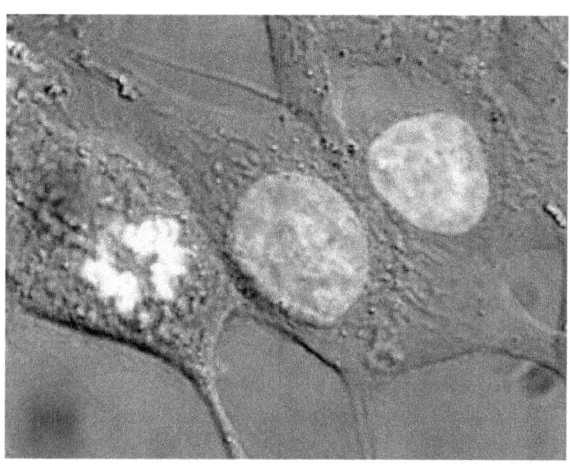

Human cancer cells with nuclei (specifically the DNA) stained blue. The central and rightmost cell are in interphase, so the entire nuclei are labeled. The cell on the left is going through mitosis and its DNA has condensed.

- **Cell nucleus**: A cell's information center, the cell nucleus is the most conspicuous organelle found in a eukaryotic cell. It houses the cell's chromosomes, and is the place where almost all DNA replication and RNA synthesis (transcription) occur. The nucleus is spherical and separated from the cytoplasm by a double membrane called the nuclear envelope. The nuclear envelope isolates and protects a cell's DNA from various molecules that could accidentally damage its structure or interfere with its processing. During processing, DNA is transcribed, or copied into a special RNA, called messenger RNA (mRNA). This mRNA is then transported out of the nucleus, where it is translated into a specific protein molecule. The nucleolus is a specialized region within the nucleus where ribosome subunits are assembled. In prokaryotes, DNA processing takes place in the cytoplasm.

- **Mitochondria and Chloroplasts**: generate energy for the cell. Mitochondria are self-replicating organelles that occur in various numbers, shapes, and sizes in the cytoplasm of all eukaryotic cells.

Respiration occurs in the cell mitochondria, which generate the cell's energy by oxidative phosphorylation, using oxygen to release energy stored in cellular nutrients (typically pertaining to glucose) to generate ATP. Mitochondria multiply by binary fission, like prokaryotes. Chloroplasts can only be found in plants and algae, and they capture the sun's energy to make carbohydrates through photosynthesis.

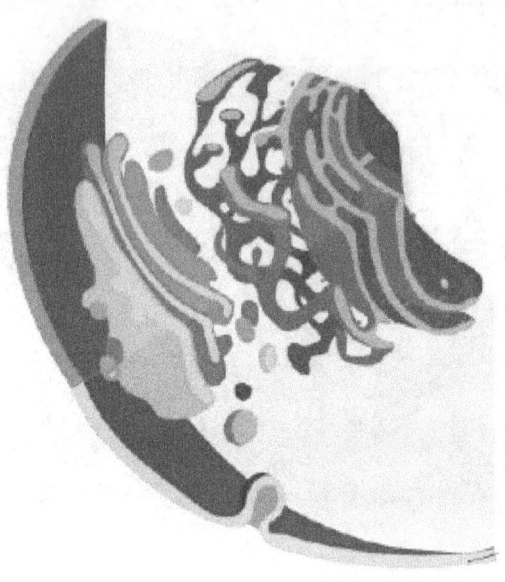

Diagram of an endomembrane system

- **Endoplasmic reticulum**: The endoplasmic reticulum (ER) is a transport network for molecules targeted for certain modifications and specific destinations, as compared to molecules that float freely in the cytoplasm. The ER has two forms: the rough ER, which has ribosomes on its surface that secrete proteins into the ER, and the smooth ER, which lacks ribosomes. The smooth ER plays a role in calcium sequestration and release.

- **Golgi apparatus**: The primary function of the Golgi apparatus is to process and package the macromolecules such as proteins and lipids that are synthesized by the cell.

- **Lysosomes and Peroxisomes**: Lysosomes contain digestive enzymes (acid hydrolases). They digest excess or worn-out organelles, food particles, and engulfed viruses or bacteria. Peroxisomes have enzymes that rid the cell of toxic peroxides. The cell could not house these destructive enzymes if they were not contained in a membrane-bound system.

- **Centrosome**: the cytoskeleton organiser: The centrosome produces the microtubules of a cell – a key component of the cytoskeleton. It directs the transport through the ER and the Golgi apparatus. Centrosomes are composed of two centrioles, which separate during cell division and help in the formation of the mitotic spindle. A single centrosome is present in the animal cells. They are also found in some fungi and algae cells.

- **Vacuoles**: Vacuoles sequester waste products and in plant cells store water. They are often described as liquid filled space and are surrounded by a membrane. Some cells, most notably *Amoeba*, have contractile vacuoles, which can pump water out of the cell if there is too much water. The vacuoles of plant cells and fungal cells are usually larger than those of animal cells.

Eukaryotic and prokaryotic

- **Ribosomes**: The ribosome is a large complex of RNA and protein molecules. They each consist of two subunits, and act as an assembly line where RNA from the nucleus is used to synthesise proteins from amino acids. Ribosomes can be found either floating freely or bound to a membrane (the rough endoplasmatic reticulum in eukaryotes, or the cell membrane in prokaryotes).[17]

3.3 Structures outside the cell membrane

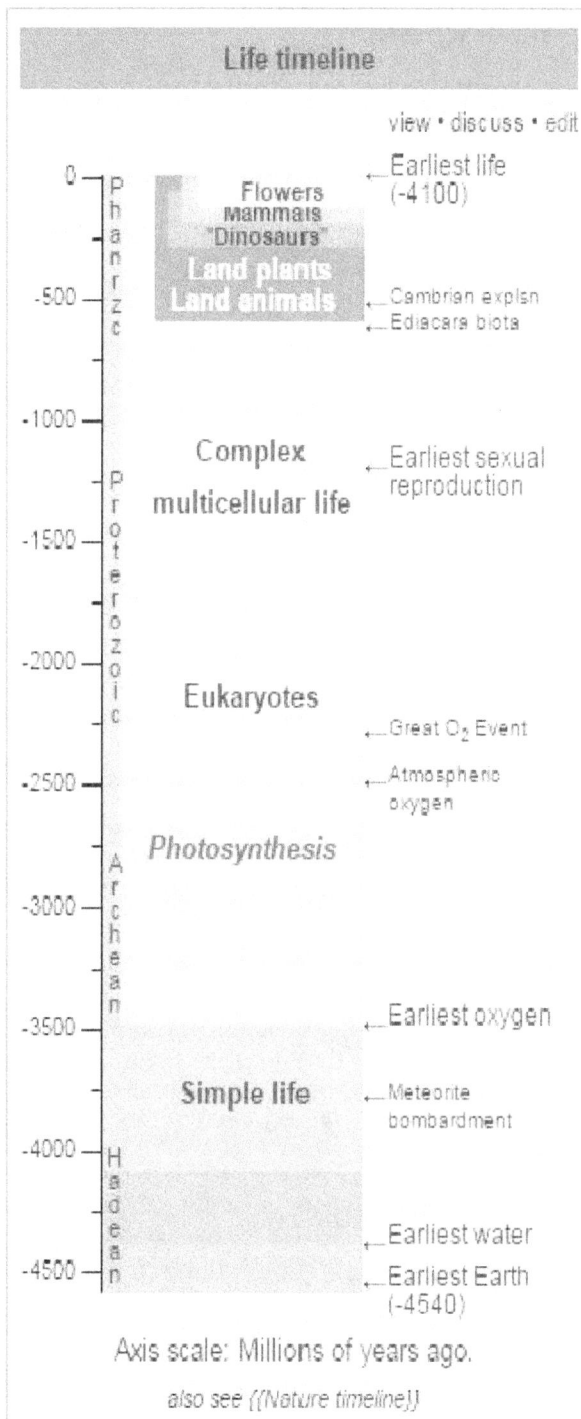

Many cells also have structures which exist wholly or partially outside the cell membrane. These structures are notable because they are not protected from the external environment by the semipermeable cell membrane. In order to assemble these structures, their components must be carried across the cell membrane by export processes.

3.3.1   Cell wall

Many types of prokaryotic and eukaryotic cells have a cell wall. The cell wall acts to protect the cell mechanically and chemically from its environment, and is an additional layer of protection to the cell membrane. Different types of cell have cell walls made up of different materials; plant cell walls are primarily made up of cellulose, fungi cell walls are made up of chitin and bacteria cell walls are made up of peptidoglycan.

3.3.2   Prokaryotic

Capsule

A gelatinous capsule is present in some bacteria outside the cell membrane and cell wall. The capsule may be polysaccharide as in pneumococci, meningococci or polypeptide as *Bacillus anthracis* or hyaluronic acid as in streptococci. Capsules are not marked by normal staining protocols and can be detected by India ink or methyl blue; which allows for higher contrast between the cells for observation.[18]:87

Flagella

Flagella are organelles for cellular mobility. The bacterial flagellum stretches from cytoplasm through the cell membrane(s) and extrudes through the cell wall. They are long and thick thread-like appendages, protein in nature. A different type of flagellum is found in archaea and a different type is found in eukaryotes.

Fimbria

A fimbria also known as a pilus is a short, thin, hair-like filament found on the surface of bacteria. Fimbriae, or pili are formed of a protein called pilin (antigenic) and are responsible for attachment of bacteria to specific receptors of human cell (cell adhesion). There are special types of specific pili involved in bacterial conjugation.

3.4 Cellular processes

3.4.1 Growth and metabolism

Main articles: Cell growth and Metabolism

Between successive cell divisions, cells grow through the functioning of cellular metabolism. Cell metabolism is the process by which individual cells process nutrient molecules. Metabolism has two distinct divisions: catabolism, in which the cell breaks down complex molecules to produce energy and reducing power, and anabolism, in which the cell uses energy and reducing power to construct complex molecules and perform other biological functions. Complex sugars consumed by the organism can be broken down into simpler sugar molecules called monosaccharides such as glucose. Once inside the cell, glucose is broken down to make adenosine triphosphate (ATP), a molecule that possesses readily available energy, through two different pathways.

3.4.2 Replication

Main article: Cell division

Cell division involves a single cell (called a *mother cell*) dividing into two daughter cells. This leads to growth in multicellular organisms (the growth of tissue) and to procreation (vegetative reproduction) in unicellular organisms. Prokaryotic cells divide by binary fission, while eukaryotic cells usually undergo a process of nuclear division, called mitosis, followed by division of the cell, called cytokinesis.

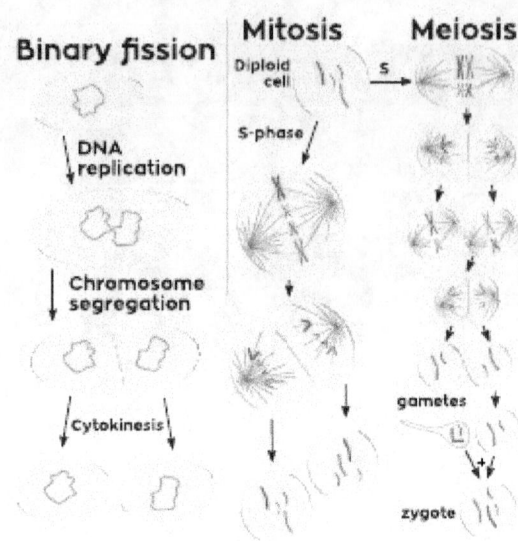

Bacteria divide by binary fission, while eukaryotes divide by mitosis or meiosis.

A diploid cell may also undergo meiosis to produce haploid cells, usually four. Haploid cells serve as gametes in multicellular organisms, fusing to form new diploid cells.

DNA replication, or the process of duplicating a cell's genome, always happens when a cell divides through mitosis or binary fission. This occurs during the S phase of the cell cycle.

In meiosis, the DNA is replicated only once, while the cell divides twice. DNA replication only occurs before meiosis I. DNA replication does not occur when the cells divide the second time, in meiosis II.[19] Replication, like all cellular activities, requires specialized proteins for carrying out the job.

3.4.3 Protein synthesis

Main article: Protein biosynthesis

Cells are capable of synthesizing new proteins, which are essential for the modulation and maintenance of cellular activities. This process involves the formation of new protein molecules from amino acid building blocks based on information encoded in DNA/RNA. Protein synthesis generally consists of two major steps: transcription and translation.

Transcription is the process where genetic information in DNA is used to produce a complementary RNA strand. This RNA strand is then processed to give messenger RNA (mRNA), which is free to migrate through the cell. mRNA molecules bind to protein-RNA complexes called

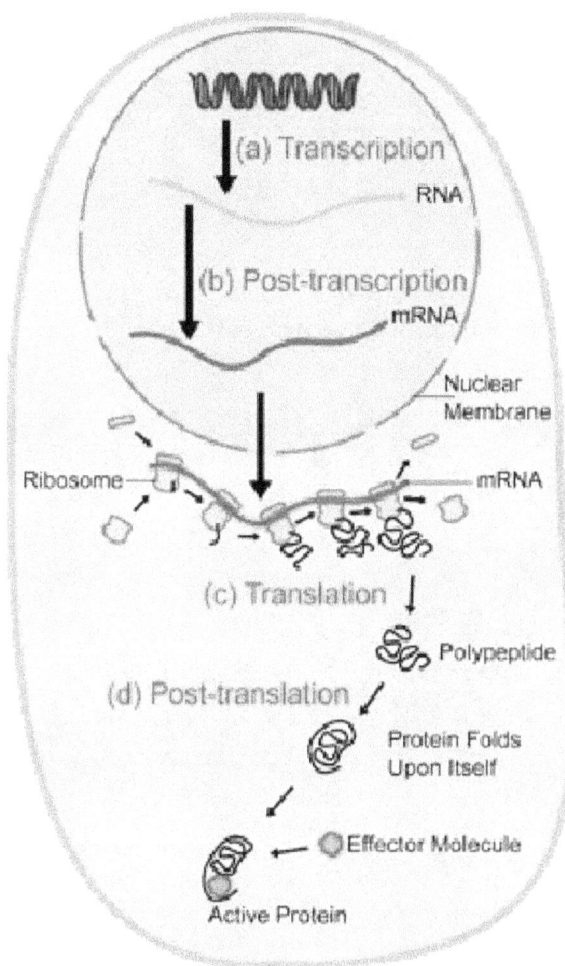

An overview of protein synthesis.

Within the nucleus of the cell (light blue), genes (DNA, dark blue) are transcribed into RNA. This RNA is then subject to post-transcriptional modification and control, resulting in a mature mRNA (red) that is then transported out of the nucleus and into the cytoplasm (peach), where it undergoes translation into a protein. mRNA is translated by ribosomes (purple) that match the three-base codons of the mRNA to the three-base anti-codons of the appropriate tRNA. Newly synthesized proteins (black) are often further modified, such as by binding to an effector molecule (orange), to become fully active.

ribosomes located in the cytosol, where they are translated into polypeptide sequences. The ribosome mediates the formation of a polypeptide sequence based on the mRNA sequence. The mRNA sequence directly relates to the polypeptide sequence by binding to transfer RNA (tRNA) adapter molecules in binding pockets within the ribosome. The new polypeptide then folds into a functional three-dimensional protein molecule.

3.4.4   Movement or motility

Main article: Motility

Unicellular organisms can move in order to find food or escape predators. Common mechanisms of motion include flagella and cilia.

In multicellular organisms, cells can move during processes such as wound healing, the immune response and cancer metastasis. For example, in wound healing in animals, white blood cells move to the wound site to kill the microorganisms that cause infection. Cell motility involves many receptors, crosslinking, bundling, binding, adhesion, motor and other proteins.[20] The process is divided into three steps – protrusion of the leading edge of the cell, adhesion of the leading edge and de-adhesion at the cell body and rear, and cytoskeletal contraction to pull the cell forward. Each step is driven by physical forces generated by unique segments of the cytoskeleton.[21][22]

3.5   Multicellularity

Main article: Multicellular organism

3.5.1   Cell specialization

Multicellular organisms are organisms that consist of more than one cell, in contrast to single-celled organisms.[23]

In complex multicellular organisms, cells specialize into different cell types that are adapted to particular functions. In mammals, major cell types include skin cells, muscle cells, neurons, blood cells, fibroblasts, stem cells, and others. Cell types differ both in appearance and function, yet are genetically identical. Cells are able to be of the same genotype but of different cell type due to the differential expression of the genes they contain.

Most distinct cell types arise from a single totipotent cell, called a zygote, that differentiates into hundreds of different cell types during the course of development. Differentiation of cells is driven by different environmental cues (such as cell–cell interaction) and intrinsic differences (such as those caused by the uneven distribution of molecules during division).

3.5.2   Origin of multicellularity

Multicellularity has evolved independently at least 25 times,[24] including in some prokaryotes,

3.6 Origins

Main article: Evolutionary history of life

The origin of cells has to do with the origin of life, which began the history of life on Earth.

3.6.1 Origin of the first cell

Stromatolites are left behind by cyanobacteria, also called blue-green algae. They are the oldest known fossils of life on Earth. This one-billion-year-old fossil is from Glacier National Park in the United States.

Further information: Abiogenesis and Evolution of cells

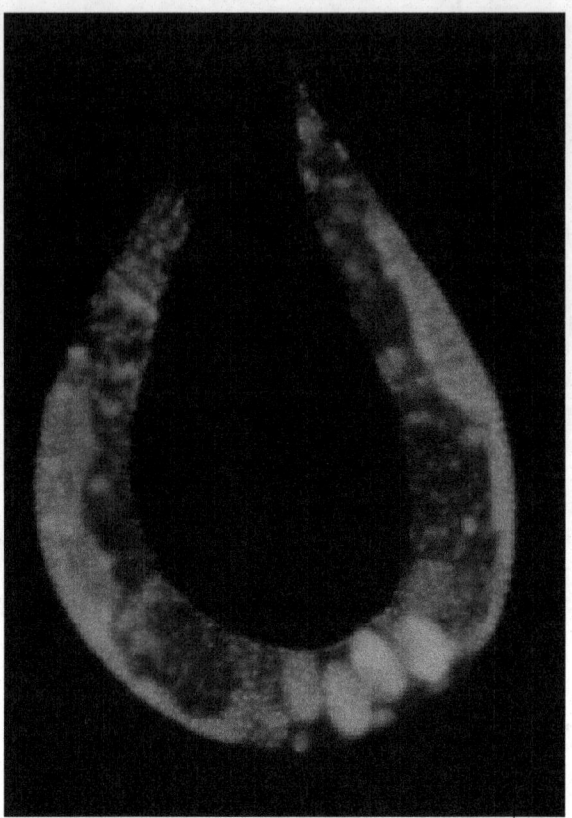

Staining of a Caenorhabditis elegans *which highlights the nuclei of its cells.*

like cyanobacteria, myxobacteria, actinomycetes, *Magnetoglobus multicellularis* or *Methanosarcina*. However, complex multicellular organisms evolved only in six eukaryotic groups: animals, fungi, brown algae, red algae, green algae, and plants.[25] It evolved repeatedly for plants (Chloroplastida), once or twice for animals, once for brown algae, and perhaps several times for fungi, slime molds, and red algae.[26] Multicellularity may have evolved from colonies of interdependent organisms, from cellularization, or from organisms in symbiotic relationships.

The first evidence of multicellularity is from cyanobacteria-like organisms that lived between 3 and 3.5 billion years ago.[24] Other early fossils of multicellular organisms include the contested Grypania spiralis and the fossils of the black shales of the Palaeoproterozoic Francevillian Group Fossil B Formation in Gabon.[27]

The evolution of multicellularity from unicellular ancestors has been replicated in the laboratory, in evolution experiments using predation as the selective pressure.[24]

There are several theories about the origin of small molecules that led to life on the early Earth. They may have been carried to Earth on meteorites (see Murchison meteorite), created at deep-sea vents, or synthesized by lightning in a reducing atmosphere (see Miller–Urey experiment). There is little experimental data defining what the first self-replicating forms were. RNA is thought to be the earliest self-replicating molecule, as it is capable of both storing genetic information and catalyzing chemical reactions (see RNA world hypothesis), but some other entity with the potential to self-replicate could have preceded RNA, such as clay or peptide nucleic acid.[28]

Cells emerged at least 3.5 billion years ago.[8][9][10] The current belief is that these cells were heterotrophs. The early cell membranes were probably more simple and permeable than modern ones, with only a single fatty acid chain per lipid. Lipids are known to spontaneously form bilayered vesicles in water, and could have preceded RNA, but the first cell membranes could also have been produced by catalytic RNA, or even have required structural proteins before they could form.[29]

3.6.2 Origin of eukaryotic cells

Further information: Evolution of sexual reproduction

The eukaryotic cell seems to have evolved from a symbiotic community of prokaryotic cells. DNA-bearing organelles like the mitochondria and the chloroplasts are descended from ancient symbiotic oxygen-breathing proteobacteria and cyanobacteria, respectively, which were endosymbiosed by an ancestral archaean prokaryote.

There is still considerable debate about whether organelles like the hydrogenosome predated the origin of mitochondria, or vice versa: see the hydrogen hypothesis for the origin of eukaryotic cells.

3.7 History of research

- 1632–1723: Antonie van Leeuwenhoek teaches himself to make lenses, constructs basic optical microscopes and draws protozoa, such as *Vorticella* from rain water, and bacteria from his own mouth.

- 1665: Robert Hooke discovers cells in cork, then in living plant tissue using an early compound microscope. He coins the term *cell* (from Latin *cella*, meaning "small room"[1]) in his book *Micrographia* (1665).[30]

- 1839: Theodor Schwann and Matthias Jakob Schleiden elucidate the principle that plants and animals are made of cells, concluding that cells are a common unit of structure and development, and thus founding the cell theory.

- 1855: Rudolf Virchow states that new cells come from pre-existing cells by cell division (*omnis cellula ex cellula*).

- 1859: The belief that life forms can occur spontaneously (*generatio spontanea*) is contradicted by Louis Pasteur (1822–1895) (although Francesco Redi had performed an experiment in 1668 that suggested the same conclusion).

- 1931: Ernst Ruska builds the first transmission electron microscope (TEM) at the University of Berlin. By 1935, he has built an EM with twice the resolution of a light microscope, revealing previously unresolvable organelles.

- 1953: Watson and Crick made their first announcement on the double helix structure of DNA on February 28.

- 1981: Lynn Margulis published *Symbiosis in Cell Evolution* detailing the endosymbiotic theory.

3.8 See also

- Cell culture
- Cellular component
- Cellular model
- Cytorrhysis
- Cytotoxicity
- Lipid raft
- Plasmolysis
- Stem cell
- Syncytium
- Topic outline of cell biology
- Vault (organelle)

3.9 References

[1] "Cell". Online Etymology Dictionary. Retrieved 31 December 2012.

[2] Cell Movements and the Shaping of the Vertebrate Body in Chapter 21 of *Molecular Biology of the Cell* fourth edition, edited by Bruce Alberts (2002) published by Garland Science.
The Alberts text discusses how the "cellular building blocks" move to shape developing embryos. It is also common to describe small molecules such as amino acids as "molecular building blocks".

[3] Alberts, p. 2.

[4] Campbell, Neil A.; Brad Williamson; Robin J. Heyden (2006). *Biology: Exploring Life*. Boston, Massachusetts: Pearson Prentice Hall. ISBN 0-13-250882-6.

[5] Karp, Gerald (19 October 2009). *Cell and Molecular Biology: Concepts and Experiments*. John Wiley & Sons. p. 2. ISBN 9780470483374. Hooke called the pores cells because they reminded him of the cells inhabited by monks living in a monastery.

[6] Alan Chong Tero (1990). *Achiever's Biology*. Allied Publishers. p. 36. ISBN 9788184243697. In 1665, an Englishman, Robert Hooke observed a thin slice of" cork under a simple microscope. (A simple microscope is a microscope with only one biconvex lens, rather like a magnifying glass). He saw many small box like structures. These reminded him of small rooms called "cells" in which Christian monks lived and meditated.

[7] Maton, Anthea (1997). *Cells Building Blocks of Life*. New Jersey: Prentice Hall. ISBN 0-13-423476-6.

[8] Schopf, JW, Kudryavtsev, AB, Czaja, AD, and Tripathi, AB. (2007). *Evidence of Archean life: Stromatolites and microfossils*. Precambrian Research 158:141-155.

[9] Schopf, JW (2006). *Fossil evidence of Archaean life*. Philos Trans R Soc Lond B Biol Sci 29;361(1470):869-85.

[10] Peter Hamilton Raven; George Brooks Johnson (2002). *Biology*. McGraw-Hill Education. p. 68. ISBN 978-0-07-112261-0. Retrieved 7 July 2013.

[11] *Campbell Biology—Concepts and Connections*. Pearson Education. 2009. p. 320.

[12] *Microbiology : Principles and Explorations* By Jacquelyn G. Black

[13] European Bioinformatics Institute, Karyn's Genomes: Borrelia burgdorferi, part of 2can on the EBI-EMBL database. Retrieved 5 August 2012

[14] Satir, Peter; Christensen, ST; Søren T. Christensen (2008-03-26). "Structure and function of mammalian cilia". *Histochemistry and Cell Biology* (Springer Berlin/Heidelberg) **129** (6): 687–693. doi:10.1007/s00418-008-0416-9. PMC 2386530. PMID 18365235. 1432-119X. Retrieved 2009-09-12.

[15] PH Raven , Evert RF, Eichhorm SE (1999) Biology of Plants, 6th edition. WH Freeman, New York

[16] Michie K, Löwe J (2006). "Dynamic filaments of the bacterial cytoskeleton". *Annu Rev Biochem* **75**: 467–92. doi:10.1146/annurev.biochem.75.103004.142452. PMID 16756499.

[17] Ménétret JF, Schaletzky J, Clemons WM, CW; Akey; et al. (December 2007). "Ribosome binding of a single copy of the SecY complex: implications for protein translocation". *Mol. Cell* **28** (6): 1083–92. doi:10.1016/j.molcel.2007.10.034. PMID 18158904.

[18] *Prokaryotes*. Newnes. Apr 11, 1996. ISBN 9780080984735.

[19] *Campbell Biology—Concepts and Connections*. Pearson Education. 2009. p. 138.

[20] Revathi Ananthakrishnan1 *, Allen Ehrlicher2 ⊕. "The Forces Behind Cell Movement". Biolsci.org. Retrieved 2009-04-17.

[21] Alberts B, Johnson A, Lewis J. et al. Molecular Biology of the Cell, 4e. Garland Science. 2002

[22] Ananthakrishnan, R; Ehrlicher, A (2007). "The Forces Behind Cell Movement". *Int J Biol Sci* **3** (5): 303–317. doi:10.7150/ijbs.3.303. PMC 1893118. PMID 17589565.

[23] Becker, Wayne M.; et al. (2009). *The world of the cell*. Pearson Benjamin Cummings. p. 480. ISBN 978-0-321-55418-5.

[24] Grosberg RK, Strathmann RR. The evolution of multicellularity: A minor major transition? Annu Rev Ecol Evol Syst. 2007;38:621–654.

[25] http://public.wsu.edu/~{}lange-m/Documnets/Teaching2011/Popper2011.pdf

[26] Bonner, John Tyler (1998). "The Origins of Multicellularity" (PDF). *Integrative Biology: Issues, News, and Reviews* **1** (1): 27–36. doi:10.1002/(SICI)1520-6602(1998)1:1<27::AID-INBI4>3.0.CO;2-6. ISSN 1093-4391. Archived from the original (PDF, 0.2 MB) on March 8, 2012.

[27] El Albani A, Bengtson S, Canfield DE, Bekker A, Macchiarelli R, Mazurier A, Hammarlund EU, Boulvais P, Dupuy JJ, Fontaine C, Fürsich FT, Gauthier-Lafaye F, Janvier P, Javaux E, Ossa FO, Pierson-Wickmann AC, Riboulleau A, Sardini P, Vachard D, Whitehouse M, Meunier A (1 July 2010). "Large colonial organisms with coordinated growth in oxygenated environments 2.1 Gyr ago". *Nature* **466** (7302): 100–104. Bibcode:2010Natur.466..100A. doi:10.1038/nature09166. ISSN 0028-0836. PMID 20596019.

[28] Orgel LE (1998). "The origin of life--a review of facts and speculations". *Trends Biochem Sci* **23** (12): 491–5. doi:10.1016/S0968-0004(98)01300-0. PMID 9868373.

[29] Griffiths G (December 2007). "Cell evolution and the problem of membrane topology". *Nature reviews. Molecular cell biology* **8** (12): 1018–24. doi:10.1038/nrm2287. PMID 17971839.

[30] "... I could exceedingly plainly perceive it to be all perforated and porous, much like a Honey-comb, but that the pores of it were not regular [..] these pores, or cells, [..] were indeed the first microscopical pores I ever saw, and perhaps, that were ever seen, for I had not met with any Writer or Person, that had made any mention of them before this. . ." – Hooke describing his observations on a thin slice of cork. Robert Hooke

- This article incorporates public domain material from the NCBI document "Science Primer".

3.10 Bibliography

- Alberts, Bruce; Johnson, Alexander; Lewis, Julian; Morgan, David; Raff, Martin; Roberts, Keith; Walter, Peter (2015). *Molecular Biology of the Cell* (6th ed.). Garland Science. p. 2. ISBN 978-0815344322.

3.11 External links

- MBInfo - Descriptions on Cellular Functions and Processes

- MBInfo - Cellular Organization

- Inside the Cell - a science education booklet by National Institutes of Health, in PDF and ePub.

- Cells Alive!

- Cell Biology in "The Biology Project" of University of Arizona.

- Centre of the Cell online

- The Image & Video Library of The American Society for Cell Biology, a collection of peer-reviewed still images, video clips and digital books that illustrate the structure, function and biology of the cell.

- HighMag Blog, still images of cells from recent research articles.

- New Microscope Produces Dazzling 3D Movies of Live Cells, March 4, 2011 - Howard Hughes Medical Institute.

- WormWeb.org: Interactive Visualization of the *C. elegans* Cell lineage - Visualize the entire cell lineage tree of the nematode *C. elegans*

3.11.1 Textbooks

- Alberts B, Johnson A, Lewis J, Raff M, Roberts K, Walter P (2002). *Molecular Biology of the Cell* (4th ed.). Garland. ISBN 0-8153-3218-1.

- Lodish H, Berk A, Matsudaira P, Kaiser CA, Krieger M, Scott MP, Zipurksy SL, Darnell J (2004). *Molecular Cell Biology* (5th ed.). WH Freeman: New York, NY. ISBN 978-0-7167-4366-8.

- Cooper GM (2000). *The cell: a molecular approach* (2nd ed.). Washington, D.C: ASM Press. ISBN 0-87893-102-3.

Chapter 4

Cellular differentiation

"Cell differentiation" redirects here. For the journal, see Cell Differentiation (journal).

In developmental biology, **cellular differentiation** is the process of a cell changing from one cell type to another.[1][2] Most commonly this is a less specialized type becoming a more specialized type, such as during cell growth. Differentiation occurs numerous times during the development of a multicellular organism as it changes from a simple zygote to a complex system of tissues and cell types. Differentiation continues in adulthood as adult stem cells divide and create fully differentiated daughter cells during tissue repair and during normal cell turnover. Some differentiation occurs in response to antigen exposure. Differentiation dramatically changes a cell's size, shape, membrane potential, metabolic activity, and responsiveness to signals. These changes are largely due to highly controlled modifications in gene expression and are the study of epigenetics. With a few exceptions, cellular differentiation almost never involves a change in the DNA sequence itself. Thus, different cells can have very different physical characteristics despite having the same genome.

A cell that can differentiate into all cell types of the adult organism is known as *pluripotent*. Such cells are called embryonic stem cells in animals and meristematic cells in higher plants. A cell that can differentiate into all cell types, including the placental tissue, is known as *totipotent*. In mammals, only the zygote and subsequent blastomeres are totipotent, while in plants many differentiated cells can become totipotent with simple laboratory techniques. In cytopathology, the level of cellular differentiation is used as a measure of cancer progression. "Grade" is a marker of how differentiated a cell in a tumor is.[3]

4.1 Mammalian cell types

See also: List of distinct cell types in the adult human body

Three basic categories of cells make up the mammalian body: germ cells, somatic cells, and stem cells. Each of the approximately 100 trillion (10^{14}) cells in an adult human has its own copy or copies of the genome except certain cell types, such as red blood cells, that lack nuclei in their fully differentiated state. Most cells are diploid; they have two copies of each chromosome. Such cells, called somatic cells, make up most of the human body, such as skin and muscle cells. Cells differentiate to specialize for different functions.[4]

Germ line cells are any line of cells that give rise to gametes—eggs and sperm—and thus are continuous through the generations. Stem cells, on the other hand, have the ability to divide for indefinite periods and to give rise to specialized cells. They are best described in the context of normal human development.

Development begins when a sperm fertilizes an egg and creates a single cell that has the potential to form an entire organism. In the first hours after fertilization, this cell divides into identical cells. In humans, approximately four days after fertilization and after several cycles of cell division, these cells begin to specialize, forming a hollow sphere of cells, called a blastocyst.[5] The blastocyst has an outer layer of cells, and inside this hollow sphere, there is a cluster of cells called the inner cell mass. The cells of the inner cell mass go on to form virtually all of the tissues of the human body. Although the cells of the inner cell mass can form virtually every type of cell found in the human body, they cannot form an organism. These cells are referred to as pluripotent.[6]

Pluripotent stem cells undergo further specialization into multipotent progenitor cells that then give rise to functional cells. Examples of stem and progenitor cells include:

- *Hematopoietic stem cells* (adult stem cells) from the bone marrow that give rise to red blood cells, white blood cells, and platelets

- *Mesenchymal stem cells* (adult stem cells) from the bone marrow that give rise to stromal cells, fat cells, and types of bone cells

- *Epithelial stem cells* (progenitor cells) that give rise to the various types of skin cells

- *Muscle satellite cells* (progenitor cells) that contribute to differentiated muscle tissue.

A pathway that is guided by the cell adhesion molecules consisting of four amino acids, arginine, glycine, asparagine, and serine, is created as the cellular blastomere differentiates from the single-layered blastula to the three primary layers of germ cells in mammals, namely the ectoderm, mesoderm and endoderm (listed from most distal (exterior) to proximal (interior)). The ectoderm ends up forming the skin and the nervous system, the mesoderm forms the bones and muscular tissue, and the endoderm forms the internal organ tissues.

4.2   Dedifferentiation

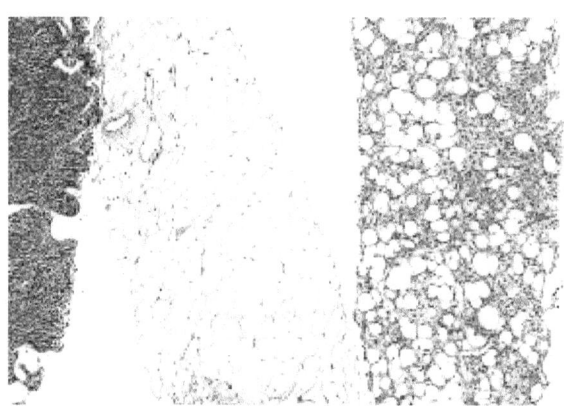

Micrograph of a liposarcoma with some dedifferentiation, that is not identifiable as a liposarcoma, (left edge of image) and a differentiated component (with lipoblasts and increased vascularity (right of image)). Fully differentiated (morphologically benign) adipose tissue (center of the image) has few blood vessels. H&E stain.

Dedifferentiation, or integration is a cellular process often seen in more basal life forms such as worms and amphibians in which a partially or terminally differentiated cell reverts to an earlier developmental stage, usually as part of a regenerative process.[7][8] Dedifferentiation also occurs in plants.[9] Cells in cell culture can lose properties they originally had, such as protein expression, or change shape. This process is also termed dedifferentiation.[10]

Some believe dedifferentiation is an aberration of the normal development cycle that results in cancer,[11] whereas others believe it to be a natural part of the immune response lost by humans at some point as a result of evolution.

A small molecule dubbed reversine, a purine analog, has been discovered that has proven to induce dedifferentiation in myotubes. These dedifferentiated cells could then redifferentiate into osteoblasts and adipocytes.[12]

Diagram exposing several methods used to revert adult somatic cells to totipotency or pluripotency.

4.3   Mechanisms

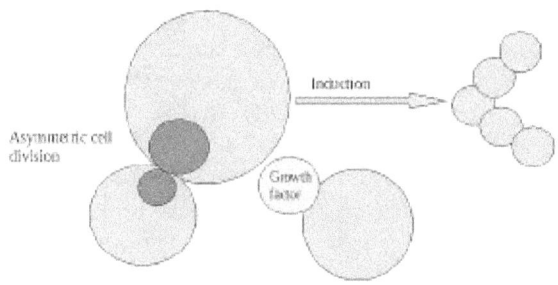

Mechanisms of cellular differentiation.

Each specialized cell type in an organism expresses a subset of all the genes that constitute the genome of that species. Each cell type is defined by its particular pattern of regulated gene expression. Cell differentiation is thus a transition of a cell from one cell type to another and it involves a switch from one pattern of gene expression to another. Cellular differentiation during development can be understood as the result of a gene regulatory network. A regulatory gene and its cis-regulatory modules are nodes in a gene regulatory network; they receive input and create output elsewhere in the network.[13] The systems biology approach to developmental biology emphasizes the importance of investigating how developmental mechanisms interact to produce predictable patterns (morphogenesis).

(However, an alternative view has been proposed recently. Based on stochastic gene expression, cellular differentiation is the result of a Darwinian selective process occurring among cells. In this frame, protein and gene networks are the result of cellular processes and not their cause. See: Cellular Darwinism)

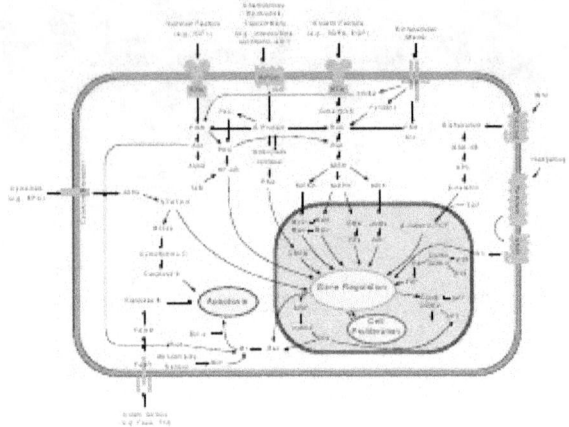

An overview of major signal transduction pathways.

A few evolutionarily conserved types of molecular processes are often involved in the cellular mechanisms that control these switches. The major types of molecular processes that control cellular differentiation involve cell signaling. Many of the signal molecules that convey information from cell to cell during the control of cellular differentiation are called growth factors. Although the details of specific signal transduction pathways vary, these pathways often share the following general steps. A ligand produced by one cell binds to a receptor in the extracellular region of another cell, inducing a conformational change in the receptor. The shape of the cytoplasmic domain of the receptor changes, and the receptor acquires enzymatic activity. The receptor then catalyzes reactions that phosphorylate other proteins, activating them. A cascade of phosphorylation reactions eventually activates a dormant transcription factor or cytoskeletal protein, thus contributing to the differentiation process in the target cell.[14] Cells and tissues can vary in competence, their ability to respond to external signals.[15]

Signal induction refers to cascades of signaling events, during which a cell or tissue signals to another cell or tissue to influence its developmental fate.[15] Yamamoto and Jeffery[16] investigated the role of the lens in eye formation in cave- and surface-dwelling fish, a striking example of induction.[15] Through reciprocal transplants, Yamamoto and Jeffery[16] found that the lens vesicle of surface fish can induce other parts of the eye to develop in cave- and surface-dwelling fish, while the lens vesicle of the cave-dwelling fish cannot.[15]

Other important mechanisms fall under the category of asymmetric cell divisions, divisions that give rise to daughter cells with distinct developmental fates. Asymmetric cell divisions can occur because of asymmetrically expressed maternal **cytoplasmic determinants** or because of signaling.[15] In the former mechanism, distinct daughter cells are created during cytokinesis because of an uneven distribution of regulatory molecules in the parent cell; the distinct cytoplasm that each daughter cell inherits results in a distinct pattern of differentiation for each daughter cell. A well-studied example of pattern formation by asymmetric divisions is body axis patterning in Drosophila. RNA molecules are an important type of intracellular differentiation control signal. The molecular and genetic basis of asymmetric cell divisions has also been studied in green algae of the genus *Volvox*, a model system for studying how unicellular organisms can evolve into multicellular organisms.[15] In *Volvox carteri*, the 16 cells in the anterior hemisphere of a 32-cell embryo divide asymmetrically, each producing one large and one small daughter cell. The size of the cell at the end of all cell divisions determines whether it becomes a specialized germ or somatic cell.[15][17]

4.3.1 Epigenetic control of cellular differentiation

Main article: Epigenetics in stem cell differentiation

Since each cell, regardless of cell type, possesses the same genome, determination of cell type must occur at the level of gene expression. While the regulation of gene expression can occur through cis- and trans-regulatory elements including a gene's promoter and enhancers, the problem arises as to how this expression pattern is maintained over numerous generations of cell division. As it turns out, epigenetic processes play a crucial role in regulating the decision to adopt a stem, progenitor, or mature cell fate. This section will focus primarily on mammalian stem cells.

In systems biology and mathematical modeling of gene regulatory networks, cell-fate determination is predicted to exhibit certain dynamics, such as attractor-convergence (the attractor can be an equilibrium point, limit cycle or strange attractor) or oscillatory.[18]

Importance of epigenetic control

The first question that can be asked is the extent and complexity of the role of epigenetic processes in the determination of cell fate. A clear answer to this question can be seen in the 2011 paper by Lister R, *et al.* [19] on aberrant epigenomic programming in human induced pluripotent stem cells. As induced pluripotent stem cells (iPSCs) are thought to mimic embryonic stem cells in their pluripotent properties, few epigenetic differences should exist between them. To test this prediction, the authors conducted whole-genome profiling of DNA methylation patterns in several human embryonic stem cell (ESC), iPSC, and progenitor cell lines.

Female adipose cells, lung fibroblasts, and foreskin fibroblasts were reprogrammed into induced pluripotent state with the OCT4, SOX2, KLF4, and MYC genes. Patterns of DNA methylation in ESCs, iPSCs, somatic cells were compared. Lister R, et al. observed significant resemblance in methylation levels between embryonic and induced pluripotent cells. Around 80% of CG dinucleotides in ESCs and iPSCs were methylated, the same was true of only 60% of CG dinucleotides in somatic cells. In addition, somatic cells possessed minimal levels of cytosine methylation in non-CG dinucleotides, while induced pluripotent cells possessed similar levels of methylation as embryonic stem cells, between 0.5 and 1.5%. Thus, consistent with their respective transcriptional activities,[19] DNA methylation patterns, at least on the genomic level, are similar between ESCs and iPSCs.

However, upon examining methylation patterns more closely, the authors discovered 1175 regions of differential CG dinucleotide methylation between at least one ES or iPS cell line. By comparing these regions of differential methylation with regions of cytosine methylation in the original somatic cells, 44-49% of differentially methylated regions reflected methylation patterns of the respective progenitor somatic cells, while 51-56% of these regions were dissimilar to both the progenitor and embryonic cell lines. In vitro-induced differentiation of iPSC lines saw transmission of 88% and 46% of hyper and hypo-methylated differentially methylated regions, respectively.

Two conclusions are readily apparent from this study. First, epigenetic processes are heavily involved in cell fate determination, as seen from the similar levels of cytosine methylation between induced pluripotent and embryonic stem cells, consistent with their respective patterns of transcription. Second, the mechanisms of de-differentiation (and by extension, differentiation) are very complex and cannot be easily duplicated, as seen by the significant number of differentially methylated regions between ES and iPS cell lines. Now that these two points have been established, we can examine some of the epigenetic mechanisms that are thought to regulate cellular differentiation.

Mechanisms of epigenetic regulation

Pioneering factors (Oct4, Sox2, Nanog) Three transcription factors, OCT4, SOX2, and NANOG – the first

two of which are used in iPSC reprogramming – are highly expressed in undifferentiated embryonic stem cells and are necessary for the maintenance of their pluripotency.[20] It is thought that they achieve this through alterations in chromatin structure, such as histone modification and DNA methylation, to restrict or permit the transcription of target genes.

Polycomb repressive complex (PRC2) In the realm of gene silencing, Polycomb repressive complex 2, one of two classes of the Polycomb group (PcG) family of proteins, catalyzes the di- and tri-methylation of histone H3 lysine 27 (H3K27me2/me3).[20][21][22] By binding to the H3K27me2/3-tagged nucleosome, PRC1 (also a complex of PcG family proteins) catalyzes the mono-ubiquitinylation of histone H2A at lysine 119 (H2AK119Ub1), blocking RNA polymerase II activity and resulting in transcriptional suppression.[20] PcG knockout ES cells do not differentiate efficiently into the three germ layers, and deletion of the PRC1 and PRC2 genes leads to increased expression of lineage-affiliated genes and unscheduled differentiation.[20] Presumably, PcG complexes are responsible for transcriptionally repressing differentiation and development-promoting genes.

Trithorax group proteins (TrxG) Alternately, upon receiving differentiation signals, PcG proteins are recruited to promoters of pluripotency transcription factors. PcG-deficient ES cells can begin differentiation but cannot maintain the differentiated phenotype.[20] Simultaneously, differentiation and development-promoting genes are activated by Trithorax group (TrxG) chromatin regulators and lose their repression.[20][22] TrxG proteins are recruited at regions of high transcriptional activity, where they catalyze the trimethylation of histone H3 lysine 4 (H3K4me3) and promote gene activation through histone acetylation.[22] PcG and TrxG complexes engage in direct competition and are thought to be functionally antagonistic, creating at differentiation and development-promoting loci what is termed a "bivalent domain" and rendering these genes sensitive to rapid induction or repression.[23]

DNA methylation Regulation of gene expression is further achieved through DNA methylation, in which the DNA methyltransferase-mediated methylation of cytosine residues in CpG dinucleotides maintains heritable repression by controlling DNA accessibility.[23] The majority of CpG sites in embryonic stem cells are unmethylated and appear to be associated with H3K4me3-carrying nucleosomes.[20] Upon differentiation, a small number of genes, including OCT4 and NANOG,[23] are methylated and their promoters repressed to prevent their further expression. Consistently, DNA methylation-deficient embryonic stem cells rapidly enter apoptosis upon in vitro differentiation.[20]

Nucleosome positioning While the DNA sequence of most cells of an organism is the same, the binding patterns of transcription factors and the corresponding gene expression patterns are different. To a large extent, differences in transcription factor binding are determined by the chromatin accessibility of their binding sites through histone modification and/or pioneer factors. In particular, it is important to know whether a nucleosome is covering a given genomic binding site or not. Recent studies have elucidated the role of nucleosome positioning during stem cell development.[24]

Role of signaling in epigenetic control

A final question to ask concerns the role of cell signaling in influencing the epigenetic processes governing differentiation. Such a role should exist, as it would be reasonable to think that extrinsic signaling can lead to epigenetic remodeling, just as it can lead to changes in gene expression through the activation or repression of different transcription factors. Interestingly, little direct data is available concerning the specific signals that influence the epigenome, and the majority of current knowledge consist of speculations on plausible candidate regulators of epigenetic remodeling.[25] We will first discuss several major candidates thought to be involved in the induction and maintenance of both embryonic stem cells and their differentiated progeny, and then turn to one example of specific signaling pathways in which more direct evidence exists for its role in epigenetic change.

The first major candidate is Wnt signaling pathway. The Wnt pathway is involved in all stages of differentiation, and the ligand Wnt3a can substitute for the overexpression of c-Myc in the generation of induced pluripotent stem cells.[25] On the other hand, disruption of ß-catenin, a component of the Wnt signaling pathway, leads to decreased proliferation of neural progenitors.

Growth factors comprise the second major set of candidates of epigenetic regulators of cellular differentiation. These morphogens are crucial for development, and include bone morphogenetic proteins, transforming growth factors (TGFs), and fibroblast growth factors (FGFs). TGFs and FGFs have been shown to sustain expression of OCT4, SOX2, and NANOG by downstream signaling to Smad proteins.[25] Depletion of growth factors promotes the differentiation of ESCs, while genes with bivalent chromatin can become either more restrictive or permissive in their transcription.[25]

Several other signaling pathways are also considered to be primary candidates. Cytokine leukemia inhibitory factors are associated with the maintenance of mouse ESCs in an undifferentiated state. This is achieved through its activation of the Jak-STAT3 pathway, which has been shown to be necessary and sufficient towards maintaining mouse ESC pluripotency.[26] Retinoic acid can induce differentiation of human and mouse ESCs,[25] and Notch signaling is involved in the proliferation and self-renewal of stem cells. Finally, Sonic hedgehog, in addition to its role as a morphogen, promotes embryonic stem cell differentiation and the self-renewal of somatic stem cells.[25]

The problem, of course, is that the candidacy of these signaling pathways was inferred primarily on the basis of their role in development and cellular differentiation. While epigenetic regulation is necessary for driving cellular differentiation, they are certainly not sufficient for this process. Direct modulation of gene expression through modification of transcription factors plays a key role that must be distinguished from heritable epigenetic changes that can persist even in the absence of the original environmental signals. Only a few examples of signaling pathways leading to epigenetic changes that alter cell fate currently exist, and we will focus on one of them.

Expression of Shh (Sonic hedgehog) upregulates the production of BMI1, a component of the PcG complex that recognizes H3K27me3. This occurs in a Gli-dependent manner, as Gli1 and Gli2 are downstream effectors of the Hedgehog signaling pathway. In culture, Bmi1 mediates the Hedgehog pathway's ability to promote human mammary stem cell self-renewal.[27] In both humans and mice, researchers showed Bmi1 to be highly expressed in proliferating immature cerebellar granule cell precursors. When Bmi1 was knocked out in mice, impaired cerebellar development resulted, leading to significant reductions in postnatal brain mass along with abnormalities in motor control and behavior.[28] A separate study showed a significant decrease in neural stem cell proliferation along with increased astrocyte proliferation in Bmi null mice.[29]

In summary, the role of signaling in the epigenetic control of cell fate in mammals is largely unknown, but distinct examples exist that indicate the likely existence of further such mechanisms.

4.4   See also

- Interbilayer Forces in Membrane Fusion

- Fusion mechanism

- Lipid bilayer fusion

- Cell-cell fusogens

- CAF-1

4.5   References

[1] Slack, J.M.W. (2013) Essential Developmental Biology. Wiley-Blackwell, Oxford.

[2] Slack, J.M.W. (2007) Metaplasia and transdifferentiation: from pure biology to the clinic. Nature Reviews Molecular Cell Biology 8, 369-378.

[3] "NCI Dictionary of Cancer Terms". National Cancer Institute. Retrieved 1 November 2013.

[4] Lodish, Harvey (2000). *Molecular Cell Biology* (4th ed.). New York: W. H. Freeman. Section 14.2. ISBN 0-7167-3136-3.

[5] Kumar, Rani (2008). *Textbook of Human Embryology*. I.K. International Publishing House. p. 22. ISBN 9788190675710.

[6] D. Binder, Marc; Hirokawa, Nobutaka; Windhorst, Uwe (2009). *Encyclopedia of Neuroscience*. Springer. ISBN 3540237356.

[7] Stocum DL (2004). "Amphibian regeneration and stem cells". *Curr. Top. Microbiol. Immunol.* Current Topics in Microbiology and Immunology 280: 1–70. doi:10.1007/978-3-642-18846-6_1. ISBN 978-3-540-02238-1. PMID 14594207.

[8] Casimir CM, Gates PB, Patient RK, Brockes JP (1988-12-01). "Evidence for dedifferentiation and metaplasia in amphibian limb regeneration from inheritance of DNA methylation". *Development* 104 (4): 657–668. PMID 3268408.

[9] Giles KL. "Dedifferentiation and Regeneration in Bryophytes: A Selective Review". *New Zealand Journal of Botany* 9: 689–94. doi:10.1080/0028825x.1971.10430231.

[10] Schnabel M, Marlovits S, Eckhoff G; et al. (January 2002). "Dedifferentiation-associated changes in morphology and gene expression in primary human articular chondrocytes in cell culture". *Osteoarthr. Cartil.* 10 (1): 62–70. doi:10.1053/joca.2001.0482. PMID 11795984.

[11] Sell S (December 1993). "Cellular origin of cancer: dedifferentiation or stem cell maturation arrest?". *Environ. Health Perspect.* 101 (Suppl 5): 15–26. doi:10.2307/3431838. JSTOR 3431838. PMC 1519468. PMID 7516873.

[12] Tsonis PA (April 2004). "Stem cells from differentiated cells". *Mol. Interv.* 4 (2): 81–3. doi:10.1124/mi.4.2.4. PMID 15087480.

[13] Ben-Tabou de-Leon S, Davidson EH. (2007). "Gene regulation: gene control network in development.". *Annu Rev Biophys Biomol Struct.* 36 (191). doi:10.1146/annurev.biophys.35.040405.102002. PMID 17291181.

[14] Knisely, Karen; Gilbert, Scott F. (2009). *Developmental Biology* (8th ed.). Sunderland, Mass: Sinauer Associates. p. 147. ISBN 0-87893-371-9.

[15] Rudel and Sommer: The evolution of developmental mechanisms. *Developmental Biology* 264, 15-37, 2003 Rudel, D.; Sommer, R. J. (2003). "The evolution of developmental mechanisms". *Developmental Biology* 264 (1): 15–37. doi:10.1016/S0012-1606(03)00353-1. PMID 14623229.

[16] Yamamoto Y and WR Jeffery: Central role for the lens in cave fish eye degeneration. *Science* 289 (5479), 631-633, 2000 Yamamoto, Y.; Jeffery, W. R. (2000). "Central Role for the Lens in Cave Fish Eye Degeneration". *Science* 289 (5479): 631–633. Bibcode:2000Sci...289..631Y. doi:10.1126/science.289.5479.631. PMID 10915628.

[17] Kirk MM, A Ransick, SE Mcrae, DL Kirk: The relationship between cell size and cell fate in *Volvox carteri*. *Journal of Cell Biology* 123, 191-208, 1993 Kirk, M. M.; Ransick, A.; McRae, S. E.; Kirk, D. L. (1993). "The relationship between cell size and cell fate in Volvox carteri". *Journal of Cell Biology* 123 (1): 191–208. doi:10.1083/jcb.123.1.191. PMC 2119814. PMID 8408198.

[18] Rabajante JF, Babierra AL (January 30, 2015). "Branching and oscillations in the epigenetic landscape of cell-fate determination". *Progress in Biophysics and Molecular Biology*. doi:10.1016/j.pbiomolbio.2015.01.006. PMID 25641423.

[19] Lister R; et al. (2011). "Hotspots of aberrant epigenomic reprogramming in human induced pluripotent stem cells". *Nature* 471 (7336): 68–73. Bibcode:2011Natur.471...68L. doi:10.1038/nature09798. PMC 3100360. PMID 21289626.

[20] Christophersen NS, Helin K (2010). "Epigenetic control of embryonic stem cell fate". *J Exp Med* 207 (11): 2287–95. doi:10.1084/jem.20101438. PMC 2964577. PMID 20975044.

[21] Zhu, J.; et al. (2013). "Genome-wide chromatin state transitions associated with developmental and environmental cues". *Cell* 152 (3): 642–654. doi:10.1016/j.cell.2012.12.033. PMID 23333102.

[22] Guenther MG, Young RA (2010). "Repressive Transcription". *Science* 329 (5988): 150–1. Bibcode:2010Sci...329..150G. doi:10.1126/science.1193995. PMC 3006433. PMID 20616255.

[23] Meissner A (2010). "Epigenetic modifications in pluripotent and differentiated cells". *Nat Biotechnol* 28 (10): 1079–88. doi:10.1038/nbt.1684. PMID 20944600.

[24] Teif VB, Vainshtein Y, Caudron-Herger M, Mallm JP, Marth C, Höfer T, Rippe K. (2012). "Genome-wide nucleosome positioning during embryonic stem cell development". *Nat Struct Mol Biol* 19 (11): 1185–92. doi:10.1038/nsmb.2419. PMID 23085715.

[25] Mohammad HP, Baylin SB (2010). "Linking cell signaling and the epigenetic machinery". *Nat Biotechnol* **28** (10): 1033–8. doi:10.1038/nbt1010-1033. PMID 20944593.

[26] Niwa H, Burdon T, Chambers I, Smith A (1998). "Self-renewal of pluripotent embryonic stem cells is mediated via activation of STAT3". *Genes Dev* **12** (13): 2048–60. doi:10.1101/gad.12.13.2048. PMC 316954. PMID 9649508.

[27] Liu S; et al. (2006). "Hedgehog Signaling and Bmi-1 Regulate Self-renewal of Normal and Malignant Human Mammary Stem Cells". *Cancer Res* **66** (12): 6063–71. doi:10.1158/0008-5472.CAN-06-0054. PMID 16778178.

[28] Leung C; et al. (2004). "Bmi1 is essential for cerebellar development and is overexpressed in human medulloblastomas". *Nature* **428** (6980): 337–41. Bibcode:2004Natur.428..337L. doi:10.1038/nature02385. PMID 15029199.

[29] Zencak D; et al. (2005). "Bmi1 loss produces an increase in astroglial cells and a decrease in neural stem cell population and proliferation". *J Neurosci* **25** (24): 5774–83. doi:10.1523/JNEUROSCI.3452-04.2005. PMID 15958744.

Chapter 5

Epigenetics in stem-cell differentiation

Embryonic stem cells are capable of self-renewing and differentiating to the desired fate depending on its position within the body. Stem cell homeostasis is maintained through epigenetic mechanisms that are highly dynamic in regulating the chromatin structure as well as specific gene transcription programs.[1] Epigenetics has been used to refer to changes in gene expression, which are heritable through modifications not affecting the DNA sequence.

The mammalian epigenome undergoes global remodeling during early stem cell development that requires commitment of cells to be restricted to the desired lineage. There has been multiple evidence suggesting that the maintenance of the lineage commitment of stem cells are controlled by epigenetic mechanisms such as DNA methylation, histone modifications and regulation of ATP-dependent remolding of chromatin structure.[1] Based on the *histone code* hypothesis, distinct covalent histone modifications can lead to functionally distinct chromatin structures that influence the fate of the cell.

This regulation of chromatin through epigenetic modifications is a molecular mechanism that will determine whether the cell will continue to differentiate into the desired fate. A research study performed by *Lee et al.* examined the effects of epigenetic modifications on the chromatin structure and the modulation of these epigenetic markers during stem cell differentiation through in vitro differentiation of murine embryonic stem (ES) cells.[2]

5.1 Experimental background

Embryonic stem cells exhibit dramatic and complex alterations to both global and site-specific chromatin structures. *Lee et al.* performed an experiment to determine the importance of deacetylation and acetylation for stem cell differentiation by looking at global acetylation and methylation levels at certain site-specific modification in histone sites *H3K9* and *H3K4*. Gene expression at these histones regulated by epigenetic modifications is critical in restricting the

embryonic stem cell to desired cell lineages and developing cellular memory.

For mammalian cells, the maintenance of cytosine methylation is catalyzed by DNA methyltransferases and any disruption to these methyltransferases will cause a lethal phenotype to the embryo. Cytosine methylation is examined at *H3K9*, which is associated with inactive heterochromatin and occurs mainly at CpG dinucleotides while global acetylation is examined at *H3K4*, which is associated with active euchromatin. The mammalian zygotic genome undergoes active and passive global cytosine demethylation following fertilization that reaches a minimal point of 20% CpG methylation at the blastocyst stage to which is then followed by a wave of methylation that reprograms the chromatin structure in order to restore global levels of CpG methylation to 60%.[2] Embryonic stem cells containing reduced or elevation levels of methylation are viable but unable to differentiate and therefore require critical regulation of cytosine methylation for mammalian development.

5.2 Effects of global histone modifications during embryonic stem cell differentiation

Histones modifications in chromatin were analyzed at various time intervals (along a 6 day period) following the initiation of in vitro embryonic stem cell differentiation. Differentiation was triggered by the removal of Leukemia inhibitory factor (LIF) which inhibits differentiation. Representative data of the histone modifications at the specific sites were assessed using Western blotting. The data confirms that strong deacetylation at the *H3K4* and *H3K9* positions of histone H3 one day after LIF removal, followed by a small increase in acetylation by day two.

The histone *H3K4* methylation also decreased after one day of LIF removal but showed a rebound between days 2-4 of differentiation, finally ending with a decrease in methylation on day five. These results indicate a decrease in the level of active euchromatin epigenetic marks upon initiation of embryonic stem cell differentiation which is then followed immediately by reprogramming of the epigenome.

Histone modifications of *H3K9* position show a decrease in di- and tri-methylation of undifferentiated embryonic stem cells and had a gradual increase in methylation during the six-day time course of in vitro differentiation, which indicated that there is a global increase of inactive heterochromatin levels at this histone mark.

As the embryonic stem cell undergoes differentiation the markers for active euchromatin (histone acetylation and *H3K4* methylation) are decreased after the removal of LIF showing that the cell is indeed becoming more differentiated. The slight rebound in each of these marks allows for further differentiation to occur by allowing another opportunity to decrease the markers once again, bringing the cell closer to its desired fate. Since there is also an increase throughout the six-day period in *H3K9me*, a marker for active heterochromatin, once differentiation occurs it is concluded that the formation of heterochromatin occurs as the cell is differentiated into its desired fate making the cell inactive to prevent further differentiation.

5.2.1 DNA methylation in differentiated versus undifferentiated cells

Global levels of cytosine methylation were compared between undifferentiated and differentiated embryonic stem cells. Global 5-methylcytosine levels have been measured prior to differentiation and after in vitro differentiation. The global cytosine methylation pattern appears to be established prior to the reprogramming of the histone code that occurs upon in vitro differentiation of embryonic stem cells.

As the embryonic stem cell undergoes differentiation the level of DNA methylation increases. This is in agreement with findings that show that there is an increase in inactive heterochromatin during differentiation.

5.3 Supplemental effects of methylation with DNMTs

In mammals, DNA methylation plays a role in regulating a key component of multipotency—the ability to rapidly self-renew. Khavari et al. discussed the fundamental mechanisms of DNA methylation and the interaction with several pathways regulating differentiation.[3] New approaches studying the genomic status of DNA methylation in vari-

ous states of differentiation have shown that methylation at CpG sites associated with putative enhancers are important in this process. DNA methylation can modulate the binding affinities of transcription factors by recruiting repressors such as *MeCP2* which display binding specificity for sequences containing methylated CpG dinucleotides. DNA methylation is controlled by certain methyltransferases, *DMNTs*, which perform different functions depending on each one. *DNMT3A* and *DNMT3B* have both been linked to a role in the establishment of DNA methylation pattern in the early development of the stem cell, whereas *DNMT1* is required to methylate a newly synthesized strand of DNA after the cell has undergone replication in order to sustain the epigenetic regulatory state. Numerous proteins can physically interact with *DNMTs* themselves, which help target *DNMT1* to hemi-methylated DNA.

Several new studies point to the central role of DNA methylation interacting with the regulation of cell cycles and DNA repair pathways in order to maintain the undifferentiated state. In embryonic stem cells, *DNMT1* depletion within the undifferentiated progenitor cell compartment led to cell cycle arrest, premature differentiation and a failure of tissue self-renewal. The loss of *DNMT1* occurred from profound effects associated with activation of differentiation genes and loss of genes promoting cell cycle progression, thus indicating that *DNMT1* and other *DNMTs* do not continuously suppress differentiation and thus maintain the pluripotent state.

These studies point to the important of the interaction of DNMTs in order to maintain stem cell states allowing for further differentiation and formation of heterochromatin to occur.

5.4 Epigenetic modifications of regulated genes during ESC differentiation

Okamoto et al. previously documented the expression level of the *Oct4* gene decreasing with embryonic stem cell differentiation.[4] *Lee et al* performed a ChIP analysis of the Oct4 promoter, associated with undifferentiated cells, region to examine the epigenetic modifications of regulated genes undergoing development during embryonic stem cell differentiation. This promoter region decreased at *H3K4* methylation and *H3K9* acetylation sites and increased at the *H3K9* methylation site during differentiation. Analysis of a CpG motif of the *Oct4* gene promoter revealed a progressive increase of DNA methylation and was completely methylated at day 10 of differentiation as previously reported in Gidekel and Bergman.[5]

These results indicate that there was a shift from the active eurchromatin to the inactive heterochromatin due to the decrease of acetylation of H3K4 and an increase of H3K9me. This means that the cell is becoming differentiated at the Oct4 gene, which is coincident with the silence of Oct4 gene expression.

Another site specific gene tested for histone modification was a *Brachyury* gene, a marker of mesoderm differentiation and is only slightly expressed in undifferentiated embryonic stem cells. "Brachyury" was induced at day five of differentiation and completely silencing by day 10, corresponding to the last day of differentiation.[6] The ChIP analysis of the "Brachyury" gene promoter revealed increase of expression in mono- and di-methylation of *H3K4* at day 0 and 5 of embryonic stem cell differentiation with a loss of gene expression at day 10. *H3K4* trimethylation coincides with the time of highest Brachyury gene expression since it only had gene expression on day 5. *H3K4* methylations in all forms are absent at day 10 of differentiation, which correlates with the silencing of Brachyury gene expression. Mono-methylation of both histones produced expression at day 0 indication a marker that is not useful for chromatin structure. Acetylation of H3K9 does not correlate to *Brachyury* gene expression since it was down regulated at the induction of differentiation. Upon examining of DNA methylation expression, there was no formation of intermediate sized bad in the Southern analysis suggesting that CpG motifs upstream of the promoter region are not methylated in the absence of cytosine methylation at this site.

It is demonstrated from these studies that both *H3K9* di-and tri-methylation correlate with the DNA methylation and gene expression while *H3K4* tri-methylation is associated the highest gene expression stage of the *Brachyury* gene. A previous report from Santos-Rosa is in agreement with these data showing that active genes are associated with *H3K4* tri-methylation in yeast.[7]

This data indicated the same results as for the *Oct4* gene, in that heterochromatin is forming as differentiation occurs again coinciding with the silence of *Brachyury* gene expression.

5.5   Effect of TSA on stem cell differentiation

Leukemia inhibitory factor (LIF) was removed from all the cell lines. LIF inhibits cell differentiation, and its removal allows the cell lines to go through cell differentiation. The cell lines were treated with Trichostatin A (TSA) - a histone deacetylase inhibitor for 6 days. One group of cell lines was treated with 10nM of TSA. The western analysis showed

the lack of initial deacetylation on Day-1 which, was observed in the control for the embryonic stem cell differentiation. The lack of histone deacetylase activity allowed the acetylation of H3K9 and histone H4. Embryonic stem cells were also analyzed morphologically to observe the formation of embryoid body formation as one of the measures of cell differentiation. The 10nM TSA treated cells failed to form the embryoid body by Day-6 as observed in the control cell line. This implies that the ES cells treated with TSA lacked the deacetylation on Day-1 and failed to differentiate after the removal of LIF. Second group,'-TSA Day4' was treated with TSA for 3days. As soon as the TSA treatment was stopped, on day 4 the deacetylation was observed and the acetylation recovered on Day-5. The morphological examination showed the formation of embryoid body formation by Day-6. In addition, "Interestingly" the embryoid body formation was faster than the control cell line. This suggests that the '-TSA Day4' lines were responding to the removal of LIF but, were unable to acquire any differentiation phenotype. They were able to acquire the differentiation phenotype after the cessation of TSA treatment and at rapid rate. The morphological examination of the third group,' 5 nM TSA' showed the intermediate effect between the control and 10nM-TSA group. The lower dose of TSA allowed the formation of some embryoid body formation. This experiment shows that TSA inhibits histone deacetylase and the activity of histone deacetylase is required for the embryonic stem cell differentiation. Without the initial deacetylation on Day-1, the ES cells cannot go through the differentiation.

5.6   Alkaline phosphatase activity

In normal stem cells, the activity of alkaline phosphatase activity is lowered upon differentiation. Trichostatin A causes the cells to maintain the activity of alkaline phosphatase. Significant increase in alkaline phosphatase extinction was observed when Trichostatin A was withdrawn after three days. Alkaline phosphatase activity correlates with the morphology changes. Initial deacetylation of histone is required for embryonic stem cell differentiation.

5.7   HDAC1, but not HDAC2 controls differentiation

Dovery et al. (2010) used HDAC knockout mice to demonstrate whether HDAC1 or HDAC2 was important for the embryonic stem cell differentiation. Examination of global histone acetylation in the absence of HDAC 1 showed an increase in acetylation. Global histone acetylation levels were unchanged by the loss of HDAC2. In order to analyze the process of HDAC knockout mouse in detail, the knockout mice embryonic stem cells were used to generate embryoid bodies. It showed that just before or during gastrulation, embryonic stem cells lacking HDAC1 acquired visible developmental defects. The continued culture of HDAC1 knockout embryonic stem cells showed that the embryoid bodies formed became irregular and reduced in size rather than uniformly spherical as in normal mice. Embryonic stem cell proliferation was unaffected by the loss of either HDAC1 or HDAC2 but the differentiation of embryonic stem cells were affected with that lack of HDAC 1. This shows that HDAC1 is required for cell fate determination during differentiation.[8]

5.8 The future

Any disturbance of a stable epigenetic regulation of gene expression mediated by DNA methylation is associated with a number of human disorders, including cancer as well as congenital diseases such as pseudohypoparathyroidism type IA, Beckwith-Wiedemann, Prader-Willi and Angelman syndromes, which are each caused by altered methylation-based imprinting at specific loci.

Perturbations of both global and gene-specific patterns of cytosine methylation are commonly observed in cancer while histone deacetylation is an important feature of nuclear reprogramming in oocytes during meiosis.[9]

Recent studies have revealed that there is an array of different pathways that cooperates with one another in order to bestow proper epigenetic regulation by DNA methylation. Future studies will be needed to further clarify the certain mechanism pathways such as DNA binding proteins, DNA repair and noncoding RNAs serve in order to regulate DNA methylation to suppress differentiation and sustain self-renewal in somatic stem cells in the epidermis and other tissues. Addressing these questions will help extend insight into these recent findings for a central role in epigenetic regulators of DNA methylation in controlling embryonic stem cell differentiation.[3]

5.9 References

[1] Zhou Y, Kim J, Yuan X, Braun T. (2011). "Epigenetic Modifications of Stem Cells-A Paradigm for the Control of Cardiac Progenitor Cells". *Circulation Research* **109**: 1067–1081. doi:10.1161/circresaha.111.243709.

[2] Lee J.H., Hart S., Skalnik D. (January 2004). "Histone Deacetylase Activity is Required for Embryonic Stem Cell Differentiation". *Genesis* **38** (1): 32–38. doi:10.1002/gene.10250. PMID 14755802.

[3] Khavari D., Sen G., Rinn J. (2010). "DNA methylation and epigenetic control of cellular differentiation". *Cell Cycle* **9** (19): 3880–3883. doi:10.4161/cc.9.19.13385.

[4] Okamoto K, Okazawa H, Okuda A, Sakai M, Murmatsu M, Hamada H. (1990). "A novel octamer binding transcription factor is differentially expressed in mouse embryonic cells". *Cell* **60**: 461–472. doi:10.1016/0092-8674(90)90597-8.

[5] Gidekel S, Bergman Y. (2002). "A unique developmental pattern of Oct3/4 DNA methylation is controlled by a cis-demodification element". *Journal of Biological Chemistry* **277**: 34521–34530. doi:10.1074/jbc.m203338200.

[6] Keller G, Kennedy M, Papayannopoulou T, Wiles M. (1993). "Hematopoietic commitment during embryonic stem cell differentiation in culture". *Molecular Cell Biology* **13**: 473–486.

[7] Santos-rosa H, Schneider R, Bannister AJ, Sherriff J, Bernstein BE, Emre NC, Schreiber SL, Mellor J, Kouzarides T. (2002). "Active genes are tri-methylated at K4 of histone H3". *Nature* **419**: 407–411. doi:10.1038/nature01080. PMID 12353038.

[8] Dovey O., Foster C., Cowley S. (2010). "Histone deacetylase 1(HDAC1), but not HDAC2, controls embryonic stem cell differentiation". *PNAS* **107** (18): 8242–8247. doi:10.1073/pnas.1000478107.

[9] Kim JM, Liu H, Tazaki M, Nagata M, Aoki F. (2003). "Changes in histone acetylation during mouse oocyte meiosis". *Journal of Cell Biology* **162**: 37–47. doi:10.1083/jcb.200303047.

Chapter 6

Embryonic stem cell

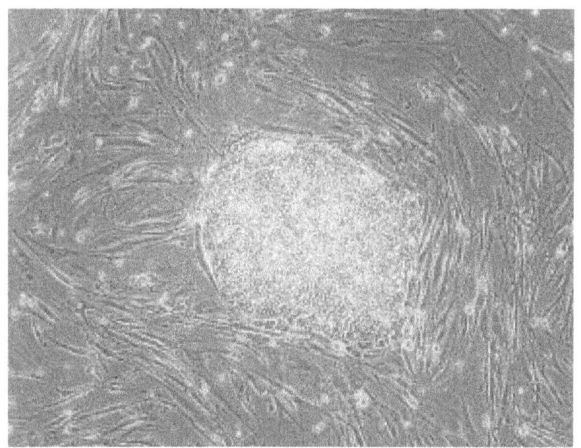

Human embryonic stem cells in cell culture

derived from the inner cell mass of a blastocyst, an early-stage preimplantation embryo.[1][2] Human embryos reach the blastocyst stage 4–5 days post fertilization, at which time they consist of 50–150 cells. Isolating the embryoblast or inner cell mass (ICM) results in destruction of the blastocyst, which raises ethical issues, including whether or not embryos at the pre-implantation stage should be considered to have the same moral or legal status as more developed human beings.[3][4]

Human ES cells measure approximately 14 μm while mouse ES cells are closer to 8 μm.[5]

6.1 Properties

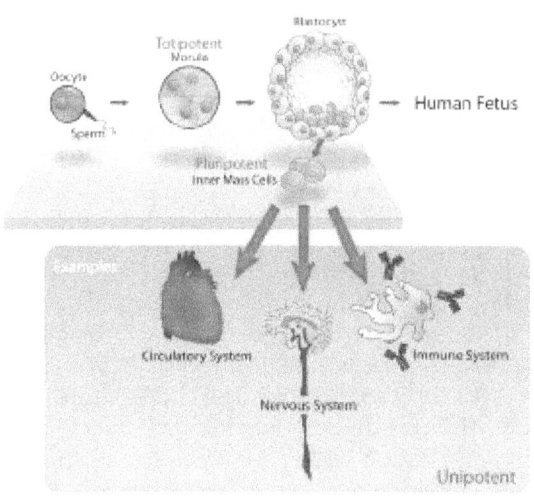

Pluripotent: Embryonic stem cells are able to develop into any type of cell, excepting those of the placenta. Only embryonic stem cells of the morula are totipotent: able to develop into any type of cell, including those of the placenta.

Embryonic stem cells (**ES cells**) are pluripotent stem cells

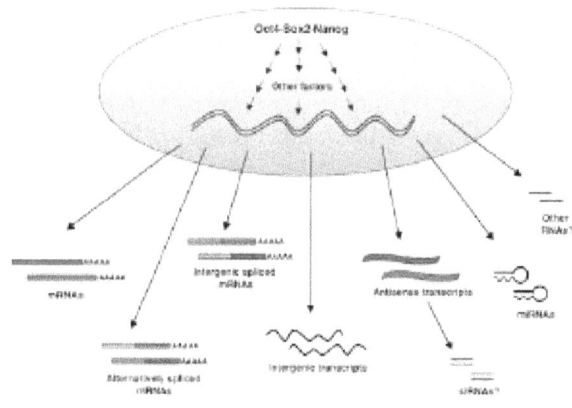

The transcriptome of embryonic stem cells

Embryonic stem cells, derived from the blastocyst stage early mammalian embryos, are distinguished by their ability to differentiate into any cell type and by their ability to propagate. Embryonic stem cell's properties include having a normal karyotype, maintaining high telomerase activity, and exhibiting remarkable long-term proliferative potential.[6]

6.1.1 Pluripotent

Embryonic stem cells of the inner cell mass are pluripotent, that is, they are able to differentiate to generate primitive ectoderm, which ultimately differentiates during gastrulation into all derivatives of the three primary germ layers: ectoderm, endoderm, and mesoderm. These include each of the more than 220 cell types in the adult body. Pluripotency distinguishes embryonic stem cells from adult stem cells found in adults; while embryonic stem cells can generate all cell types in the body, adult stem cells are multipotent and can produce only a limited number of cell types. If the pluripotent differentiation potential of embryonic stem cells could be harnessed in vitro, it might be a means of deriving cell or tissue types virtually to order. This would provide a radical new treatment approach to a wide variety of conditions where age, disease, or trauma has led to tissue damage or dysfunction.

6.1.2 Propagation

Additionally, under defined conditions, embryonic stem cells are capable of propagating themselves indefinitely in an undifferentiated state and have the capacity when provided with the appropriate signals to differentiate, presumably via the formation of precursor cells, to almost all mature cell phenotypes.[7] This allows embryonic stem cells to be employed as useful tools for both research and regenerative medicine, because they can produce limitless numbers of themselves for continued research or clinical use.

6.1.3 Usefulness

Because of their plasticity and potentially unlimited capacity for self-renewal, embryonic stem cell therapies have been proposed for regenerative medicine and tissue replacement after injury or disease. Diseases that could potentially be treated by pluripotent stem cells include a number of blood and immune-system related genetic diseases, cancers, and disorders; juvenile diabetes; Parkinson's disease; blindness and spinal cord injuries. Besides the ethical concerns of stem cell therapy (see *stem cell controversy*), there is a technical problem of graft-versus-host disease associated with allogeneic stem cell transplantation. However, these problems associated with histocompatibility may be solved using autologous donor adult stem cells, therapeutic cloning. The therapeutic cloning done by a method called somatic cell nuclear transfer (SCNT) may be advantageous against mitochondrial DNA (mtDNA) mutated diseases.[8] Stem cell banks or more recently by reprogramming of somatic cells with defined factors (e.g. induced pluripotent stem cells). Embryonic stem cells provide hope that it will

be possible to overcome the problems of donor tissue shortage and also, by making the cells immunocompatible with the recipient. Other potential uses of embryonic stem cells include investigation of early human development, study of genetic disease and as in vitro systems for toxicology testing.[6]

6.2 Utilizations

6.2.1 Potential clinical use

According to a 2002 article in *PNAS*, "Human embryonic stem cells have the potential to differentiate into various cell types, and, thus, may be useful as a source of cells for transplantation or tissue engineering."[9]

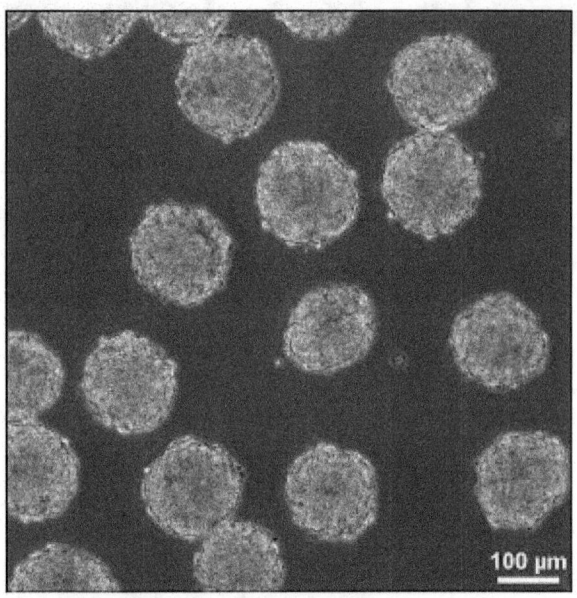

Embryoid bodies 24 hours after formation.

Current research focuses on differentiating ES into a variety of cell types for eventual use as cell replacement therapies (CRTs). Some of the cell types that have or are currently being developed include cardiomyocytes (CM), neurons, hepatocytes, bone marrow cells, islet cells and endothelial cells.[10] However, the derivation of such cell types from ESs is not without obstacles and hence current research is focused on overcoming these barriers. For example, studies are underway to differentiate ES in to tissue specific CMs and to eradicate their immature properties that distinguish them from adult CMs.[11]

Besides in the future becoming an important alternative to organ transplants, ES are also being used in field of toxicology and as cellular screens to uncover new chemical entities (NCEs) that can be developed as small molecule drugs. Studies have shown that cardiomyocytes derived from ES are validated in vitro models to test drug responses and predict toxicity profiles.[10] ES derived cardiomyocytes have been shown to respond to pharmacological stimuli and hence can be used to assess cardiotoxicity like *Torsades de Pointes*.[12]

ES-derived hepatocytes are also useful models that could be used in the preclinical stages of drug discovery. However, the development of hepatocytes from ES has proven to be challenging and this hinders the ability to test drug metabolism. Therefore, current research is focusing on establishing fully functional ES-derived hepatocytes with stable phase I and II enzyme activity.[13]

Researchers have also differentiated ES into dopamine-producing cells with the hope that these neurons could be used in the treatment of Parkinson's disease.[14][15] Recently, the development of ESC after Somatic Cell Nuclear Transfer (SCNT) of Olfactory ensheathing cells (OEC's) to a healthy Oocyte has been recommended for Neurodegenerative diseases.[16] The same group (*Baig et al.*) also have advocated the use of Olfactory ensheathing cells for demyelinating diseases like Multiple Sclerosis.[8] ESs have also been differentiated to natural killer (NK) cells and bone tissue.[17] Studies involving ES are also underway to provide an alternative treatment for diabetes. For example, D'Amour et al. were able to differentiate ES into insulin producing cells[18] and researchers at Harvard University were able to produce large quantities of pancreatic beta cells from ES.[19]

6.2.2 Human embryonic stem cells as models of genetic disorders

Several new studies have started to address this issue. This has been done either by genetically manipulating the cells, or more recently by deriving diseased cell lines identified by prenatal genetic diagnosis (PGD). This approach may very well prove invaluable at studying disorders such as Fragile-X syndrome, Cystic fibrosis, and other genetic maladies that have no reliable model system.

Yury Verlinsky, a Russian-American medical researcher who specialized in embryo and cellular genetics (genetic cytology), developed prenatal diagnosis testing methods to determine genetic and chromosomal disorders a month and a half earlier than standard amniocentesis. The techniques are now used by many pregnant women and prospective parents, especially those couples with a history of genetic abnormalities or where the woman is over the age of 35, when the risk of genetically related disorders is higher. In addition, by allowing parents to select an embryo without genetic disorders, they have the potential of saving the lives of siblings that already had similar disorders and diseases

using cells from the disease free offspring.[20]

Scientists have discovered a new technique for deriving human embryonic stem cell (ESC). Normal ESC lines from different sources of embryonic material including morula and whole blastocysts have been established. These findings allows researchers to construct ESC lines from embryos that acquire different genetic abnormalities; therefore, allowing for recognition of mechanisms in the molecular level that are possibly blocked that could impede the disease progression. The ESC lines originating from embryos with genetic and chromosomal abnormalities provide the data necessary to understand the pathways of genetic defects.[21]

A donor patient acquires one defective gene copy and one normal, and only one of these two copies is used for reproduction. By selecting egg cell derived from embryonic stem cells that have two normal copies, researchers can find variety of treatments for various diseases. To test this theory Dr. McLaughlin and several of his colleagues looked at whether parthenogenetic embryonic stem cells can be used in a mouse model that has thalassemia intermedia. This disease is described as an inherited blood disorder in which there is a lack of hemoglobin leading to anemia. The mouse model used, had one defective gene copy. Embryonic stem cells from an unfertilized egg of the diseased mice were gathered and those stem cells that contained only healthy hemoglobin genes were identified. The healthy embryonic stem cell lines were then converted into cells transplanted into the carrier mice. After five weeks, the test results from the transplant illustrated that these carrier mice now had a normal blood cell count and hemoglobin levels.[22]

6.2.3 Repair of DNA damage

Differentiated somatic cells and ES cells use different strategies for dealing with DNA damage. For instance, human foreskin fibroblasts, one type of somatic cell, use non-homologous end joining (NHEJ), an error prone DNA repair process, as the primary pathway for repairing double-strand breaks (DSBs) during all cell cycle stages.[23] Because of its error-prone nature, NHEJ tends to produce mutations in a cell's clonal descendants.

ES cells use a different strategy to deal with DSBs.[24] Because ES cells give rise to all of the cell types of an organism including the cells of the germ line, mutations arising in ES cells due to faulty DNA repair are a more serious problem than in differentiated somatic cells. Consequently, robust mechanisms are needed in ES cells to repair DNA damages accurately, and if repair fails, to remove those cells with un-repaired DNA damages. Thus, mouse ES cells predominantly use high fidelity homologous recombinational repair (HRR) to repair DSBs.[24] This type of repair depends on the interaction of the two sister chromosomes formed during S phase and present together during the G2 phase of the cell cycle. HRR can accurately repair DSBs in one sister chromosome by using intact information from the other sister chromosome. Cells in the G1 phase of the cell cycle (i.e. after metaphase/cell division but prior the next round of replication) have only one copy of each chromosome (i.e. sister chromosomes aren't present). Mouse ES cells lack a G1 checkpoint and do not undergo cell cycle arrest upon acquiring DNA damage.[25] Rather they undergo programmed cell death (apoptosis) in response to DNA damage.[26] Apoptosis can be used as a fail-safe strategy to remove cells with un-repaired DNA damages in order to avoid mutation and progression to cancer.[27] Consistent with this strategy, mouse ES stem cells have a mutation frequency about 100-fold lower than that of isogenic mouse somatic cells.[28]

6.3 Adverse effect

The major concern with the possible transplantation of ESC into patients as therapies is their ability to form tumors including teratoma.[29] Safety issues prompted the FDA to place a hold on the first ESC clinical trial (see below), however no tumors were observed.

The main strategy to enhance the safety of ESC for potential clinical use is to differentiate the ESC into specific cell types (e.g. neurons, muscle, liver cells) that have reduced or eliminated ability to cause tumors. Following differentiation, the cells are subjected to sorting by flow cytometry for further purification. ESC are predicted to be inherently safer than IPS cells because they are not genetically modified with genes such as c-Myc that are linked to cancer. Nonetheless, ESC express very high levels of the iPS inducing genes and these genes including Myc are essential for ESC self-renewal and pluripotency,[30] and potential strategies to improve safety by eliminating Myc expression are unlikely to preserve the cells' "stemness".

6.4 History

In 1964, researchers isolated a single type of cell from a teratocarcinoma, a tumor now known to be derived from a germ cell. These cells isolated from the teratocarcinoma replicated and grew in cell culture as a stem cell and are now known as embryonal carcinoma (EC) cells.[31] Although similarities in morphology and differentiating potential (pluripotency) led to the use of EC cells as the *in vitro* model for early mouse development,[32] EC cells harbor genetic mutations and often abnormal karyotypes that accumulated during the development of the teratocarcinoma. These genetic aberrations further emphasized the need to

be able to culture pluripotent cells directly from the inner cell mass.

Martin Evans revealed a new technique for culturing the mouse embryos in the uterus to allow for the derivation of ES cells from these embryos.

In 1981, embryonic stem cells (ES cells) were independently first derived from mouse embryos by two groups. Martin Evans and Matthew Kaufman from the Department of Genetics, University of Cambridge published first in July, revealing a new technique for culturing the mouse embryos in the uterus to allow for an increase in cell number, allowing for the derivation of ES cells from these embryos.[33] Gail R. Martin, from the Department of Anatomy, University of California, San Francisco, published her paper in December and coined the term "Embryonic Stem Cell".[34] She showed that embryos could be cultured *in vitro* and that ES cells could be derived from these embryos. In 1998, a breakthrough occurred when researchers, led by James Thomson at the University of Wisconsin-Madison, first developed a technique to isolate and grow human embryonic stem cells in cell culture.[35]

6.4.1   First clinical trial

On January 23, 2009. Phase I clinical trials for transplantation of oligodendrocytes (a cell type of the brain and spinal cord) derived from human ES cells into spinal cord-injured individuals received approval from the U.S. Food and Drug Administration (FDA). marking it the world's first human ES cell human trial.[36] The study leading to this scientific advancement was conducted by Hans Keirstead and colleagues at the University of California, Irvine and supported by Geron Corporation of Menlo Park, CA. founded by Michael D. West, PhD. A previous experiment had shown an improvement in locomotor recovery in spinal cord-injured rats after a 7-day delayed transplantation of human ES cells that had been pushed into an oligodendrocytic lineage.[37] The phase I clinical study was designed to enroll about eight to ten paraplegics who have had their injuries no longer than two weeks before the trial begins, since the cells must be injected before scar tissue is able to form. The researchers emphasized that the injections were not expected to fully cure the patients and restore all mobility. Based on the results of the rodent trials, researchers speculated that restoration of myelin sheathes and an increase in mobility might occur. This first trial was primarily designed to test the safety of these procedures and if everything went well, it was hoped that it would lead to future studies that involve people with more severe disabilities.[38] The trial was put on hold in August 2009 due to FDA concerns regarding a small number of microscopic cysts found in several treated rat models but the hold was lifted on July 30, 2010.[39]

In October 2010 researchers enrolled and administered ESTs to the first patient at Shepherd Center in Atlanta.[40] The makers of the stem cell therapy, Geron Corporation, estimated that it would take several months for the stem cells to replicate and for the GRNOPC1 therapy to be evaluated for success or failure.

In November 2011 Geron announced it was halting the trial and dropping out of stem cell research for financial reasons, but would continue to monitor existing patients, and was attempting to find a partner that could continue their research.[41] In 2013 BioTime (NYSE MKT: BTX). led by CEO Dr. Michael D. West, acquired all of Geron's stem cell assets, with the stated intention of restarting Geron's embryonic stem cell-based clinical trial for spinal cord injury research.[42]

6.5   Techniques and conditions for embryonic stem cell derivation and culture

6.5.1   Derivation of human embryonic stem cells

In vitro fertilization generates multiple embryos. The surplus of embryos is not clinically used or is unsuitable for implantation into the patient, and therefore may be donated by the donor with consent. Human embryonic stem cells can be derived from these donated embryos or additionally they can also be extracted from cloned embryos using a cell from a patient and a donated egg.[43] The inner cell mass (cells of interest), from the blastocyst stage of the embryo, is separated from the trophectoderm, the cells that would differentiate into extra-embryonic tissue. Immunosurgery, the process in which antibodies are bound to the trophectoderm and removed by another solution, and mechanical dissection are performed to achieve separation. The resulting inner cell mass cells are plated onto cells that will supply support. The inner cell mass cells attach and expand further to form a human embryonic cell line, which are undifferentiated. These cells are fed daily and are enzymatically or mechanically separated every four to seven days. For differentiation to occur, the human embryonic stem cell line is removed from the supporting cells to form embryoid bodies, is co-cultured with a serum containing necessary signals, or is grafted in a three-dimensional scaffold to result.[44]

6.5.2   Derivation of embryonic stem cells from other animals

Embryonic stem cells are derived from the inner cell mass of the early embryo, which are harvested from the donor mother animal. Martin Evans and Matthew Kaufman reported a technique that delays embryo implantation, allowing the inner cell mass to increase. This process includes removing the donor mother's ovaries and dosing her with progesterone, changing the hormone environment, which causes the embryos to remain free in the uterus. After 4–6 days of this intrauterine culture, the embryos are harvested and grown in *in vitro* culture until the inner cell mass forms "egg cylinder-like structures," which are dissociated into single cells, and plated on fibroblasts treated with mitomycin-c (to prevent fibroblast mitosis). Clonal cell lines are created by growing up a single cell. Evans and Kaufman showed that the cells grown out from these cultures could form teratomas and embryoid bodies, and differentiate *in vitro*, all of which indicating that the cells are pluripotent.[33]

Gail Martin derived and cultured her ES cells differently. She removed the embryos from the donor mother at approximately 76 hours after copulation and cultured them overnight in a medium containing serum. The following day, she removed the inner cell mass from the late blastocyst using microsurgery. The extracted inner cell mass was cultured on fibroblasts treated with mitomycin-c in a medium containing serum and conditioned by ES cells. After approximately one week, colonies of cells grew out. These cells grew in culture and demonstrated pluripotent characteristics, as demonstrated by the ability to form teratomas, differentiate *in vitro*, and form embryoid bodies. Martin referred to these cells as ES cells.[34]

It is now known that the feeder cells provide leukemia inhibitory factor (LIF) and serum provides bone morphogenetic proteins (BMPs) that are necessary to prevent ES cells from differentiating.[45][46] These factors are extremely important for the efficiency of deriving ES cells. Furthermore, it has been demonstrated that different mouse strains have different efficiencies for isolating ES cells.[47] Current uses for mouse ES cells include the generation of transgenic mice, including knockout mice. For human treatment, there is a need for patient specific pluripotent cells. Generation of human ES cells is more difficult and faces ethical issues. So, in addition to human ES cell research, many groups are focused on the generation of induced pluripotent stem cells (iPS cells).[48]

6.5.3 Potential method for new cell line derivation

On August 23, 2006, the online edition of *Nature* scientific journal published a letter by Dr. Robert Lanza (medical director of Advanced Cell Technology in Worcester, MA) stating that his team had found a way to extract embryonic stem cells without destroying the actual embryo.[49] This technical achievement would potentially enable scientists to work with new lines of embryonic stem cells derived using public funding in the USA, where federal funding was at the time limited to research using embryonic stem cell lines derived prior to August 2001. In March, 2009, the limitation was lifted.[50]

6.5.4 Induced pluripotent stem cells

Main article: Induced pluripotent stem cell

In 2007 it was shown that pluripotent stem cells highly similar to embryonic stem cells can be generated by the delivery of three genes (*Oct4*, *Sox2*, and *Klf4*) to differentiated cells.[51] The delivery of these genes "reprograms" differentiated cells into pluripotent stem cells, allowing for the generation of pluripotent stem cells without the embryo. Because ethical concerns regarding embryonic stem cells typically are about their derivation from terminated embryos, it is believed that reprogramming to these "induced

pluripotent stem cells" (iPS cells) may be less controversial. Both human and mouse cells can be reprogrammed by this methodology, generating both human pluripotent stem cells and mouse pluripotent stem cells without an embryo.[52]

This may enable the generation of patient specific ES cell lines that could potentially be used for cell replacement therapies. In addition, this will allow the generation of ES cell lines from patients with a variety of genetic diseases and will provide invaluable models to study those diseases.

However, as a first indication that the induced pluripotent stem cell (iPS) cell technology can in rapid succession lead to new cures, it was used by a research team headed by Rudolf Jaenisch of the Whitehead Institute for Biomedical Research in Cambridge, Massachusetts, to cure mice of sickle cell anemia, as reported by *Science* journal's online edition on December 6, 2007.[53][54]

On January 16, 2008, a California-based company, Stemagen, announced that they had created the first mature cloned human embryos from single skin cells taken from adults. These embryos can be harvested for patient matching embryonic stem cells.[55]

6.5.5 Contamination by reagents used in cell culture

The online edition of *Nature Medicine* published a study on January 24, 2005, which stated that the human embryonic stem cells available for federally funded research are contaminated with non-human molecules from the culture medium used to grow the cells.[56] It is a common technique to use mouse cells and other animal cells to maintain the pluripotency of actively dividing stem cells. The problem was discovered when non-human sialic acid in the growth medium was found to compromise the potential uses of the embryonic stem cells in humans, according to scientists at the University of California, San Diego.[57]

However, a study published in the online edition of *Lancet Medical Journal* on March 8, 2005 detailed information about a new stem cell line that was derived from human embryos under completely cell- and serum-free conditions. After more than 6 months of undifferentiated proliferation, these cells demonstrated the potential to form derivatives of all three embryonic germ layers both *in vitro* and in teratomas. These properties were also successfully maintained (for more than 30 passages) with the established stem cell lines.[58]

6.6 See also

- Embryoid body

- Embryonic Stem Cell Research Oversight Committees

- Fetal tissue implant

- Stem cell controversy

- Induced stem cells

6.7 References

[1] Thomson et. al; Itskovitz-Eldor, J; Shapiro, SS; Waknitz, MA; Swiergiel, JJ; Marshall, VS; Jones, JM (1998). "Blastocysts Embryonic Stem Cell Lines Derived from Human". *Science* **282** (5391): 1145–1147. doi:10.1126/science.282.5391.1145. PMID 9804556.

[2] "NIH Stem Cell Basics. What are embryonic stem cells?".

[3] Baldwing A (2009). "Morality and human embryo research. Introduction to the Talking Point on morality and human embryo research.". *EMBO reports* **10** (4): 299–300. doi:10.1038/embor.2009.37. PMC 2672902. PMID 19337297.

[4] Nakaya, Andrea C. (August 1, 2011). *Biomedical ethics*. San Diego, CA: ReferencePoint Press. p. 96. ISBN 160152157X.

[5] Thomson, James A.; Zwaka (10 February 2003). "Homologous recombination in human embryonic stem cells". *Nature Biotechnology* **21** (3): 319–321. doi:10.1038/nbt788. PMID 12577066.

[6] Thomson, J. A.; Itskovitz-Eldor, J; Shapiro, S. S.; Waknitz, M. A.; Swiergiel, J. J.; Marshall, V. S.; Jones, J. M. (1998). "Embryonic Stem Cell Lines Derived from Human Blastocysts". *Science* **282** (5391): 1145–7. doi:10.1126/science.282.5391.1145. PMID 9804556.

[7] Ying et. al; Nichols, J; Chambers, I; Smith, A (2003). "BMP Induction of Id Proteins Suppresses Differentiation and Sustains Embryonic Stem Cell Self-Renewal in Collaboration with STAT3". *Cell* **115** (3): 281–292. doi:10.1016/S0092-8674(03)00847-X. PMID 14636556.

[8] Mannan Baig, Abdul (2014). "Cloned Microglias with novel delivery systems in Multiple Sclerosis". *J Stem Cell Res Ther* **4** (11): 11. doi:10.4172/2157-7633.1000252.

[9] Levenberg, S. (2002). "Endothelial cells derived from human embryonic stem cells". *Proceedings of the National Academy of Sciences* **99** (7): 4391–4396. doi:10.1073/pnas.032074999.

[10] Davila, JC; Cezar, GG; Thiede, M; Strom, S; Miki, T; Trosko, J (2004). "Use and application of stem cells in toxicology". *Toxicological sciences : an official journal of the Society of Toxicology* **79** (2): 214–23. doi:10.1093/toxsci/kfh100. PMID 15014205.

[11] Siu, CW; Moore, JC; Li, RA (2007). "Human embryonic stem cell-derived cardiomyocytes for heart therapies". *Cardiovascular & hematological disorders drug targets* **7** (2): 145–52. doi:10.2174/187152907780830851. PMID 17584049.

[12] Jensen, J; Hyllner, J; Björquist, P (2009). "Human embryonic stem cell technologies and drug discovery". *Journal of cellular physiology* **219** (3): 513–9. doi:10.1002/jcp.21732. PMID 19277978.

[13] Söderdahl, T; Küppers-Munther, B; Heins, N; Edsbagge, J; Björquist, P; Cotgreave, I; Jernström, B (2007). "Glutathione transferases in hepatocyte-like cells derived from human embryonic stem cells". *Toxicology in vitro : an international journal published in association with BIBRA* **21** (5): 929–37. doi:10.1016/j.tiv.2007.01.021. PMID 17346923.

[14] Perrier, A. L. (2004). "Derivation of midbrain dopamine neurons from human embryonic stem cells". *Proceedings of the National Academy of Sciences* **101** (34): 12543–12548. doi:10.1073/pnas.0404700101.

[15] Parish, CL; Arenas, E (2007). "Stem-cell-based strategies for the treatment of Parkinson's disease". *Neuro-degenerative diseases* **4** (4): 339–47. doi:10.1159/000101892. PMID 17627139.

[16] Abdul Mannan Baig. Designer's Microglia with Novel delivery system in Neurodegenerative Diseases. Medical Hypotheses (Impact Factor: 1.18). 08/2014; DOI: 10.1016/j. May. 2014.08.003

[17] Waese, EY; Kandel, RA; Stanford, WL (2008). "Application of stem cells in bone repair". *Skeletal radiology* **37** (7): 601–8. doi:10.1007/s00256-007-0438-8. PMID 18193216.

[18] d'Amour, KA; Bang, AG; Eliazer, S; Kelly, OG; Agulnick, AD; Smart, NG; Moorman, MA; Kroon, E; Carpenter, MK; Baetge, EE (2006). "Production of pancreatic hormone-expressing endocrine cells from human embryonic stem cells". *Nature Biotechnology* **24** (11): 1392–401. doi:10.1038/nbt1259. PMID 17053790.

[19] Colen, B.D. (9 October 2014) Giant leap against diabetes The Harvard Gazette. Retrieved 24 November 2014

[20] "Dr. Yury Verlinsky, 1943–2009: Expert in reproductive technology" *Chicago Tribune*. July 20, 2009

[21] Verlinsky, Y; Strelchenko, N; Kukharenko, V; Rechitsky, S; Verlinsky, O; Galat, V; Kuliev, A (2005). "Human embryonic stem cell lines with genetic disorders". *Reproductive biomedicine online* **10** (1): 105–10. doi:10.1016/S1472-6483(10)60810-3. PMID 15705304.

[22] Embryonic Stem Cells Help Deliver 'Good Genes' In A Model Of Inherited Blood Disorder. ScienceDaily (February 13, 2011).

[23] Mao Z, Bozzella M, Seluanov A, Gorbunova V (September 2008). "DNA repair by nonhomologous end joining and homologous recombination during cell cycle in human cells". *Cell Cycle* **7** (18): 2902–6. doi:10.4161/cc.7.18.6679. PMC 2754209. PMID 18769152.

[24] Tichy ED, Pillai R, Deng L; et al. (November 2010). "Mouse embryonic stem cells, but not somatic cells, predominantly use homologous recombination to repair double-strand DNA breaks". *Stem Cells Dev.* **19** (11): 1699–711. doi:10.1089/scd.2010.0058. PMC 3128311. PMID 20446816.

[25] Hong Y, Stambrook PJ (October 2004). "Restoration of an absent G1 arrest and protection from apoptosis in embryonic stem cells after ionizing radiation". *Proc. Natl. Acad. Sci. U.S.A.* **101** (40): 14443–8. doi:10.1073/pnas.0401346101. PMC 521944. PMID 15452351.

[26] Aladjem MI, Spike BT, Rodewald LW; et al. (January 1998). "ES cells do not activate p53-dependent stress responses and undergo p53-independent apoptosis in response to DNA damage". *Curr. Biol.* **8** (3): 145–55. doi:10.1016/S0960-9822(98)70061-2. PMID 9443911.

[27] Bernstein C, Bernstein H, Payne CM, Garewal H (June 2002). "DNA repair/pro-apoptotic dual-role proteins in five major DNA repair pathways: fail-safe protection against carcinogenesis". *Mutat. Res.* **511** (2): 145–78. doi:10.1016/S1383-5742(02)00009-1. PMID 12052432.

[28] Cervantes RB, Stringer JR, Shao C, Tischfield JA, Stambrook PJ (March 2002). "Embryonic stem cells and somatic cells differ in mutation frequency and type". *Proc. Natl. Acad. Sci. U.S.A.* **99** (6): 3586–90. doi:10.1073/pnas.062527199. PMC 122567. PMID 11891338.

[29] Knoepfler, Paul S. (2009). "Deconstructing Stem Cell Tumorigenicity: A Roadmap to Safe Regenerative Medicine". *Stem Cells* **27** (5): 1050–6. doi:10.1002/stem.37. PMC 2733374. PMID 19415771.

[30] Varlakhanova, Natalia V.; Cotterman, Rebecca F.; Devries, Wilhelmine N.; Morgan, Judy; Donahue, Leah Rae; Murray, Stephen; Knowles, Barbara B.; Knoepfler, Paul S. (2010). "Myc maintains embryonic stem cell pluripotency and self-renewal". *Differentiation* **80** (1): 9–19. doi:10.1016/j.diff.2010.05.001. PMC 2916696. PMID 20537458.

[31] Andrews P, Matin M, Bahrami A, Damjanov I, Gokhale P, Draper J (2005). "Embryonic stem (ES) cells and embryonal carcinoma (EC) cells: opposite sides of the same coin" (PDF). *Biochem Soc Trans* **33** (Pt 6): 1526–30. doi:10.1042/BST20051526. PMID 16246161.

[32] Martin GR (1980). "Teratocarcinomas and mammalian embryogenesis". *Science* **209** (4458): 768–76. doi:10.1126/science.6250214. PMID 6250214.

[33] Evans M, Kaufman M (1981). "Establishment in culture of pluripotent cells from mouse embryos". *Nature* **292** (5819): 154–6. doi:10.1038/292154a0. PMID 7242681.

[34] Martin G (1981). "Isolation of a pluripotent cell line from early mouse embryos cultured in medium conditioned by teratocarcinoma stem cells". *Proc Natl Acad Sci USA* **78** (12): 7634–8. doi:10.1073/pnas.78.12.7634. PMC 349323. PMID 6950406.

[35] Thomson J, Itskovitz-Eldor J, Shapiro S, Waknitz M, Swiergiel J, Marshall V, Jones J (1998). "Embryonic stem cell lines derived from human blastocysts". *Science* **282** (5391): 1145–7. doi:10.1126/science.282.5391.1145. PMID 9804556.

[36] "FDA approves human embryonic stem cell study - CNN.com". January 23, 2009. Retrieved May 1, 2010.

[37] Keirstead HS, Nistor G, Bernal G; et al. (2005). "Human embryonic stem cell-derived oligodendrocyte progenitor cell transplants remyelinate and restore locomotion after spinal cord injury". *J. Neurosci.* **25** (19): 4694–705. doi:10.1523/JNEUROSCI.0311-05.2005. PMID 15888645.

[38] Reinberg, Steven (2009-01-23) FDA OKs 1st Embryonic Stem Cell Trial. *The Washington Post*

[39] Geron comments on FDA hold on spinal cord injury trial http://www.geron.com/media/pressview.aspx?id=1188

[40] Vergano, Dan (11 October 2010). "Embryonic stem cells used on patient for first time". *USA Today*. Retrieved 12 October 2010.

[41] Brown, Eryn (November 15, 2011). "Geron exits stem cell research". LA Times. Retrieved 2011-11-15.

[42] "Great news: BioTime Subsidiary Asterias Acquires Geron Embryonic Stem Cell Program". *iPScell.com*. October 1, 2013.

[43] Mountford, JC (2008). "Human embryonic stem cells: origins, characteristics and potential for regenerative therapy". *Transfus Med* **18** (1): 1–12. doi:10.1111/j.1365-3148.2007.00807.x. PMID 18279188.

[44] Thomson JA, Itskovitz-Eldor J, Shapiro SS, Waknitz MA, Swiergiel JJ, Marshall VS, Jones JM, J. A.; Itskovitz-Eldor, J; Shapiro, SS; Waknitz, MA; Swiergiel, JJ; Marshall, VS; Jones, JM (1998). "Embryonic stem cell lines derived from human blastocysts". *Science* **282** (5391): 1145–1147. doi:10.1126/science.282.5391.1145. PMID 9804556.

[45] Smith AG, Heath JK, Donaldson DD, Wong GG, Moreau J, Stahl M, Rogers D (1988). "Inhibition of pluripotential embryonic stem cell differentiation by purified polypeptides". *Nature* **336** (6200): 688–690. doi:10.1038/336688a0. PMID 3143917.

[46] Williams RL, Hilton DJ, Pease S, Willson TA, Stewart CL, Gearing DP, Wagner EF, Metcalf D, Nicola NA, Gough NM (1988). "Myeloid leukaemia inhibitory factor maintains the developmental potential of embryonic stem cells". *Nature* **336** (6200): 684–687. doi:10.1038/336684a0. PMID 3143916.

[47] Ledermann B, Bürki K (1991). "Establishment of a germ-line competent C57BL/6 embryonic stem cell line". *Exp Cell Res* **197** (2): 254–258. doi:10.1016/0014-4827(91)90430-3. PMID 1959560.

[48] Takahashi K, Tanabe K, Ohnuki M, Narita M, Ichisaka T, Tomoda K, Yamanaka S. (2007). "Induction of pluripotent stem cells from adult human fibroblasts by defined factors". *Cell* **131** (5): 861–872. doi:10.1016/j.cell.2007.11.019. PMID 18035408.

[49] Klimanskaya I, Chung Y, Becker S, Lu SJ, Lanza R. (2006). "Human embryonic stem cell lines derived from single blastomeres". *Nature* **444** (7118): 481–5. doi:10.1038/nature05142. PMID 16929302.

[50] US scientists relieved as Obama lifts ban on stem cell research. The Guardian, 10 March 2009

[51] Wernig, Marius; Meissner, Alexander; Foreman, Ruth; Brambrink, Tobias; Ku, Manching; Hochedlinger, Konrad; Bernstein, Bradley E.; Jaenisch, Rudolf (2007-07-19). "In vitro reprogramming of fibroblasts into a pluripotent ES-cell-like state". *Nature* **448** (7151): 318–324. doi:10.1038/nature05944. ISSN 1476-4687. PMID 17554336.

[52] "Embryonic stem cells made without embryos". Reuters. 2007-11-21.

[53] Weiss, Rick (2007-12-07). "Scientists Cure Mice Of Sickle Cell Using Stem Cell Technique: New Approach Is From Skin, Not Embryos". The Washington Post. pp. A02.

[54] Hanna, J.; Wernig, M.; Markoulaki, S.; Sun, C.-W.; Meissner, A.; Cassady, J. P.; Beard, C.; Brambrink, T.; Wu, L.-C.; Townes, T. M.; Jaenisch, R. (2007). "Treatment of Sickle Cell Anemia Mouse Model with iPS Cells Generated from Autologous Skin". *Science* **318** (5858): 1920. doi:10.1126/science.1152092. PMID 18063756.

[55] Helen Briggs (2008-01-17). "US team makes embryo clone of men". BBC. pp. A01.

[56] Ebert, Jessica (24 January 2005). "Human stem cells trigger immune attack". *News from "Nature"* (London: Nature Publishing Group). doi:10.1038/news050124-1. Retrieved 2009-02-27.

[57] Martin MJ, Muotri A, Gage F, Varki A (2005). "Human embryonic stem cells express an immunogenic non-human sialic acid". *Nat. Med.* **11** (2): 228–32. doi:10.1038/nm1181. PMID 15685172.

[58] Klimanskaya I, Chung Y, Meisner L, Johnson J, West MD, Lanza R (2005). "Human embryonic stem cells derived without feeder cells". *Lancet* **365** (9471): 1636–41. doi:10.1016/S0140-6736(05)66473-2. PMID 15885296.

6.8   External links

- Understanding Stem Cells: A View of the Science and Issues from the National Academies

- National Institutes of Health

- University of Oxford practical workshop on pluripotent stem cell technology

- Fact sheet on embryonic stem cells

- Fact sheet on ethical issues in embryonic stem cell research

- Information & Alternatives to Embryonic Stem Cell Research

- A blog focusing specifically on ES cells and iPS cells including research, biotech, and patient-oriented issues

Chapter 7

Embryo

For other uses, see Embryo (disambiguation).
See also: Human embryogenesis and Prenatal development

An **embryo** is a multicellular diploid eukaryote in an early stage of embryogenesis, or development. In general, in organisms that reproduce sexually, an embryo develops from a zygote, the single cell resulting from the fertilization of the female egg cell by the male sperm cell. The zygote possesses half the DNA of each of its two parents. In plants, animals, and some protists, the zygote will begin to divide by mitosis to produce a multicellular organism. The result of this process is an embryo.

In humans, a pregnancy is generally considered to be in the embryonic stage of development between the fifth and the eleventh weeks after fertilization,[1] and is considered a fetus from the twelfth week on.

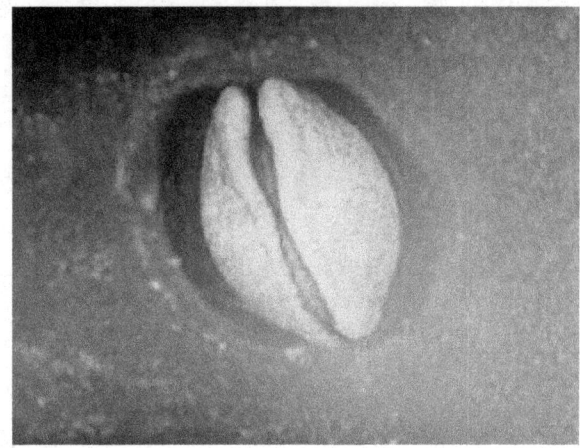

Embryonic development of salamander, circa the 1920s.

7.1 Etymology

First attested in English in the mid-14c., the word *embryon* derives from Medieval Latin *embryo*, itself from Greek ἔμβρυον (*embruon*), lit. "young one",[2] which is the neuter of ἔμβρυος (*embruos*), lit. "growing in",[3] from ἐν (*en*), "in"[4] and βρύω (*bruō*), "swell, be full";[5] the proper Latinized form of the Greek term would be *embryum*.

7.2 Development

In animals, the development of the zygote into an embryo proceeds through specific recognizable stages of blastula, gastrula, and organogenesis. The blastula stage typically features a fluid-filled cavity, the blastocoel, surrounded by a sphere or sheet of cells, also called blastomeres. In a placental mammal, an ovum is fertilized in a fallopian tube through which it travels into the uterus. An embryo is called a fetus at a more advanced stage of development and up until birth or hatching. In humans, this is from the eleventh

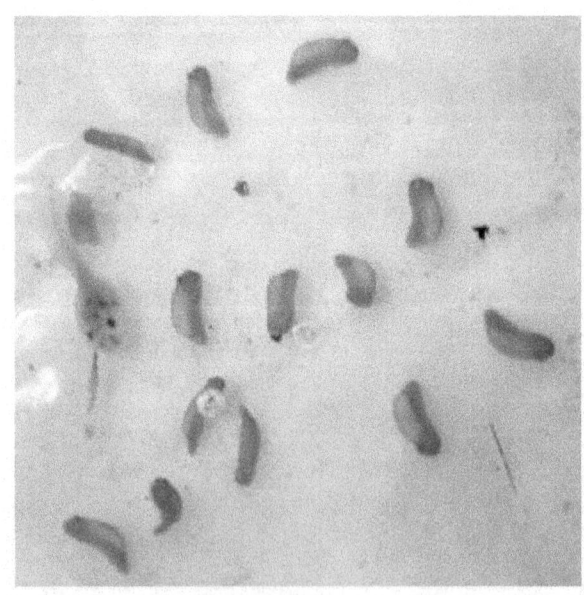

Embryos (and one tadpole) of the wrinkled frog (Rana rugosa)

week of gestation. However, animals which develop in eggs outside the mother's body, are usually referred to as embryos throughout development; e.g. one would refer to a chick embryo, not a "chick fetus," even at later stages.

During gastrulation the cells of the blastula undergo coordinated processes of cell division, invasion, and/or migration to form two (diploblastic) or three (triploblastic) tissue layers. In triploblastic organisms, the three germ layers are called endoderm, ectoderm, and mesoderm. The position and arrangement of the germ layers are highly species-specific, however, depending on the type of embryo produced. In vertebrates, a special population of embryonic cells called the neural crest has been proposed as a "fourth germ layer", and is thought to have been an important novelty in the evolution of head structures.

During organogenesis, molecular and cellular interactions between germ layers, combined with the cells' developmental potential, or competence to respond, prompt the further differentiation of organ-specific cell types. For example, in neurogenesis, a subpopulation of ectoderm cells is set aside to become the brain, spinal cord, and peripheral nerves. Modern developmental biology is extensively probing the molecular basis for every type of organogenesis, including angiogenesis (formation of new blood vessels from pre-existing ones), chondrogenesis (cartilage), myogenesis (muscle), osteogenesis (bone), and many others.

Generally, if a structure pre-dates another structure in evolutionary terms, then it often appears earlier than the second in an embryo; this general observation is sometimes summarized by the phrase "ontogeny recapitulates phylogeny".[6] For example, the backbone is a common structure among all vertebrates such as fish, reptiles, and mammals, and the backbone also appears as one of the earliest structures laid out in all vertebrate embryos. The cerebrum in humans, which is the most sophisticated part of the brain, develops last. This sequencing rule is not absolute, but it is recognized as being partly applicable to development of the human embryo.

7.2.1   Plant embryos

Further information: Sporophyte
In botany, a seed plant *embryo* is part of a seed, consisting of precursor tissues for the leaves, stem (see **hypocotyl**), and root (see **radicle**), as well as one or more **cotyledons**. Once the embryo begins to germinate — grow out from the seed — it is called a **seedling** (plantlet).

Bryophytes and ferns also produce an embryo, but do not produce seeds. In these plants, the embryo begins its existence attached to the inside of the archegonium on a parental gametophyte from which the egg cell was generated. The inner wall of the archegonium lies in close contact with the "foot" of the developing embryo; this "foot" consists of a bulbous mass of cells at the base of the embryo

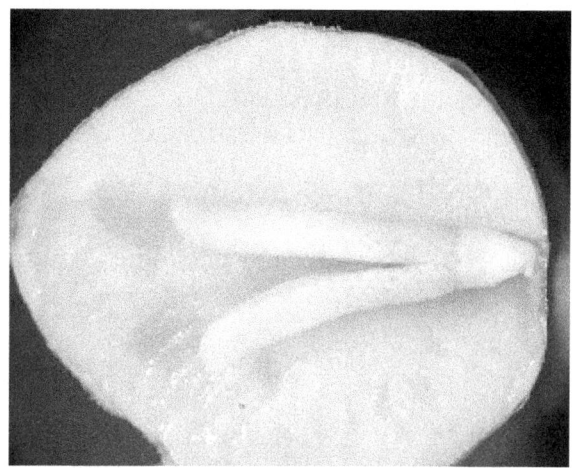

The inside of a Ginkgo seed, showing the embryo

which may receive nutrition from its parent gametophyte. The structure and development of the rest of the embryo varies by group of plants. Once the embryo has expanded beyond the enclosing archegonium, it is no longer termed an embryo.

7.3   Research and technology

Embryos are used in various fields of research and in techniques of assisted reproductive technology. An egg may be fertilized in vitro and the resulting embryo may be frozen for later use. The potential of embryonic stem cell research, reproductive cloning, and germline engineering are currently being explored. Prenatal diagnosis or preimplantation diagnosis enables testing embryos for diseases or conditions.

The embryos of *Arabidopsis thaliana* have been used as a model to understand gene activation, patterning, and organogenesis of seed plants.[7]

In regards to research using human embryos, the ethics and legalities of this application continue to be debated.[8][9][10]

7.3.1   Fossilized embryos

Main article: Fossil embryos

Fossilized animal embryos are known from the Precambrian, and are found in great numbers during the Cambrian period. Even fossilized dinosaur embryos have been discovered.[11]

7.4 Abortion

See also: Miscarriage and Abortion
Some embryos do not survive to the next stage of develop-

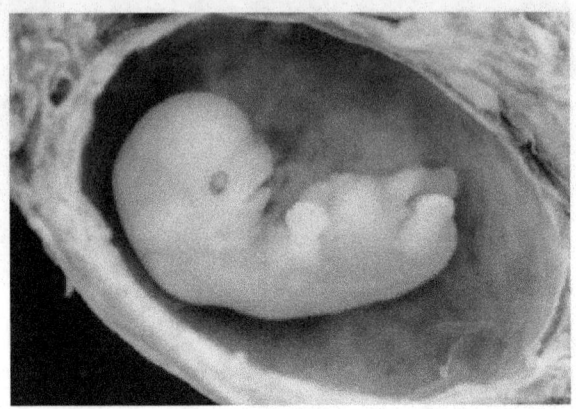

A complete spontaneous abortion at about 6 weeks from conception, i.e. 8 weeks from LMP

ment. When this happens naturally, it is called spontaneous abortion. There are many reasons why this may occur. The most common natural cause of abortion is chromosomal abnormality in animals[12] or genetic load in plants.[13]

In species which produce multiple embryos at the same time, abortion of some embryos can provide the remaining embryos with a greater share of maternal resources. Genetic strains which spontaneously abort their embryos are the source of commercial seedless fruits.

Abortion can also be artificially induced through pharmaceutical and surgical methods.

7.5 See also

- In vitro fertilization

- Plant embryogenesis

- Pregnancy

- Prenatal development

- Proembryo

7.6 Notes

[1] "embryo". *Mayo*. Retrieved 21 February 2016.

[2] ἔμβρυον, Henry George Liddell, Robert Scott, *A Greek-English Lexicon*, on Perseus

[3] ἔμβρυος, Henry George Liddell, Robert Scott, *A Greek-English Lexicon*, on Perseus

[4] ἐν, Henry George Liddell, Robert Scott, *A Greek-English Lexicon*, on Perseus

[5] βρύω, Henry George Liddell, Robert Scott, *A Greek-English Lexicon*, on Perseus

[6] Gould, Stephen. Ontogeny and Phylogeny, p. 206 (1977): "recapitulation was not 'disproved'; it could not be, for too many well-established cases fit its expectations."

[7] Boscá, S; Knauer, S; Laux, T (2011). "Embryonic development in Arabidopsis thaliana: from the zygote division to the shoot meristem". *Frontiers in Plant Science*. doi:10.3389/fpls.2011.00093. PMID 22639618.

[8] Freedman, Jeri. "America Debates Stem Cell Research." New York, NY: The Rosen Publishing Group, 2008.

[9] Sandel, Michael J. "The Case Against Perfection." Michael J. Sandel, 2007.

[10] Zavos, Panayiotis. "Reproductive Cloning is Moral." Ed. James Woodward. The Ethics of Human Cloning: At Issue. Farmington Hills, MI: Greenhaven, 2005. 14–24.

[11] Morelle, Rebecca. "Dinosaur embryo fossils reveal life inside the egg". BBC News. Retrieved 8 August 2015.

[12] Conrad Stöppler, Melissa. Shiel, Jr., William C., ed. "Miscarriage (Spontaneous Abortion)". *MedicineNet.com*. Retrieved 2009-04-07.

[13] Kärkkäinen, Katri; Koski, Veikko (1999). "Why do plants abort so many developing seeds: bad offspring or bad maternal genotypes?". *Evolutionary Ecology* **13** (3): 305–317. doi:10.1023/A:1006746900736. |first2= missing |last2= in Authors list (help)

7.7 External links

- UNSW Embryology - Educational website

- A Comparative Embryology Gallery

- 4-H Embryology, University of Nebraska-Lincoln Extension in Lancaster County

- Video with embryo of a small-spotted catshark inside the egg on YouTube

Chapter 8

Germ layer

See also: Germ cell

A **germ layer** is a primary layer of cells that form during embryogenesis.[1] The three germ layers in vertebrates are particularly pronounced; however, all eumetazoans, (animals more complex than the sponge) produce two or three primary germ layers. Animals with radial symmetry, like cnidarians, produce two germ layers (the ectoderm and endoderm) making them diploblastic. Animals with bilateral symmetry produce a third layer (the mesoderm), between these two layers, making them triploblastic. Germ layers eventually give rise to all of an animal's tissues and organs through the process of organogenesis.

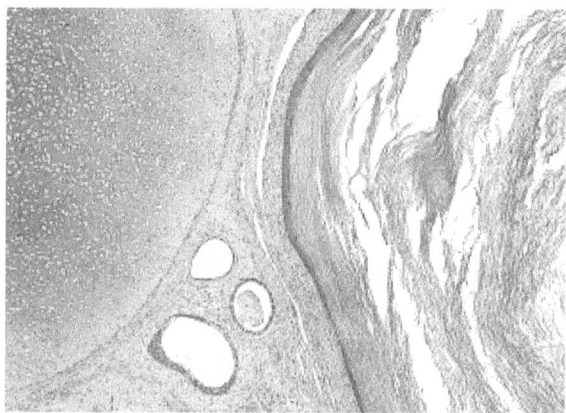

*Micrograph of a teratoma, a tumour that characteristically has tissue from all three **germ layers**. The image shows tissue derived from the mesoderm (immature cartilage - left-upper corner of image), endoderm (gastrointestinal glands - center-bottom of image) and ectoderm (epidermis - right of image). H&E stain.*

8.1 Germ layers

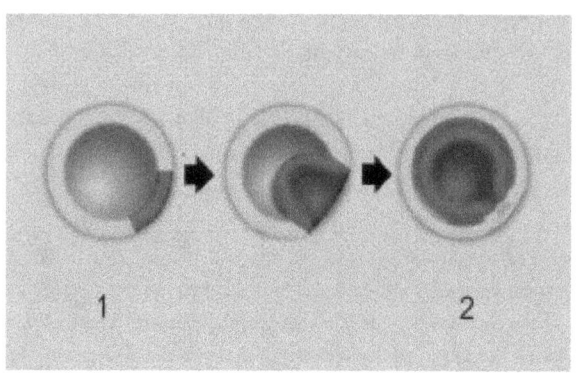

Gastrulation of a diploblast: *The formation of germ layers from a (1) blastula to a (2) gastrula. Some of the ectoderm cells (orange) move inward forming the endoderm (red).*

Caspar Friedrich Wolff observed organization of the early embryo in leaf-like layers. In 1817, Heinz Christian Pander discovered three primordial germ layers while studying chick embryos. Between 1850 and 1855, Robert Remak had further refined the germ cell layer concept, and introduced into English were the terms "mesoderm" by Huxley

in 1871 and "ectoderm" and "endoderm" by Lankester in 1873.

Among animals, sponges show the simplest organization, having a single germ layer. Although they have differentiated cells (e.g. collar cells), they lack true tissue coordination. Diploblastic animals, Cnidaria and Ctenophora, show an increase in complexity, having two germ layers, the endoderm and ectoderm. Diploblastic animals are organized into recognisable tissues. All higher animals (from flatworms to humans) are triploblastic, possessing a mesoderm in addition to the germ layers found in Diploblasts. Triploblastic animals develop recognisable organs.

8.1.1 Development

Fertilization leads to the formation of a zygote. During the next stage, cleavage, mitotic cell divisions transform the zygote into a hollow ball of cells, a blastula. This early embryonic form undergoes gastrulation, forming a gastrula with either two or three layers (the germ layers). In all vertebrates, these progenitor cells differentiate into all adult tissues and organs.[2]

In humans, after about three days, the zygote forms a solid mass of cells by mitotic division, called a morula. This then changes to a blastocyst, consisting of an outer layer called a trophoblast, and an inner cell mass called the embryoblast. Filled with uterine fluid, the blastocyst breaks out of the zona pellucida and undergoes implantation. The inner cell mass initially has two layers: the hypoblast and epiblast. At the end of the second week, a primitive streak appears. The epiblast in this region moves towards the primitive streak, dives down into it, and forms a new layer, called the endoderm, pushing the hypoblast out of the way (this goes on to form the amnion.) The epiblast keeps moving and forms a second layer, the mesoderm. The top layer is now called the ectoderm.[3]

8.1.2 Endoderm

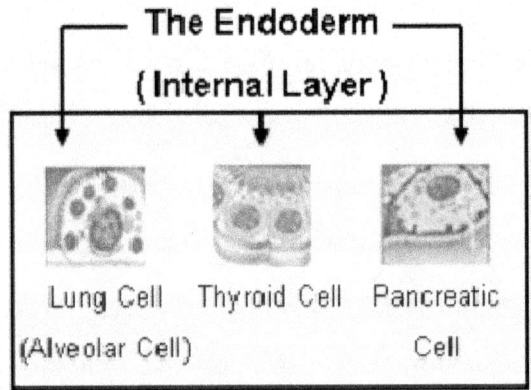

The endoderm produces tissue within the lungs, thyroid, and pancreas.

Main article: Endoderm

The **endoderm** is one of the germ layers formed during animal embryogenesis. Cells migrating inward along the archenteron form the inner layer of the gastrula, which develops into the endoderm.

The endoderm consists at first of flattened cells, which subsequently become columnar. It forms the epithelial lining of the whole of the digestive tube except part of the mouth and pharynx and the terminal part of the rectum (which are lined by involutions of the ectoderm). It also forms the lining cells of all the glands which open into the digestive tube, including those of the liver and pancreas; the epithelium of the auditory tube and tympanic cavity; the trachea, bronchi,

and air cells of the lungs; the urinary bladder and part of the urethra; and the follicle lining of the thyroid gland and thymus.

The endoderm forms: the stomach, the colon, the liver, the pancreas, the urinary bladder, the epithelial parts of trachea, the lungs, the pharynx, the thyroid, the parathyroid, and the intestines.

8.1.3 Mesoderm

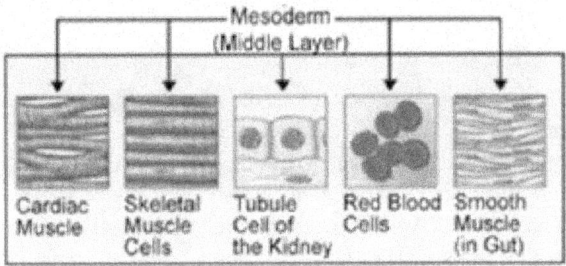

The mesoderm aids in the production of cardiac muscle, skeletal muscle, smooth muscle, tissues within the kidneys, and red blood cells.

Main article: Mesoderm

The **mesoderm** germ layer forms in the embryos of triploblastic animals. During gastrulation, some of the cells migrating inward contribute to the mesoderm, an additional layer between the endoderm and the ectoderm. The formation of a mesoderm leads to the development of a coelom. Organs formed inside a coelom can freely move, grow, and develop independently of the body wall while fluid cushions and protects them from shocks.

The mesoderm has several components which develop into tissues: intermediate mesoderm, paraxial mesoderm, lateral plate mesoderm, and chorda-mesoderm. The chorda-mesoderm develops into the notochord. The intermediate mesoderm develops into kidneys and gonads. The paraxial mesoderm develops into cartilage, skeletal muscle, and dermis. The lateral plate mesoderm develops into the circulatory system (including the heart and spleen), the wall of the gut, and wall of the human body.[4]

Through cell signaling cascades and interactions with the ectodermal and endodermal cells, the mesodermal cells begin the process of differentiation.[5]

The mesoderm forms: muscle (smooth and striated), bone, cartilage, connective tissue, adipose tissue, circulatory system, lymphatic system, dermis, genitourinary system, serous membranes, and notochord.

8.1.4   Ectoderm

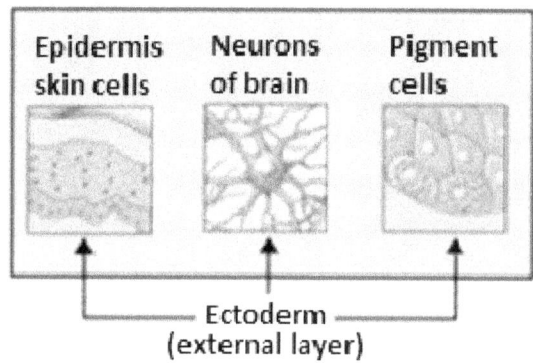

The ectoderm produces tissues within the epidermis, aids in the formation of neurons within the brain, and constructs melanocytes.

Main article: Ectoderm

The **ectoderm** generates the outer layer of the embryo, and it forms from the embryo's epiblast.[6] The ectoderm develops into the surface ectoderm, neural crest, and the neural tube.[7]

The surface ectoderm develops into: epidermis, hair, nails, lens of the eye, sebaceous glands, cornea, tooth enamel, the epithelium of the mouth and nose.

The neural crest of the ectoderm develops into: peripheral nervous system, adrenal medulla, melanocytes, facial cartilage, dentin of teeth.

The neural tube of the ectoderm develops into: brain, spinal cord, posterior pituitary, motor neurons, retina.

Note: The anterior pituitary develops from the ectodermal tissue of Rathke's pouch.

8.1.5   Neural crest

Because of its great importance, the neural crest is sometimes considered a fourth germ layer.[8] It is, however, derived from the ectoderm.

8.2   See also

- Histogenesis

- Neurulation

8.3   References

[1] Gilbert, Scott F (2003). "The Epidermis and the Origin of Cutaneous Structures". *Developmental Biology*. Sinauer Associates.

[2] Gilbert, Scott F (2000). "Comparative Embryology". *Developmental Biology*. Sinauer Associates.

[3] Gilbert, Scott F (2000). "Early Mammalian Development". *Developmental Biology*. Sinauer Associates.

[4] Gilbert, Scott F (2003). "Paraxial and Intermediate Mesoderm". *Developmental Biology*. Sinauer Associates.

[5] Brand, Thomas (1 June 2003). "Heart development: molecular insights into cardiac specification and early morphogenesis". *Developmental Biology* **258** (1): 1–19. doi:10.1016/S0012-1606(03)00112-X.

[6] Gilbert, Scott F (2003). "Early Mammalian Development". *Developmental Biology*. Sinauer Associates.

[7] Gilbert, Scott F (2003). "The Central Nervous System and The Epidermis". *Developmental Biology*. Sinauer Associates.

[8] Hall BK (2000). "The neural crest as a fourth germ layer and vertebrates as quadroblastic not triploblastic". *Evolution & Development 2, 3-5* **2**: 3–5. doi:10.1046/j.1525-142x.2000.00032.x. PMID 11256415.

Chapter 9

Inner cell mass

In early embryogenesis of most eutherian mammals, the **inner cell mass** (abbreviated **ICM** and also known as the **embryoblast** or pluriblast, the latter term being applicable to all mammals) is the mass of cells inside the primordial embryo that will eventually give rise to the definitive structures of the fetus. This structure forms in the earliest steps of development, before implantation into the endometrium of the uterus has occurred. The ICM lies within the blastocoele (more correctly termed "blastocyst cavity," as it is not strictly homologous to the blastocoele of anamniote vertebrates) and is entirely surrounded by the single layer of cells called trophoblast.

9.1   Further development

The physical and functional separation of the inner cell mass from the trophectoderm (TE) is a special feature of mammalian development and is the first cell lineage specification in these embryos. Following fertilization in the oviduct, the mammalian embryo undergoes a relatively slow round of cleavages to produce an eight cell morula. Each cell of the morula, called a blastomere, increases surface contact with its neighbors in a process called compaction. This results in a polarization of the cells within the morula, and further cleavage yields a blastocyst of roughly 32 cells.[1] In mice, about 12 internal cells comprise the new inner cell mass and 20 – 24 cells comprise the surrounding trophectoderm.[2][3] There is variation between species of mammals as to number of cells at compaction with bovine embryos showing differences related to compaction as early as 9-15 cells and in rabbits not until after 32 cells.[4] There is also interspecies variation in gene expression patterns in early embryos [5]

The ICM and the TE will generate distinctly different cell types as implantation starts and embryogenesis continues. Trophectoderm cells form extraembryonic tissues, which act in a supporting role for the embryo proper. Furthermore, these cells pump fluid into the interior of the blastocyst, causing the formation of a polarized blastocyst with the ICM attached to the trophectoderm at one end (see figure). This difference in cellular localization causes the ICM cells exposed to the fluid cavity to adopt a primitive endoderm (or hypoblast) fate, while the remaining cells adopt a primitive ectoderm (or epiblast) fate. The hypoblast contributes to extraembryonic membranes and the epiblast will give rise to the ultimate embryo proper as well as some extraembryonic tissues.[1]

9.2   Regulation of cellular specification

Since segregation of pluripotent cells of the inner cell mass from the remainder of the blastocyst is integral to mammalian development, considerable research has been performed to elucidate the corresponding cellular and molecular mechanisms of this process. There is primary interest in which transcription factors and signaling molecules direct blastomere asymmetric divisions leading to what are known as inside and outside cells and thus cell lineage specification. However, due to the variability and regulative nature of mammalian embryos, experimental evidence for establishing these early fates remains incomplete.[2]

At the transcription level, the transcription factors Oct4, Nanog, Cdx2, and Tead4 have all been implicated in establishing and reinforcing the specification of the ICM and the TE in early mouse embryos.[2]

- Oct4: *Oct4* is expressed in the ICM and participate in maintaining its pluripotency, a role that has been recapitulated in ICM derived mouse embryonic stem cells.[6] *Oct4* genetic knockout cells both in vivo and in culture display TE morphological characteristics. It has been shown that one transcriptional target of Oct4 is the *Fgf4* gene. This gene normally encodes a ligand secreted by the ICM, which induces proliferation in the adjacent polar TE.[6]

- Nanog: *Nanog* is also expressed in the ICM and par-

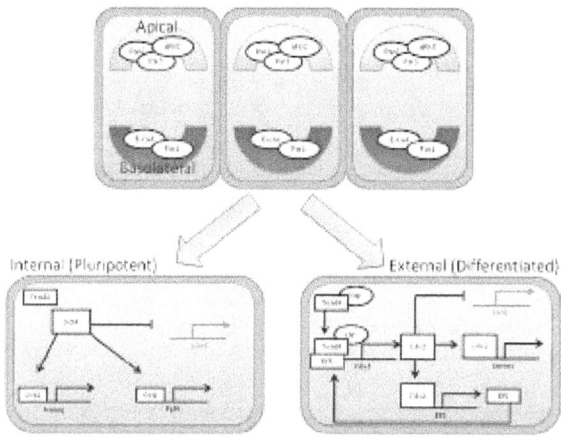

Early embryo apical and basolateral polarization is established at the 8-16 cell stage following compaction. This initial difference in environment strengthens a transcriptional feedback loop in either an internal or external direction. Inside cells express high levels of Oct4 which maintains pluripotency and suppresses Cdx2. Outside cells express high levels of Cdx2 which causes TE differentiation and suppresses Oct4.

ticipates in maintaining its pluripotency. In contrast with *Oct4*, studies of *Nanog*-null mice do not show the reversion of the ICM to a TE-like morphology, but demonstrate that loss of *Nanog* prevents the ICM from generating primitive endoderm.[7]

- Cdx2: *Cdx2* is strongly expressed in the TE and is required for maintaining its specification. Knockout mice for the *Cdx2* gene undergo compaction, but lose the TE epithelial integrity during the late blastocyst stage. Furthermore, *Oct4* expression is subsequently raised in these TE cells, indicating Cdx2 plays a role in suppressing *Oct4* in this cell lineage. Moreover, embryonic stem cells can be generated from *Cdx2*-null mice, demonstrating that Cdx2 is not essential for ICM specification.[8]

- Tead4: Like *Cdx2*, *Tead4* is required for TE function, although the transcription factor is expressed ubiquitously. *Tead4*-null mice similarly undergo compaction, but fail to generate the blastocoel cavity. Like *Cdx2*-null embryos, the Tead4-null embryos can yield embryonic stem cells, indicating that Tead4 is dispensable for ICM specification.[9] Recent work has shown that *Tead4* may help to upregulate Cdx2 in the TE and its transcriptional activity depends on the coactivator Yap. Yap's nuclear localization in outside cells allows it to contribute to TE specificity, whereas inside cells sequester Yap in the cytoplasm through a phosphorylation event.[10]

Together these transcription factors function in a positive feedback loop that strengthens the ICM to TE cellular allocation. Initial polarization of blastomeres occurs at the 8-16 cell stage. An apical-basolateral polarity is visible through the visualization of apical markers such as Par3, Par6, and aPKC as well as the basal marker E-Cadherin.[2] The establishment of such a polarity during compaction is thought to generate an environmental identity for inside and outside cells of the embryo. Consequently, stochastic expression of the above transcription factors is amplified into a feedback loop that specifies outside cells to a TE fate and inside cells to an ICM fate. In the model, an apical environment turns on *Cdx2*, which upregulates its own expression through a downstream transcription factor, Elf5. In concert with a third transcription factor, Eomes, these genes act to suppress pluripotency genes like *Oct4* and *Nanog* in the outside cells.[2][8] Thus, TE becomes specified and differentiates. Inside cells, however, do not turn on the *Cdx2* gene, and express high levels of *Oct4*, *Nanog*, and *Sox2*.[2][3] These genes suppress *Cdx2* and the inside cells maintain pluripotency generate the ICM and eventually the rest of the embryo proper.

Although this dichotomy of genetic interactions is clearly required to divide the blastomeres of the mouse embryo into both the ICM and TE identities, the initiation of these feedback loops remains under debate. Whether they are established stochastically or through an even earlier asymmetry is unclear, and current research seeks to identify earlier markers of asymmetry. For example, some research correlates the first two cleavages during embryogenesis with respect to the prospective animal and vegetal poles with ultimate specification. The asymmetric division of epigenetic information during these first two cleavages, and the orientation and order in which they occur, may contribute to a cell's position either inside or outside the morula..[11][12]

9.3   Stem cells

Blastomeres isolated from the ICM of mammalian embryos and grown in culture are known as embryonic stem (ES) cells. These pluripotent cells, when grown in a carefully coordinated media, can give rise to all three germ layers (ectoderm, endoderm, and mesoderm) of the adult body.[13] For example, the transcription factor LIF4 is required for mouse ES cells to be maintained in vitro.[14] Blastomeres are dissociated from an isolated ICM in an early blastocyst, and their transcriptional code governed by *Oct4*, *Sox2*, and *Nanog* helps maintain an undifferentiated state.

One benefit to the regulative nature in which mammalian embryos develop is the manipulation of blastomeres of the ICM to generate knockout mice. In mouse, mutations in a gene of interest can be introduced retrovirally into cultured ES cells, and these can be reintroduced into the ICM of an intact embryo. The result is a chimeric mouse, which develops with a portion of its cells containing the ES cell genome. The aim of such a procedure is to incorporate the mutated gene into the germ line of the mouse such that its progeny will be missing one or both alleles of the gene of interest. Geneticists widely take advantage of this ICM manipulation technique in studying the function of genes in the mammalian system,.[1][13]

9.4 Additional images

- Blastodermic vesicle of Vespertilio murinus.

- Section through embryonic disk of Vespertilio murinus.

9.5 See also

- Homeobox genes

9.6 References

[1] Wolpert, Lewis. Principles of Development: Third Edition. 2007. Oxford University Press.

[2] Marikawa, Yusuke, et al. Establishment of Trophectoderm and Inner Cell Mass Lineages in the Mouse Embryo. Molecular Reproduction & Development 76:1019–1032 (2009)

[3] Suwinska A, Czołowska R, Ozdze_nski W, Tarkowski AK. 2008. Blastomeres of the mouse embryo lose totipotency after the fifth cleavage division: Expression of Cdx2 and Oct4 and developmental potential of inner and outer blastomeres of 16- and 32-cell embryos. Dev Biol 322:133–144.

[4] Koyama *et al* Analysis of Polarity of Bovine and Rabbit Embryos by Scanning Electron Microscopy Biol of Reproduction, 50, 163-170 1994

[5] Kuijk. *et al* Validation of reference genes for quantitative RT-PCR studies in porcine oocytes and preimplantation embryos BMC Developmental Biology 2007, 7:58 doi: 10.1186/1471-213X-7-58

[6] Nichols J, Zevnik B, Anastassiadis K, Niwa H, Klewe-Nebenius D, Chambers I, Sch€oler H, Smith A. 1998. Formation of pluripotent stem cells in the mammalian embryo depends on the POU transcription factor Oct4. Cell 95:379–391.

[7] Rodda DJ, Chew JL, Lim LH, Loh YH, Wang B, Ng HH, Robson P. 2005. Transcriptional regulation of nanog by OCT4 and SOX2. J Biol Chem 280:24731–24737.

[8] Strumpf D, Mao CA, Yamanaka Y, Ralston A, Chawengsaksophak K, Beck F, Rossant J. 2005. Cdx2 is required for correct cell fate specification and differentiation of trophectoderm in the mouse blastocyst. Development 132:2093–2102.

[9] Nishioka N, Yamamoto S, Kiyonari H, Sato H, Sawada A, Ota M, Nakao K, Sasaki H. 2008. *Tead4* is required for specification of trophectoderm in pre-implantation mouse embryos. Mech Dev 125:270–283.

[10] Nishioka N, et al. 2009. The Hippo signaling pathway components Lats and Yap pattern Tead4 activity to distinguish mouse trophectoderm from inner cell mass. Dev Cell 16: 398–410.

[11] Bischoff, Marcus, et al. Formation of the embryonic-abembryonic axis of the mouse blastocyst: relationships between orientation of early cleavage divisions and pattern of symmetric/asymmetric divisions. Development 135, 953-962 (2008)

[12] Jedrusik, Agnieszka, et al. Role of Cdx2 and cell polarity in cell allocation and specification of trophectoderm and inner cell mass in the mouse embryo. Genes Dev. 2008 22: 2692-2706

[13] Robertson, Elizabeth , et al. Germ-line transmission of genes introduced into cultured pluripotential cells by retroviral vector. Nature 323, 445 - 448 (2 October 1986)

[14] Smith AG, Heath JK, Donaldson DD, Wong GG, Moreau J, Stahl M and Rogers D (1988) Inhibition of pluripotential embryonic stem cell differentiation by purified polypeptides. Nature, 336, 688–690

9.7 External links

- Thomas A. Marino, Ph.D. - Embryology Lectures, Temple university (archive)

- Week 1: Implantation

Chapter 10

Stem-cell line

A **stem-cell line** is a group of stem cells that is cultured in vitro and can be propagated indefinitely. Stem-cell lines are derived from either animal or human tissues and come from one of three sources: embryonic stem cells, adult stem cells, or induced stem cells. They are commonly used in research and regenerative medicine.

10.1   Properties

Main article: Stem cell

By definition, stem cells possess two properties: (1) they can self-renew, which means that they can divide indefinitely while remaining in an undifferentiated state; and (2) they are pluripotent or multipotent, which means that they can differentiate to form specialized cell types. Due to the self-renewal capacity of stem cells, a stem cell line can be cultured *in vitro* indefinitely.

A stem-cell line is distinctly different from an immortalized cell line, such as the HeLa line. While stem cells can propagate indefinitely in culture due to their inherent properties, immortalized cells would not normally divide indefinitely but have gained this ability due to mutation. Immortalized cell lines can be generated from cells isolated from tumors, or mutations can be introduced to make the cells immortal.[1]

A stem cell line is also distinct from primary cells. Primary cells are cells that have been isolated and then used immediately. Primary cells cannot divide indefinitely and thus cannot be cultured for long periods of time in vitro.

10.2   Types and methods of derivation

10.2.1   Embryonic stem cell line

Main article: Embryonic stem cell

An embryonic stem cell line is created from cells derived from the inner cell mass of a blastocyst, an early stage, pre-implantation embryo.[2] In humans, the blastocyst stage occurs 4–5 days post fertilization. To create an embryonic stem cell line, the inner cell-mass is removed from the blastocyst, separated from the trophoectoderm, and cultured on a layer of supportive cells in vitro. In the derivation of human embryonic stem cell lines, embryos leftover from in vitro fertilization (IVF) procedures are used. The fact that the embryo is destroyed during the process has raised controversy and ethical concerns.

Embryonic stem cells are pluripotent, meaning they can differentiate to form all cell types in the body. In vitro, embryonic stem cells can be cultured under defined conditions to keep them in their pluripotent state, or they can be stimulated with biochemical and physical cues to differentiate them to different cell types.

10.2.2   Adult stem cell line

Main article: Adult stem cell

Adult stem cells are found in juvenile or adult tissues. Adult stem cells are multipotent: they can generate a limited number of differentiated cell types (unlike pluripotent embryonic stem cells). Types of adult stem cells include hematopoietic stem cells and mesenchymal stem cells. Hematopoetic stem cells are found in the bone marrow and generate all cells of the immune system all blood cell types. Mesenchymal stem cells are found in umbilical cord blood, amniotic fluid, and adipose tissue and can generate a number of cell types, including osteoblasts, chondrocytes, and adipocytes. In medicine, adult stem cells are mostly commonly used in bone marrow transplants to treat many bone and blood cancers as well as some autoimmune diseases.[3]

(See Hematopoietic stem cell transplantation)

Of the types of adult stem cells have successfully been isolated and identified, only mesenchymal stem cells can successfully be grown in culture for long periods of time. Other adult stem cell types, such as hematopoietic stem cells, are difficult to grow and propagate in vitro.[4] Identifying methods for maintaining hematopoietic stem cells in vitro is an active area of research. Thus, while mesenchymal stem cell lines exist, other types of adult stem cells that are grown in vitro can better be classified as primary cells.

10.2.3 Induced pluripotent stem-cell (iPSC) line

Main articles: Induced pluripotent stem cells and Induced stem cells

Induced pluripotent stem cell (iPSC) lines are pluripotent stem cells that have been generated from adult/somatic cells. The method of generating iPSCs was developed by Shinya Yamanaka's lab in 2006; his group demonstrated that the introduction of four specific genes could induce somatic cells to revert to a pluripotent stem cell state.[5]

Compared to embryonic stem-cell lines, iPSC lines are also pluripotent in nature but can be derived without the use of human embryos—a process that has raised ethical concerns. Furthermore, patient-specific iPSC cell lines can be generated—that is, cell lines that are genetically matched to an individual. Patient-specific iPSC lines have been generated for the purposes of studying diseases[6] and for developing patient-specific medical therapies.

10.3 Methods of culture

Main article: Cell culture

Stem-cell lines are grown and maintained at specific temperature and atmospheric conditions (37 degrees Celsius and 5% CO_2) in incubators. Culture conditions such as the cell growth medium and surface on which cells are grown vary widely depending on the specific stem cell line. Different biochemical factors can be added to the medium to control the cell phenotype—for example to keep stem cells in a pluripotent state or to differentiate them to a specific cell type.

10.4 Uses

Stem-cell lines are used in research and regenerative medicine. They can be used to study stem-cell biology and early human development. In the field of regenerative medicine, it has been proposed that stem cells be used in cell-based therapies to replace injured or diseased cells and tissues. Examples of conditions that researchers are working to develop stem-cell-based treatments for include neurodegenerative diseases, diabetes, and spinal cord injuries.

10.5 Ethical issues

Main article: Stem cell controversy

There is controversy associated with the derivation and use of human embryonic stem cell lines. This controversy stems from the fact that derivation of human embryonic stem cells requires the destruction of a blastocyst-stage, pre-implantation human embryo. There is a wide range of viewpoints regarding the moral consideration that blastocyst-stage human embryos should be given.[7][8]

10.6 Access to human embryonic stem-cell lines

10.6.1 United States

In the United States, Executive Order 13505 established that federal money can be used for research in which approved human embryonic stem-cell (hESC) lines are used, but it cannot be used to derive new lines.[9] The National Institutes of Health (NIH) Guidelines on Human Stem Cell Research, effective July 7, 2009, implemented the Executive Order 13505 by establishing criteria which hESC lines must meet to be approved for funding.[10] The NIH Human Embryonic Stem Cell Registry can be accessed online and has updated information on cell lines eligible for NIH funding.[11] There are 279 approved lines as of April 2014.

Studies have found that approved hESC lines are not uniformly used in the US data from cell banks and surveys of researchers indicate that only a handful of the available hESC lines are routinely used in research. Access and utility are cited as the two primary factors influencing what hESC lines scientists choose to work with.[12]

A 2011 survey of stem cell scientists in the US who use hESC lines in their research found that 54% of respondents used two or fewer lines and 75% used three or fewer lines.[13]

Another study tracked cell-line requests fulfilled from the largest US repositories, the National Stem Cell Bank (NSCB) and the Harvard Stem Cell Institute (HSCI; Cambridge, MA, USA), for the periods March 1999 – December 2008 (for NSCB) and April 2004 – December 2008 (for HSCI).[14] For NSCB, out of twenty-one approved cell lines, 77% of requests were for two of the lines (H1 and H9). For HSCI, out of the 17 lines requested more than once, 24.7% of requests were for the two most commonly requested lines.

10.7 See also

- Stem cell
- Embryonic stem cell
- Induced pluripotent stem cell
- Induced stem cells
- Adult stem cell
- Cell culture
- Immortalised cell line
- Stem-cell controversy
- Stem-cell treatments

10.8 References

[1] Irfan Maqsood, M; M. M.; Bahrami, A. R.; Ghasroldasht, M. M. (2013). "Immortality of cell lines: Challenges and advantages of establishment". *Cell Biology International* **37** (10): 1038–1045. doi:10.1002/cbin.10137. PMID 23723166.

[2] Thomson, JA; Itskovitz-Eldor J; Shapiro SS; Waknitz MA; Swiergiel JJ; Marshall VS; Jones JM (November 6, 1998). "Embryonic stem cell lines derived from human blastocysts". *Science*. 282 **6** (5391): 1145–1147. doi:10.1126/science.282.5391.1145. PMID 9804556.

[3] http://stemcells.nih.gov/info/basics/pages/basics4.aspx. Missing or empty |title= (help)

[4] Walasek, MA; van Os R; de Haan G (August 2012). "Hematopoietic stem cell expansion: challenges and opportunities". *Ann N Y Acad Sci* **1266**: 138–150. doi:10.1111/j.1749-6632.2012.06549.x.

[5] Takahashi, Katzutoshi; Shinya Yamanaka (August 25, 2006). "Induction of Pluripotent Stem Cells from Mouse Embryonic and Adult Fibroblast Cultures by Defined Factors". *Cell* **126** (4): 663–676. doi:10.1016/j.cell.2006.07.024. PMID 16904174.

[6] Park, IH; Arora, N; Huo, H; Maherali, N; Ahfeldt, T; Shimamura, A; Lensch, MW; Cowan, C; Hochedlinger, K; Daley, GQ (September 5, 2008). "Disease-specific induced pluripotent stem cells". *Cell* **134** (5): 877–886. doi:10.1016/j.cell.2008.07.041. PMC 2633781. PMID 18691744.

[7] George, Robert P; Alfonso Gomez-Lobo (2005). "The Moral Status of the Human Embryo". *Perspectives in Biology and Medicine* **48** (2): 201–210. doi:10.1353/pbm.2005.0052.

[8] Cohen, Cynthia B (June 25, 2007). *Renewing the Stuff of Life: Stem Cells, Ethics, and Public Policy*. Oxford University Press. ISBN 9780195305241.

[9] "Executive Order: Removing barriers to responsible scientific research involving human stem cells". The White House.

[10] "National Institutes of Health Guidelines on Human Stem Cell Research". Retrieved 24 April 2014.

[11] "NIH Human Embryonic Stem Cell Registry". Retrieved 24 April 2014.

[12] Levine, Aaron D (December 2011). "Access to human embryonic stem cell lines". *Nature Biotechnology* **29** (12): 1079–1081. doi:10.1038/nbt.2029.

[13] Levine, Aaron D (December 2011). "Access to human embryonic stem cell lines". *Nature Biotechnology* **29** (12): 1079–1081. doi:10.1038/nbt.2029.

[14] Christopher, Thomas Scott; Jennifer B. McCormick; Jason Owen-Smith (August 2009). "And then there were two: use of hESC lines". *Nature Biotechnology* **27** (8): 696–697. doi:10.1038/nbt0809-696.

Chapter 11

Adult stem cell

Adult stem cells are undifferentiated cells, found throughout the body after development, that multiply by cell division to replenish dying cells and regenerate damaged tissues. Also known as **somatic stem cells** (from Greek Σωματικός, meaning *of the body*), they can be found in juvenile as well as adult animals and human bodies.

Scientific interest in adult stem cells is centered on their ability to divide or *self-renew* indefinitely, and generate all the cell types of the organ from which they originate, potentially regenerating the entire organ from a few cells. Unlike embryonic stem cells, the use of human adult stem cells in research and therapy is not considered to be controversial, as they are derived from adult tissue samples rather than human embryos designated for scientific research. They have mainly been studied in humans and model organisms such as mice and rats.

11.1 Defining properties

A stem cell possesses two properties:

- *Self-renewal*, which is the ability to go through numerous cycles of cell division while still maintaining its undifferentiated state.

- *multipotency* or *multidifferentiative potential*, which is the ability to generate progeny of several distinct cell types, (for example glial cells and neurons) as opposed to unipotency, which is the term for cells that are restricted to producing a single-cell type. However, some researchers do not consider multipotency to be essential, and believe that unipotent self-renewing stem cells can exist.[1] These properties can be illustrated with relative ease *in vitro*, using methods such as clonogenic assays, where the progeny of a single cell is characterized. However, it is known that *in vitro* cell culture conditions can alter the behavior of cells, proving that a particular subpopulation of cells possesses stem cell properties *in vivo* is challenging, and so con-

siderable debate exists as to whether some proposed stem cell populations in the adult are indeed stem cells.

11.2 Lineage

To ensure the safety of others, stem cells undergo two types of cell division (see *Stem cell division and differentiation* diagram). Symmetric division gives rise to two identical daughter cells, both endowed with stem cell properties, whereas asymmetric division produces only one of those stem cells and a progenitor cell with limited self-renewal potential. Progenitors can go through several rounds of cell division before finally differentiating into a mature cell. It is believed that the molecular distinction between symmetric and asymmetric divisions lies in differential segregation of cell membrane proteins (such as receptors) between the daughter cells.

11.3 Multidrug resistance

Adult stem cells express transporters of the ATP-binding cassette family that actively pump a diversity of organic molecules out of the cell.[2] Many pharmaceuticals are exported by these transporters conferring multidrug resistance onto the cell. This complicates the design of drugs, for instance neural stem cell targeted therapies for the treatment of clinical depression.

11.4 Signaling pathways

Adult stem cell research has been focused on uncovering the general molecular mechanisms that control their self-renewal and differentiation.

- **Notch**

71

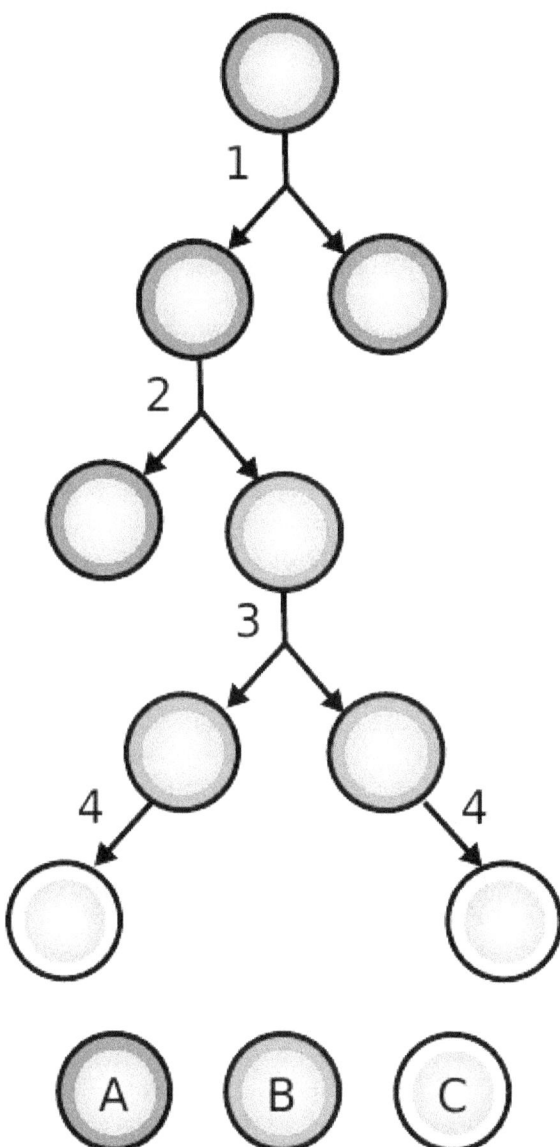

Stem cell division and differentiation. A – stem cells; B – progenitor cell; C – differentiated cell; 1 – symmetric stem cell division; 2 – asymmetric stem cell division; 3 – progenitor division; 4 – terminal differentiation

The Notch pathway has been known to developmental biologists for decades. Its role in control of stem cell proliferation has now been demonstrated for several cell types including haematopoietic, neural, and mammary[3] stem cells.

• **Wnt**

These developmental pathways are also strongly implicated as stem cell regulators.[4]

• **TGFβ**

11.5 Plasticity/Adult stem cell pluripotency

Discoveries in recent years have suggested that adult stem cells might have the ability to differentiate into cell types from different germ layers. For instance, neural stem cells from the brain, which are derived from ectoderm, can differentiate into ectoderm, mesoderm, and endoderm.[5] Stem cells from the bone marrow, which is derived from mesoderm, can differentiate into liver, lung, GI tract and skin, which are derived from endoderm and mesoderm.[6] This phenomenon is referred to as stem cell transdifferentiation or plasticity. It can be induced by modifying the growth medium when stem cells are cultured *in vitro* or transplanting them to an organ of the body different from the one they were originally isolated from. There is yet no consensus among biologists on the prevalence and physiological and therapeutic relevance of stem cell plasticity. More recent findings suggest that pluripotent stem cells may reside in blood and adult tissues in a dormant state.[7] These cells are referred to as "Blastomere Like Stem Cells" (Am Surg. 2007 Nov;73:1106-10) and "very small embryonic like" - "VSEL" stem cells, and display pluripotency in vitro.[7] As BLSC's and VSEL cells are present in virtually all adult tissues, including lung, brain, kidneys, muscles, and pancreas[8] Co-purification of BLSC's and VSEL cells with other populations of adult stem cells may explain the apparent pluripotency of adult stem cell populations. However, recent studies have shown that both human and murine VSEL cells lack stem cell characteristics and are not pluripotent.[9][10][11][12]

11.6 Aging

Stem cell function becomes impaired with age, and this contributes to progressive deterioration of tissue maintenance and repair.[13] A likely important cause of increasing stem cell dysfunction is age-dependent accumulation of DNA damage in both stem cells and the cells that comprise the stem cell environment.[13] (See also DNA damage theory of aging.)

11.7 Types

11.7.1 Hematopoietic stem cells

[14]

Main article: Hematopoietic stem cell

Hematopoietic stem cells are found in the bone marrow and umbilical cord blood and give rise to all the blood cell types.

11.7.2 Mammary stem cells

Mammary stem cells provide the source of cells for growth of the mammary gland during puberty and gestation and play an important role in carcinogenesis of the breast.[15] Mammary stem cells have been isolated from human and mouse tissue as well as from cell lines derived from the mammary gland. Single such cells can give rise to both the luminal and myoepithelial cell types of the gland, and have been shown to have the ability to regenerate the entire organ in mice.[15]

11.7.3 Intestinal stem cells

Intestinal stem cells divide continuously throughout life and use a complex genetic program to produce the cells lining the surface of the small and large intestines.[16] Intestinal stem cells reside near the base of the stem cell niche, called the crypts of Lieberkuhn. Intestinal stem cells are probably the source of most cancers of the small intestine and colon.[17]

11.7.4 Mesenchymal stem cells

Main article: Mesenchymal stem cell

Mesenchymal stem cells (MSCs) are of stromal origin and may differentiate into a variety of tissues. MSCs have been isolated from placenta, adipose tissue, lung, bone marrow and blood, Wharton's jelly from the umbilical cord,[18] and teeth (perivascular niche of dental pulp and periodontal ligament).[19] MSCs are attractive for clinical therapy due to their ability to differentiate, provide trophic support, and modulate innate immune response.[18]

11.7.5 Endothelial stem cells

Main article: Endothelial stem cell

Endothelial stem cells are one of the three types of multipotent stem cells found in the bone marrow. They are a rare and controversial group with the ability to differentiate into endothelial cells, the cells that line blood vessels.

11.7.6 Neural stem cells

Main article: Neural stem cell

The existence of stem cells in the adult brain has been postulated following the discovery that the process of neurogenesis, the birth of new neurons, continues into adulthood in rats.[20] The presence of stem cells in the mature primate brain was first reported in 1967.[21] It has since been shown that new neurons are generated in adult mice, songbirds and primates, including humans. Normally, adult neurogenesis is restricted to two areas of the brain – the subventricular zone, which lines the lateral ventricles, and the dentate gyrus of the hippocampal formation.[22] Although the generation of new neurons in the hippocampus is well established, the presence of true self-renewing stem cells there has been debated.[23] Under certain circumstances, such as following tissue damage in ischemia, neurogenesis can be induced in other brain regions, including the neocortex.

Neural stem cells are commonly cultured *in vitro* as so called neurospheres – floating heterogeneous aggregates of cells, containing a large proportion of stem cells.[24] They can be propagated for extended periods of time and differentiated into both neuronal and glia cells, and therefore behave as stem cells. However, some recent studies suggest that this behaviour is induced by the culture conditions in progenitor cells, the progeny of stem cell division that normally undergo a strictly limited number of replication cycles *in vivo*.[25] Furthermore, neurosphere-derived cells do not behave as stem cells when transplanted back into the brain.[26]

Neural stem cells share many properties with haematopoietic stem cells (HSCs). Remarkably, when injected into the blood, neurosphere-derived cells differentiate into various cell types of the immune system.[27]

11.7.7 Olfactory adult stem cells

Olfactory adult stem cells have been successfully harvested from the human olfactory mucosa cells, which are found in the lining of the nose and are involved in the sense of smell.[28] If they are given the right chemical environment these cells have the same ability as embryonic stem cells to develop into many different cell types. Olfactory stem cells hold the potential for therapeutic applications and, in contrast to neural stem cells, can be harvested with ease without harm to the patient. This means they can be easily obtained from all individuals, including older patients who might be most in need of stem cell therapies.

11.7.8 Neural crest stem cells

Hair follicles contain two types of stem cells, one of which appears to represent a remnant of the stem cells of the embryonic neural crest. Similar cells have been found in the gastrointestinal tract, sciatic nerve, cardiac outflow tract and spinal and sympathetic ganglia. These cells can generate neurons, Schwann cells, myofibroblast, chondrocytes and melanocytes.[29][30]

11.7.9 Testicular cells

Multipotent stem cells with a claimed equivalency to embryonic stem cells have been derived from spermatogonial progenitor cells found in the testicles of laboratory mice by scientists in Germany[31][32][33] and the United States,[34][35][36][37] and, a year later, researchers from Germany and the United Kingdom confirmed the same capability using cells from the testicles of humans.[38] The extracted stem cells are known as human adult germline stem cells (GSCs)[39]

Multipotent stem cells have also been derived from germ cells found in human testicles.[40]

11.8 Adult stem cell therapies

Main article: Stem cell treatments

The therapeutic potential of adult stem cells is the focus of much scientific research, due to their ability to be harvested from the patient.[41][42][43] In common with embryonic stem cells, adult stem cells have the ability to differentiate into more than one cell type, but unlike the former they are often restricted to certain types or "lineages". The ability of a differentiated stem cell of one lineage to produce cells of a different lineage is called transdifferentiation. Some types of adult stem cells are more capable of transdifferentiation than others, but for many there is no evidence that such a transformation is possible. Consequently, adult stem therapies require a stem cell source of the specific lineage needed, and harvesting and/or culturing them up to the numbers required is a challenge.[44][45] Additionally, cues from the immediate environment (including how stiff or porous the surrounding structure/extracellular matrix is) can alter or enhance the fate and differentiation of the stem cells.[46]

11.8.1 Sources

Pluripotent stem cells, i.e. cells that can give rise to

any fetal or adult cell type, can be found in a number of tissues, including umbilical cord blood.[47] Using genetic reprogramming, pluripotent stem cells equivalent to embryonic stem cells have been derived from human adult skin tissue.[48][49][50][51][52] Other adult stem cells are multipotent, meaning they are restricted in the types of cell they can become, and are generally referred to by their tissue origin (such as mesenchymal stem cell, adipose-derived stem cell, endothelial stem cell, etc.).[53][54] A great deal of adult stem cell research has focused on investigating their capacity to divide or self-renew indefinitely, and their potential for differentiation.[55] In mice, pluripotent stem cells can be directly generated from adult fibroblast cultures.[56]

11.8.2 Clinical applications

Adult stem cell treatments have been used for many years to successfully treat leukemia and related bone/blood cancers utilizing bone marrow transplants.[57] The use of adult stem cells in research and therapy is not considered as controversial as the use of embryonic stem cells, because the production of adult stem cells does not require the destruction of an embryo.

Early regenerative applications of adult stem cells has focused on intravenous delivery of blood progenitors known as Hematopetic Stem Cells (HSC's). CD34+ hematopoietic Stem Cells have been clinically applied to treat various diseases including spinal cord injury,[58] liver cirrhosis [59] and Peripheral Vascular disease.[60] Research has shown that CD34+ hematopoietic Stem Cells are relatively more numerous in men than in women of reproductive age group among spinal cord Injury victims.[61] Other early commercial applications have focused on Mesenchymal Stem Cells (MSCs). For both cell lines, direct injection or placement of cells into a site in need of repair may be the preferred method of treatment, as vascular delivery suffers from a "pulmonary first pass effect" where intravenous injected cells are sequestered in the lungs.[62] Clinical case reports in orthopedic applications have been published. Wakitani has published a small case series of nine defects in five knees involving surgical transplantation of mesenchymal stem cells with coverage of the treated chondral defects.[63] Centeno et al. have reported high field MRI evidence of increased cartilage and meniscus volume in individual human clinical subjects as well as a large n=227 safety study.[64][65][66][67] Many other stem cell based treatments are operating outside the US, with much controversy being reported regarding these treatments as some feel more regulation is needed as clinics tend to exaggerate claims of success and minimize or omit risks.[68]

11.8.3 First transplanted human organ grown from adult stem cells

In 2008 the first full transplant of a human organ grown from adult stem cells was carried out by Paolo Macchiarini, at the Hospital Clínic of Barcelona on Claudia Castillo, a Colombian female adult whose trachea had collapsed due to tuberculosis. Researchers from the University of Padua, the University of Bristol, and Politecnico di Milano harvested a section of trachea from a donor and stripped off the cells that could cause an immune reaction, leaving a grey trunk of cartilage. This section of trachea was then "seeded" with stem cells taken from Ms. Castillo's bone marrow and a new section of trachea was grown in the laboratory over four days. The new section of trachea was then transplanted into the left main bronchus of the patient.[69][70][71][72][73] Because the stem cells were harvested from the patient's own bone marrow Professor Macchiarini did not think it was necessary for her to be given anti-rejection (immunosuppressive) medication and when the procedure was reported four months later in *The Lancet*, the patient's immune system was showing no signs of rejecting the transplant.[74]

11.9 Adult stem cells and cancer

In recent years, acceptance of the concept of adult stem cells has increased. There is now a hypothesis that stem cells reside in many adult tissues and that these unique reservoirs of cells not only are responsible for the normal reparative and regenerative processes but are also considered to be a prime target for genetic and epigenetic changes, culminating in many abnormal conditions including cancer.[75][76]

11.10 See also

- induced somatic stem cells

11.11 News and external links

- NIH Stem Cell Information Resource, resource for stem cell research

- Nature Reports Stem Cells Background information, research advances and debates about stem cell science

- UMDNJ Stem Cell and Regenerative Medicine, provides educational materials and research resources

- Stem Cell Research at Johns Hopkins University

11.12 References

[1] Mlsna, Lucas J. (2010). "Stem Cell Based Treatments and Novel Considerations for Conscience Clause Legislation". *Indiana Health Law Review* (United States: Indiana University Robert H. McKinney School of Law) **8** (2): 471–496. ISSN:1549-3199. LCCN:2004212209. OCLC:OCLC 54703225.

[2] Chaudhary PM, Roninson IB (July 1991). "Expression and activity of P-glycoprotein, a multidrug efflux pump, in human hematopoietic stem cells". *Cell* **66** (1): 85–94. doi:10.1016/0092-8674(91)90141-K. PMID 1712673.

[3] Dontu G, Jackson KW, McNicholas E, Kawamura MJ, Abdallah WM, Wicha MS (2004). "Role of Notch signaling in cell-fate determination of human mammary stem/progenitor cells". *Breast Cancer Research* **6** (6): R605–15. doi:10.1186/bcr920. PMC 1064073. PMID 15535842.

[4] Beachy PA, Karhadkar SS, Berman DM; Karhadkar; Berman (November 2004). "Tissue repair and stem cell renewal in carcinogenesis". *Nature* **432** (7015): 324–31. Bibcode:2004Natur.432..324B. doi:10.1038/nature03100. PMID 15549094.

[5] Clarke, D. L.; Johansson, CB; Wilbertz, J; Veress, B; Nilsson, E; Karlström, H; Lendahl, U; Frisén, J (2000). "Generalized Potential of Adult Neural Stem Cells". *Science* **288** (5471): 1660–3. Bibcode:2000Sci...288.1660C. doi:10.1126/science.288.5471.1660. PMID 10834848.

[6] Krause, Diane S.; Theise, Neil D.; Collector, Michael I.; Henegariu, Octavian; Hwang, Sonya; Gardner, Rebekah; Neutzel, Sara; Sharkis, Saul J. (2001). "Multi-Organ, Multi-Lineage Engraftment by a Single Bone Marrow-Derived Stem Cell". *Cell* **105** (3): 369–77. doi:10.1016/S0092-8674(01)00328-2. PMID 11348593.

[7] Kucia, M; Reca, R; Campbell, F R; Zuba-Surma, E; Majka, M; Ratajczak, J; Ratajczak, M Z (2006). "A population of very small embryonic-like (VSEL) CXCR4+SSEA-1+Oct-4+ stem cells identified in adult bone marrow". *Leukemia* **20** (5): 857–69. doi:10.1038/sj.leu.2404171. PMID 16498386.

[8] Zuba-Surma, Ewa K.; Kucia, Magdalena; Wu, Wan; Klich, Izabela; Lillard, James W.; Ratajczak, Janina; Ratajczak, Mariusz Z. (2008). "Very small embryonic-like stem cells are present in adult murine organs: ImageStream-based morphological analysis and distribution studies". *Cytometry Part A* **73A** (12): 1116–1127. doi:10.1002/cyto.a.20667.

[9] Danova-Alt, Ralitza; Heider, Andreas; Egger, Dietmar; Cross, Michael; Alt, Rüdiger; Ivanovic, Zoran (2 April 2012). Ivanovic, Zoran, ed. "Very Small Embryonic-Like Stem Cells Purified from Umbilical Cord Blood Lack Stem Cell Characteristics". *PLoS ONE* **7** (4): e34899. Bibcode:2012PLoSO...734899D. doi:10.1371/journal.pone.0034899. PMC 3318011. PMID 22509366.

[10] Szade, Krzysztof; Bukowska-Strakova, Karolina; Nowak, Witold Norbert; Szade, Agata; Kachamakova-Trojanowska, Neli; Zukowska, Monika; Jozkowicz, Alicja; Dulak, Jozef; Asakura, Atsushi (16 May 2013). Asakura, Atsushi, ed. "Murine Bone Marrow Lin–Sca-1+CD45– Very Small Embryonic-Like (VSEL) Cells Are Heterogeneous Population Lacking Oct-4A Expression". PLoS ONE 8 (5): e63329. Bibcode:2013PLoSO...863329S. doi:10.1371/journal.pone.0063329. PMC 3656957. PMID 23696815.

[11] Miyanishi M, Mori Y, Seita J, Chen JY, Karten S, Chan CKF, et al. Stem Cell Reports. Stem Cell Reports. 2013 Jul 23;:1–11. http://www.cell.com/stem-cell-reports/abstract/S2213-6711(13)00050-7?fb_action_ids=10201558251787555&fb_action_types=og.likes

[12] Miyanishi, Masanori; Mori, Yasuo; Seita, Jun; Chen, James Y.; Karten, Seth; Chan, Charles K.F.; Nakauchi, Hiromitsu; Weissman, Irving L. (31 July 2013). "Do Pluripotent Stem Cells Exist in Adult Mice as Very Small Embryonic Stem Cells?". Stem Cell Reports 1 (2): 198–208. doi:10.1016/j.stemcr.2013.07.001. PMC 3757755. PMID 24052953.

[13] Behrens A, van Deursen JM, Rudolph KL, Schumacher B (2014). "Impact of genomic damage and ageing on stem cell function". Nat. Cell Biol. 16 (3): 201–7. doi:10.1038/ncb2928. PMC 4214082. PMID 24576896.

[14] http://stemcells.nih.gov/info/scireport/pages/chapter5.aspx

[15] Liu S, Dontu G, Wicha MS (2005). "Mammary stem cells, self-renewal pathways, and carcinogenesis". Breast Cancer Research 7 (3): 86–95. doi:10.1186/bcr1021. PMC 1143566. PMID 15987436.

[16] Van Der Flier, L. G.; Clevers, H. (2009). "Stem Cells, Self-Renewal, and Differentiation in the Intestinal Epithelium". Annual Review of Physiology 71: 241–260. doi:10.1146/annurev.physiol.010908.163145. PMID 18808327.

[17] Barker, N.; Ridgway, R. A.; Van Es, J. H.; Van De Wetering, M.; Begthel, H.; Van Den Born, M.; Danenberg, E.; Clarke, A. R.; Sansom, O. J.; Clevers, H. (2008). "Crypt stem cells as the cells-of-origin of intestinal cancer". Nature 457 (7229): 608–611. Bibcode:2009Natur.457..608B. doi:10.1038/nature07602. PMID 19092804.

[18] Phinney DG, Prockop DJ (November 2007). "Concise review: mesenchymal stem/multipotent stromal cells: the state of transdifferentiation and modes of tissue repair—current views". Stem Cells 25 (11): 2896–902. doi:10.1634/stemcells.2007-0637. PMID 17901396.

[19] Shi S, Bartold PM, Miura M, Seo BM, Robey PG, Gronthos S (August 2005). "The efficacy of mesenchymal stem cells to regenerate and repair dental structures". Orthod Craniofac Res 8 (3): 191–9. doi:10.1111/j.1601-6343.2005.00331.x. PMID 16022721.

[20] Altman J, Das GD (June 1965). "Autoradiographic and histological evidence of postnatal hippocampal neurogenesis in rats". The Journal of Comparative Neurology 124 (3): 319–35. doi:10.1002/cne.901240303. PMID 5861717.

[21] Lewis PD (March 1968). "Mitotic activity in the primate subependymal layer and the genesis of gliomas". Nature 217 (5132): 974–5. Bibcode:1968Natur.217..974L. doi:10.1038/217974a0. PMID 4966809.

[22] Alvarez-Buylla A, Seri B, Doetsch F (April 2002). "Identification of neural stem cells in the adult vertebrate brain". Brain Research Bulletin 57 (6): 751–8. doi:10.1016/S0361-9230(01)00770-5. PMID 12031271.

[23] Bull ND, Bartlett PF (November 2005). "The adult mouse hippocampal progenitor is neurogenic but not a stem cell". The Journal of Neuroscience 25 (47): 10815–21. doi:10.1523/JNEUROSCI.3249-05.2005. PMID 16306394.

[24] Reynolds BA, Weiss S; Weiss (March 1992). "Generation of neurons and astrocytes from isolated cells of the adult mammalian central nervous system". Science 255 (5052): 1707–10. Bibcode:1992Sci...255.1707R. doi:10.1126/science.1553558. PMID 1553558.

[25] Doetsch F, Petreanu L, Caille I, Garcia-Verdugo JM, Alvarez-Buylla A (December 2002). "EGF converts transit-amplifying neurogenic precursors in the adult brain into multipotent stem cells". Neuron 36 (6): 1021–34. doi:10.1016/S0896-6273(02)01133-9. PMID 12495619.

[26] Marshall GP, Laywell ED, Zheng T, Steindler DA, Scott EW (March 2006). "In vitro-derived "neural stem cells" function as neural progenitors without the capacity for self-renewal". Stem Cells 24 (3): 731–8. doi:10.1634/stemcells.2005-0245. PMID 16339644.

[27] Bjornson CR, Rietze RL, Reynolds BA, Magli MC, Vescovi AL; Rietze; Reynolds; Cristina Magli; Vescovi (January 1999). "Turning brain into blood: a hematopoietic fate adopted by adult neural stem cells in vivo". Science 283 (5401): 534–7. Bibcode:1999Sci...283..534B. doi:10.1126/science.283.5401.534. PMID 9915700.

[28] Murrell W, Féron F, Wetzig A, et al. (June 2005). "Multipotent stem cells from adult olfactory mucosa". Developmental Dynamics 233 (2): 496–515. doi:10.1002/dvdy.20360. PMID 15782416.

[29] Sieber-Blum M, Hu Y (December 2008). "Epidermal neural crest stem cells (EPI-NCSC) and pluripotency". Stem Cell Rev 4 (4): 256–60. doi:10.1007/s12015-008-9042-0. PMID 18712509.

[30] Kruger GM, Mosher JT, Bixby S, Joseph N, Iwashita T, Morrison SJ (August 2002). "Neural Crest Stem Cells Persist in the Adult Gut but Undergo Changes in Self-Renewal, Neuronal Subtype Potential, and Factor Responsiveness". Neuron 35 (4): 657–69. doi:10.1016/S0896-6273(02)00827-9. PMC 2728576. PMID 12194866.

[31] "Testicle cells may aid research". BBC. 25 March 2006.

[32] CBS/Associated Press (24 March 2006). "Study: Mice Testes Act Like Stem Cells". CBS.

[33] Rick Weiss (25 March 2006). "Embryonic Stem Cell Success". *Washington Post*.

[34] "Promising New Source Of Stem Cells: Mouse Testes Produce Wide Range Of Tissue Types". Science Daily. 24 September 2007.

[35] Barbara Miller (20 September 2007). "Testicles yield stem cells in science breakthrough". Australian Broadcasting Corporation.

[36] J.R. Minkel (19 September 2007). "Testes May Prove Fertile Source of Stem Cells". Scientific American.

[37] "Stem Cells in Adult Testes Provide Alternative to Embryonic Stem Cells for Organ Regeneration". Cornell University. 20 September 2007.

[38] Rob Waters (8 October 2008). "Testicle Stem Cells Become Bone, Muscle in German Experiments". Bloomberg.

[39] Nora Schultz (9 October 2008). "A Source of Men's Stem Cells – Stem cells from human testes could be used for personalized medicine.". Technology Review.

[40] Maggie Fox (Reuters) (2 April 2006). "U.S. Firm Says It Made Stem Cells From Human Testes". *Washington Post*.

[41] Liao, YH; Verchere, CB; Warnock, GL (April 2007). "Adult stem or progenitor cells in treatment for type 1 diabetes: current progress.". *Canadian Journal of Surgery* **50** (2): 137–42. PMC 2384257. PMID 17550719.

[42] Mimeault, M; Hauke, R; Batra, S K (1 August 2007). "Stem Cells: A Revolution in Therapeutics—Recent Advances in Stem Cell Biology and Their Therapeutic Applications in Regenerative Medicine and Cancer Therapies". *Clinical Pharmacology & Therapeutics* **82** (3): 252–264. doi:10.1038/sj.clpt.6100301. PMID 17671448.

[43] Christoforou, N; Gearhart, JD (May–Jun 2007). "Stem cells and their potential in cell-based cardiac therapies.". *Progress in cardiovascular diseases* **49** (6): 396–413. doi:10.1016/j.pcad.2007.02.006. PMID 17498520.

[44] Raff, M (2003). "Adult stem cell plasticity: Fact or Artifact?". *Annual Review of Cell and Developmental Biology* **19**: 1–22. doi:10.1146/annurev.cellbio.19.111301.143037. PMID 14570561.

[45] Smith, S; Neaves, W; Teitelbaum, S; Prentice, D. A.; Tarne, G. (8 June 2007). "Adult Versus Embryonic Stem Cells: Treatments". *Science* **316** (5830): 1422–1423. doi:10.1126/science.316.5830.1422b. PMID 17556566.

[46] Huang, C; et al. (2015). "Environmental physical cues determine the lineage specification of mesenchymal stem cells". *Biochim Biophys Acta* **1850**: 1261–6. doi:10.1016/j.bbagen.2015.02.011. PMID 25727396.

[47] Ratajczak MZ, Machalinski B, Wojakowski W, Ratajczak J, Kucia M (2007). "A hypothesis for an embryonic origin of pluripotent Oct-4(+) stem cells in adult bone marrow and other tissues". *Leukemia* **21** (5): 860–7. doi:10.1038/sj.leu.2404630. PMID 17344915.

[48] "Me too, too – How to make human embryonic stem cells without destroying human embryos". *The Economist*. 22 November 2007.

[49] Gina Kolata (22 November 2007). "Man Who Helped Start Stem Cell War May End It". *New York Times*.

[50] Gina Kolata (21 November 2007). "Scientists Bypass Need for Embryo to Get Stem Cells". *New York Times*.

[51] Anne McIlroy (21 November 2007). "Stem-cell method hailed as 'massive breakthrough'". *Globe and Mail*. Canada.

[52] Alice Park (20 November 2007). "A Breakthrough on Stem Cells". *Time Magazine*.

[53] Barrilleaux B, Phinney DG, Prockop DJ, O'Connor KC (2006). "Review: ex vivo engineering of living tissues with adult stem cells". *Tissue Eng.* **12** (11): 3007–19. doi:10.1089/ten.2006.12.3007. PMID 17518617.

[54] Gimble JM, Katz AJ, Bunnell BA (2007). "Adipose-derived stem cells for regenerative medicine". *Circ. Res.* **100** (9): 1249–60. doi:10.1161/01.RES.0000265074.83288.09. PMID 17495232.

[55] Gardner RL (March 2002). "Stem cells: potency, plasticity and public perception". *Journal of Anatomy* **200** (Pt 3): 277–82. doi:10.1046/j.1469-7580.2002.00029.x. PMC 1570679. PMID 12033732.

[56] Takahashi K, Yamanaka S (2006). "Induction of pluripotent stem cells from mouse embryonic and adult fibroblast cultures by defined factors". *Cell* **126** (4): 663–76. doi:10.1016/j.cell.2006.07.024. PMID 16904174.

[57] Bone Marrow Transplant Retrieved on 21 November 2008

[58] Srivastava A, Bapat M, Ranade S, Srinivasan V, Murugan P, Manjunath S, Thamaraikannan P, Abraham S (2010). "Autologous Multiple Injections of in Vitro Expanded Autologous Bone Marrow Stem Cells For Cervical Level Spinal Cord Injury - A Case Report". *Journal of Stem Cells and Regenerative Medicine*.

[59] Terai S, Ishikawa T, Omori K, Aoyama K, Marumoto Y, Urata Y, Yokoyama Y,Uchida K, Yamasaki T, Fujii Y, Okita K, Sakaida I (2006). "Improved liver function in patients with liver cirrhosis after autologous bone marrow cell infusion therapy". *Stem Cells* **24** (10): 2292–8. doi:10.1634/stemcells.2005-0542. PMID 16778155.

[60] Subrammaniyan R, Amalorpavanathan J, Shankar R, Rajkumar M, Baskar S,Manjunath SR, Senthilkumar R, Murugan P, Srinivasan VR, Abraham S (2011). "Application of autologous bone marrow mononuclear cells in six patients with advanced chronic critical limb ischemia as a result

of diabetes: our experience". *Cytotherapy* **13** (8): 993–9. doi:10.3109/14653249.2011.579961. PMID 21671823.

[61] Dedeepiya V, Rao Y Y, Jayakrishnan G, Parthiban JKBC, Baskar S, Manjunath S, Senthilkumar R and Abraham S (2012). "Index of CD34+ cells and mononuclear cells in the bone marrow of spinal cord injury patients of different age groups- A comparative analysis". *Bone Marrow Research* **2012**: 1–8. doi:10.1155/2012/787414. PMC 3398573. PMID 22830032.

[62] Fischer UM, Harting MT, Jimenez F, et al. (June 2009). "Pulmonary passage is a major obstacle for intravenous stem cell delivery: the pulmonary first-pass effect". *Stem Cells and Development* **18** (5): 683–92. doi:10.1089/scd.2008.0253. PMC 3190292. PMID 19099374.

[63] Wakitani S, Nawata M, Tensho K, Okabe T, Machida H, Ohgushi H (2007). "Repair of articular cartilage defects in the patello-femoral joint with autologous bone marrow mesenchymal cell transplantation: three case reports involving nine defects in five knees". *Journal of Tissue Engineering and Regenerative Medicine* **1** (1): 74–9. doi:10.1002/term.8. PMID 18038395.

[64] >>Centeno; et al. "Regeneration of meniscus cartilage in a knee treated with percutaneously implanted autologous mesenchymal stem cells, platelate lysate, and dexamethasome".

[65] Centeno CJ, Busse D, Kisiday J, Keohan C, Freeman M, Karli D (December 2008). "Regeneration of meniscus cartilage in a knee treated with percutaneously implanted autologous mesenchymal stem cells". *Medical Hypotheses* **71** (6): 900–8. doi:10.1016/j.mehy.2008.06.042. PMID 18786777.

[66] Centeno CJ, Busse D, Kisiday J, Keohan C, Freeman M, Karli D (2008). "Increased knee cartilage volume in degenerative joint disease using percutaneously implanted, autologous mesenchymal stem cells". *Pain Physician* **11** (3): 343–53. PMID 18523506.

[67] Centeno CJ, Schultz JR, Cheever M, Robinson B, Freeman M, Marasco W (2010). "Safety and complications reporting on the re-implantation of culture-expanded mesenchymal stem cells using autologous platelet lysate technique". *Curr Stem Cell Res Ther* **5** (1): 81–93. doi:10.2174/157488810790442796. PMID 19951252.

[68] >PR Newswire. "The International Society for Stem Cell Research Releases New Guidelines to Shape Future of Stem Cell Therapy Regulation needed as new study reveals clinics exaggerate claims and omit risks".

[69] "Claudia Castillo: The pioneer's story". The Independent(United Kingdom). 19 November 2008.

[70] Michael Kahn (18 November 2008). "Woman gets first trachea transplant without drugs". Reuters.

[71] Kate Devlin (18 November 2008). "British doctors help perform world's first transplant of a whole organ grown in lab". *The Daily Telegraph*. United Kingdom.

[72] Tanya Thompson (19 November 2008). "World first as woman gets organ made from stem cells". *The Scotsman*. UK.

[73] Jeremy Laurance (19 November 2008). "The medical miracle". The Independent(United Kingdom).

[74] Rose, David (19 November 2008). "Claudia Castillo gets windpipe tailor-made from her own stem cells". *The Times*. London: The Times Newspapers Ltd. Retrieved 20 November 2008.

[75] Bioinfobank FAQ:Stem cells in adult tissues Retrieved on 21 November 2008 Archived 27 September 2007 at the Wayback Machine.

[76] Cogle CR, Guthrie SM, Sanders RC, Allen WL, Scott EW, Petersen BE (August 2003). "An overview of stem cell research and regulatory issues". *Mayo Clinic Proceedings. Mayo Clinic* **78** (8): 993–1003. doi:10.4065/78.8.993. PMID 12911047.

Chapter 12

Progenitor cell

For the term meaning a person or thing from which others are descended or originate, see Progenitor.

A **progenitor cell** is a biological cell that, like a stem

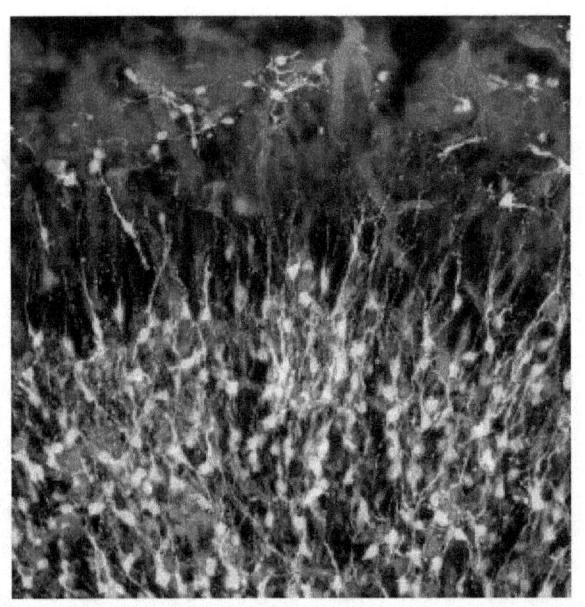

Neural progenitors (green) in olfactory bulb with astrocytes (blue).

cell, has a tendency to differentiate into a specific type of cell, but is already more specific than a stem cell and is pushed to differentiate into its "target" cell. The most important difference between stem cells and progenitor cells is that stem cells can replicate indefinitely, whereas progenitor cells can divide only a limited number of times. Controversy about the exact definition remains and the concept is still evolving.[1]

The terms "progenitor cell" and "stem cell" are sometimes equated.[2]

12.1 Properties

Most progenitors are described as oligopotent. In this point of view, they may be compared to adult stem cells. But progenitors are said to be in a further stage of cell differentiation. They are in the "center" between stem cells and fully differentiated cells. The kind of potency they have depends on the type of their "parent" stem cell and also on their niche. Some progenitor cells were found during research, and were isolated. After their marker was found, it was proven that these progenitor could move through the body and migrate towards the tissue where they are needed. Many properties are shared by adult stem cells and progenitor cells.

Progenitor cells are found in adult organisms and they act as a repair system for the body. They replenish special cells, but also maintain the blood, skin and intestinal tissues. They can also be found in developing embryonic pancreatic tissue.

12.2 Function

The majority of progenitor cells lie dormant or possess little activity in the tissue in which they reside. They exhibit slow growth and their main role is to replace cells lost by normal attrition. In case of tissue injury, damaged or dead cells, progenitor cells can be activated. Growth factors or cytokines are two substances that trigger the progenitors to mobilize toward the damaged tissue. At the same time, they start to differentiate into the target cells. Not all progenitors are mobile and are situated near the tissue of their target differentiation. When the cytokines, growth factors and other cell division enhancing stimulators take on the progenitors, a higher rate of cell division is introduced. It leads to the recovery of the tissue.

12.3   Examples

The characterization or the defining principle of progenitor cells, in order to separate them from others, is based on the different cell markers rather than their morphological appearance.[3]

- Satellite cells found in muscles. They play a major role in muscle cell differentiation and injury recoveries.

- Intermediate progenitor cells formed in the subventricular zone.[4] Some of these transit amplifying neural progenitors migrate via rostral migratory stream to the olfactory bulb and differentiate further into specific types of neural cells.

- Bone marrow stromal cells, basal cell of epidermis have 10% of progenitor cell, although they are often classed as stem cells due to their high plasticity and potential for unlimited capacity for self-renewal.

- Periosteum contains progenitor cells that develop into osteoblasts and chondroblasts.

- Pancreatic progenitor cells are among the most-studied progenitors.[5] They are used in research to develop a cure against diabetes type-1.

- Angioblasts or endothelial progenitor cells (EPC). These are very important for research on fracture and wounds healing.[6]

- Blast cells are involved in generation of B- and T-lymphocytes, which participate in immune responses.[7][8]

12.4   Development of the human cerebral cortices

Before embryonic day 40 (E40), progenitor cells generate other progenitor cells; after that period, progenitor cells produce only dissimilar mesenchymal stem cell daughters. The cells from a single progenitor cell form a proliferative unit that creates one cortical column; these columns contain a variety of neurons with different shapes.

12.5   See also

- Induced progenitor stem cells

- Endothelial progenitor cell

12.6   References

[1] Seaberg, R. M.; Van Der Kooy, D. (2003). "Stem and progenitor cells: The premature desertion of rigorous definitions". *Trends in Neurosciences* **26** (3): 125–131. doi:10.1016/S0166-2236(03)00031-6. PMID 12591214.

[2] "progenitor cell" at *Dorland's Medical Dictionary*

[3] Morgan, JE; Partridge, TA (August 2003). "Muscle satellite cells.". *The international journal of biochemistry & cell biology* **35** (8): 1151–6. doi:10.1016/s1357-2725(03)00042-6. PMID 12757751.

[4] Noctor SC, Martínez-Cerdeño V, Kriegstein AR (May 2007). "Contribution of intermediate progenitor cells to cortical histogenesis". *Arch. Neurol.* **64** (5): 639–42. doi:10.1001/archneur.64.5.639. PMID 17502462.

[5] Awong, G. (2011). "Thymus-bound: The many features of T cell progenitors". *Frontiers in Bioscience*: 961. doi:10.2741/200.

[6] Barber, C. L.; Iruela-Arispe, M. L. (2006). "The Ever-Elusive Endothelial Progenitor Cell: Identities, Functions and Clinical Implications". *Pediatric Research* **59** (4 Pt 2): 26R–32R. doi:10.1203/01.pdr.0000203553.46471.18. PMID 16549545.

[7] Carotta, S.; Nutt, S. L. (2008). "Losing B cell identity". *BioEssays* **30** (3): 203–207. doi:10.1002/bies.20725. PMID 18293359.

[8] Awong, G. (2011). "Thymus-bound: The many features of T cell progenitors". *Frontiers in Bioscience*: 961. doi:10.2741/200.

Chapter 13

Cancer stem cell

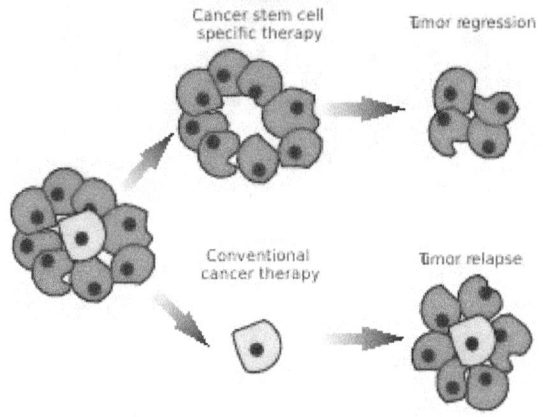

Figure 1: Stem cell specific and conventional cancer therapies

Cancer stem cells (CSCs) are cancer cells (found within tumors or hematological cancers) that possess characteristics associated with normal stem cells, specifically the ability to give rise to all cell types found in a particular cancer sample. CSCs are therefore tumorigenic (tumor-forming), perhaps in contrast to other non-tumorigenic cancer cells. CSCs may generate tumors through the stem cell processes of self-renewal and differentiation into multiple cell types. Such cells are hypothesized to persist in tumors as a distinct population and cause relapse and metastasis by giving rise to new tumors. Therefore, development of specific therapies targeted at CSCs holds hope for improvement of survival and quality of life of cancer patients, especially for patients with metastatic disease.

Existing cancer treatments have mostly been developed based on animal models, where therapies able to promote tumor shrinkage were deemed effective. However, animals do not provide a complete model of human disease. In particular, in mice, whose life spans do not exceed two years, tumor relapse is difficult to study.

The efficacy of cancer treatments is, in the initial stages of testing, often measured by the ablation fraction of tumor mass (fractional kill). As CSCs form a small proportion of the tumor, this may not necessarily select for drugs that

act specifically on the stem cells. The theory suggests that conventional chemotherapies kill differentiated or differentiating cells, which form the bulk of the tumor but do not generate new cells. A population of CSCs, which gave rise to it, could remain untouched and cause relapse.

Cancer stem cells were first identified by John Dick in acute myeloid leukemia in the late 1990s. Since the early 2000s they have been an intense cancer research focus.[1]

13.1 Tumor propagation models

In different tumor subtypes, cells within the tumor population exhibit functional heterogeneity, and tumors are formed from cells with various proliferative and differentiation capacities.[2] This functional heterogeneity among cancer cells has led to the creation of multiple propagation models to account for heterogeneity and differences in tumor-regenerative capacity: the cancer stem cell (CSC) and clonal evolution models[3]

The cancer stem cell model refers to a subset of tumor cells that have the ability to self-renew and are able to generate the diverse tumor cells.[3] These cells have been termed cancer stem cells to reflect their stem-like properties. One implication of the CSC model and the existence of CSCs is that the tumor population is hierarchically arranged with CSCs lying at the apex[4] (Fig. 3).

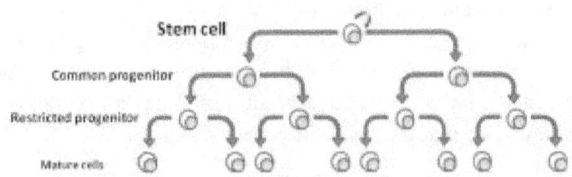

Figure 2: A normal cellular hierarchy comprising stem cells at the apex, which generate common and more restricted progenitor cells and ultimately the mature cell types that constitute particular tissues.

The clonal evolution model postulates that mutant tumor cells with a growth advantage outproliferate others. Cells

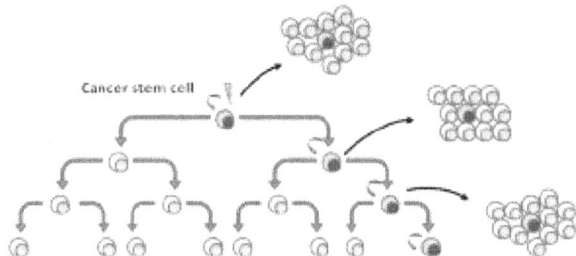

Figure 3. *In the cancer stem cell (CSC) model, only the CSCs have the ability to generate a tumor, based on their self-renewal properties and proliferative potential.*

in the dominant population have a similar potential for initiating tumor growth[5] (Fig. 4).

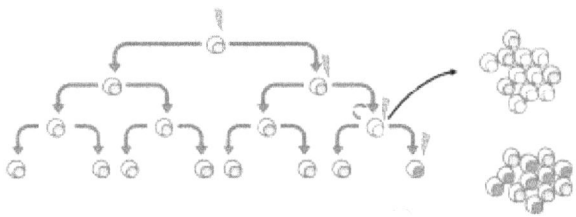

Figure 4: *In the clonal evolution model, all undifferentiated cells have similar possibility to change into a tumorigenic cell.*

[6] These two models are not mutually exclusive, as CSCs themselves undergo clonal evolution. Thus, the secondary more dominant CSCs may emerge, if a mutation confers more aggressive properties[7] (Fig. 5).

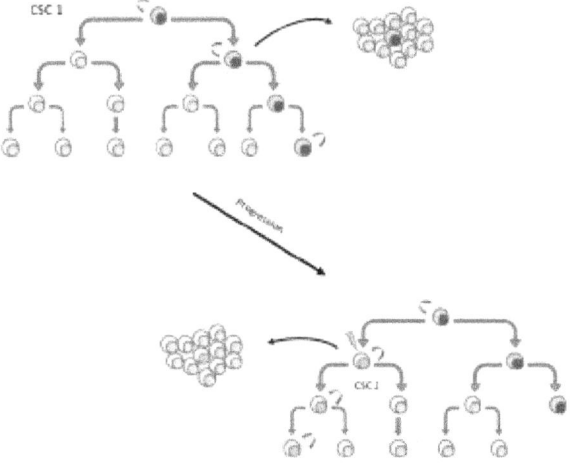

Figure 5: *Both tumor models may play a role in the maintenance of a tumor. Initially, tumor growth is assured with a specific CSC (CSC1). With tumor progression, another CSC (CSC 2) may arise due the clonal selection. The development of a new more aggressive CSC may result from the acquisition of an additional mutation or epigenetic modification.*

13.2 Debate

The existence of CSCs is under debate, because many studies have failed to discover their specific characteristics.[8] Cancer cells must be capable of continuous proliferation and self-renewal in order to retain the many mutations required for carcinogenesis, and to sustain the growth of a tumor, since differentiated cells (constrained by the Hayflick Limit[9]) cannot divide indefinitely. If most tumor cells are endowed with stem cell properties, targeting tumor size directly remains a valid strategy. If they are a small minority, targeting may be more effective. Another debate is over the origin of CSCs - whether from disregulation of normal stem cells or from a more specialized population that acquired the ability to self-renew (which is related to the issue of stem cell plasticity).

13.3 Evidence

The first conclusive evidence for CSCs came in 1997. Bonnet and Dick isolated a subpopulation of leukemia cells that expressed surface marker CD34, but not CD38.[10] The authors established that the CD34+/CD38− subpopulation is capable of initiating tumors in NOD/SCID mice that were histologically similar to the donor. The first evidence of a solid tumor cancer stem-like cell followed in 2002 with the discovery of a clonogenic, sphere-forming cell isolated and characterized from human brain gliomas. Human cortical glial tumors contain neural stem-like cells expressing astroglial and neuronal markers *in vitro*.[11]

In cancer research experiments, tumor cells are sometimes injected into an experimental animal to establish a tumor. Disease progression is then followed in time and novel drugs can be tested for their efficacy. Tumor formation requires thousands or tens of thousands of cells to be introduced. Classically, this was explained by poor methodology (i.e., the tumor cells lose their viability during transfer) or the critical importance of the microenvironment, the particular biochemical surroundings of the injected cells. Supporters of the CSC paradigm argue that only a small fraction of the injected cells, the CSCs, have the potential to generate a tumor. In human acute myeloid leukemia the frequency of these cells is less than 1 in 10,000.[10]

Further evidence comes from histology. Many tumors are heterogeneous and contain multiple cell types native to the host organ. Heterogeneity is commonly retained by tumor metastases. This suggests that the cell that produced them had the capacity to generate multiple cell types, a classical hallmark of stem cells.[10]

The existence of leukemia stem cells prompted research into other cancers. CSCs have recently been identified in several solid tumors, including:

- Brain[12]

- Breast[13]

- Colon[14]

- Ovary[15][16][17]

- Pancreas[18]

- Prostate[19][20]

- Melanoma[21][22][23][24]

- Multiple Myeloma[25][26]

- Non-melanoma skin cancer[27][28]

13.3.1 Mechanistic and mathematical models

Once the pathways to cancer are hypothesized, it is possible to develop predictive mathematical models,[29] e.g., based on the cell compartment method. For instance, the growths of abnormal cells can be denoted with specific mutation probabilities. Such a model predicted that repeated insult to mature cells increases the formation of abnormal progeny and the risk of cancer.[30] The clinical efficacy of such models[31] remains unestablished.

13.4 Origin

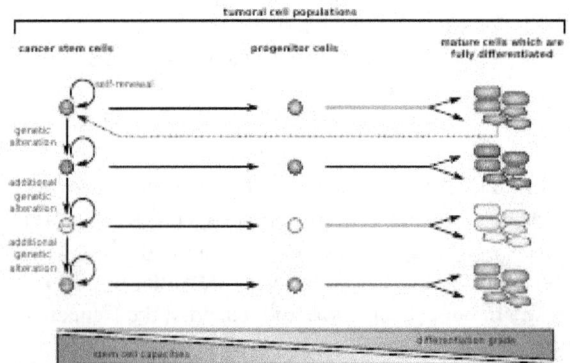

Figure 6: Hierarchical organisation of a tumour according to the CSC model

The origin of CSCs is an active research area. The answer may depend on the tumor type and phenotype. So far the hypothesis that tumors originate from a single "cell of origin" cannot be demonstrated using the cancer stem cell model. This is because cancer stem cells are not present in end-stage tumors. Therefore, describing a cancer stem cell as a cell of origin is often an inaccurate claim, even though a cancer stem cell is capable of initiating tumor formation.

Origin hypotheses include mutants in developing stem or progenitor cells, mutants in adult stem cells or adult progenitor cells, and mutant differentiated cells that acquire stem-like attributes. These theories often focus on a tumor's cell of origin.

13.4.1 Hypotheses

Stem cell mutation

The "mutation in stem cell niche populations during development" hypothesis claims that these developing stem populations are mutated and then reproduce so that the mutation is shared by many descendants. These daughter cells are much closer to becoming tumors, and their numbers increase the chance of a cancerous mutation.[32]

Adult stem cells

Another theory associates adult stem cells with tumor formation. This is most often associated with tissues with a high rate of cell turnover (such as the skin or gut). In these tissues, adult stem cells are candidates because of their frequent cell divisions (compared to most adult stem cells) in conjunction with the long lifespan of adult stem cells. This combination creates the ideal set of circumstances for mutations to accumulate: mutation accumulation is the primary factor that drives cancer initiation. Evidence shows that the association represents an actual phenomenon, although it is impossible to link a specific cancer to a specific cause.[33][34]

De-differentiation

De-differentiation of mutated cells may create stem cell-like characteristics, suggesting that any cell might become a cancer stem cell.

13.4.2 Hierarchy

The concept of tumor hierarchy claims that a tumor is a heterogeneous population of mutant cells, all of which share some mutations but vary in specific phenotype. A tumor hosts several types of stem cells, one optimal to the specific environment and other less successful lines. These secondary lines may be more successful in other environments, allowing the tumor to adapt, including adaptation to therapeutic intervention. If correct, this concept impacts cancer stem cell-specific treatment regimes.[35] Such a hierarchy would complicate attempts to pinpoint the cancer stem cell's origin.

13.5   Cancer stem cell isolation

CSCs, now reported in most human tumors, are commonly identified and enriched using strategies for identifying normal stem cells that are similar across studies.[36] These procedures include fluorescence-activated cell sorting (FACS), with antibodies directed at cell-surface markers and functional approaches including SP analysis (side population assay) or Aldefluor assay.[37] The CSC-enriched result is then implanted, at various doses, in immune-deficient mice to assess its tumor development capacity. This *in vivo* assay is called limiting dilution assay. The tumor cell subsets that can initiate tumor development at low cell numbers are further tested for self-renewal capacity in serial tumor studies.[38]

CSC can also be identified by efflux of incorporated Hoechst dyes via multidrug resistance (MDR) and ATP-binding cassette (ABC) Transporters.[37]

Another approach is sphere-forming assays. Many normal stem cells such as hematopoietics or stem cells from tissues, under special culture conditions, form three-dimensional spheres that can differentiate. As with normal stem cells, the CSCs isolated from brain or prostate tumors also have the ability to form anchor-independent spheres.[39]

13.6   Heterogeneity (CSC markers)

CSCs have been identified in various solid tumors. Markers specific for normal stem cells are commonly used for isolating CSCs from solid and hematological tumors. Cell surface markers have proved useful for isolation of CSC-enriched populations including CD133 (also known as PROM1), CD44, CD24, EpCAM (epithelial cell adhesion molecule, also known as epithelial specific antigen, ESA), THY1, ATP-binding cassette B5 (ABCB5),.[40] and CD200.

CD133 (prominin 1) is a five-transmembrane domain glycoprotein expressed on CD34+ stem and progenitor cells, in endothelial precursors and fetal neural stem cells. It has been detected using its glycosylated epitope known as AC133.

EpCAM (epithelial cell adhesion molecule, ESA, TROP1) is hemophilic Ca^{2+}-independent cell adhesion molecule expressed on the basolateral surface of most epithelial cells.

CD90   (THY1)   is   a   glycosylphosphatidylinositol

glycoprotein anchored in the plasma membrane and involved in signal transduction. It may also mediate adhesion between thymocytes and thymic stroma.

CD44 (PGP1) is an adhesion molecule that has pleiotropic roles in cell signaling, migration and homing. It has multiple isoforms, including CD44H, which exhibits high affinity for hyaluronate, and CD44V which has metastatic properties.

CD24   (HSA)   is   a   glycosylated glycosylphosphatidylinositol-anchored adhesion molecule, which has co-stimulatory role in B and T cells.

CD200 (OX-2) is a type 1 membrane glycoprotein, which delivers an inhibitory signal to immune cells including T cells, NK cells and macrophages.

ALDH is a ubiquitous aldehyde dehydrogenase family of enzymes, which catalyzes the oxidation of aromatic aldehydes to carboxyl acids. For instance, it has role in conversion of retinol to retinoic acid, which is essential for survival.[41][42]

The first solid malignancy from which CSCs were isolated and identified was breast cancer. Therefore, these CSCs are the most intensely studied. Breast CSCs have been enriched in CD44+CD24−/low,[40] SP,[43] ALDH+ subpopulations.[44][45] Breast CSCs are apparently phenotypically diverse. CSC marker expression in breast cancer cells is apparently heterogeneous and breast CSC populations vary across tumors.[46] Both CD44+CD24− and CD44+CD24+ cell populations are tumor initiating cells; however, CSC are most highly enriched using the marker profile CD44+CD49f^{hi}CD133/2^{hi}.[47]

CSCs have been reported in many brain tumors. Stem-like tumor cells have been identified using cell surface markers including CD133,[48] SSEA-1 (stage-specific embryonic antigen-1),[49] EGFR[50] and CD44.[51] The use of CD133 for identification of brain tumor stem-like cells may be problematic because tumorigenic cells are found in both CD133+ and CD133− cells in some gliomas, and some CD133+ brain tumor cells may not possess tumor-initiating capacity.[50]

CSCs were reported in human colon cancer.[52] For their identification, cell surface markers such as CD133,[52] CD44[53] and ABCB5,[54] functional analysis including clonal analysis [55] and Aldefluor assay were used.[56] Using CD133 as a positive marker for colon CSCs generated conflicting results. The AC133 epitope, but not the CD133 protein, is specifically expressed in colon CSCs and its expression is lost upon differentiation.[57] In addition, CD44+ colon cancer cells and additional sub-fractionation of CD44+EpCAM+ cell population with CD166 enhance the success of tumor engraftments.[53]

Multiple CSCs have been reported in prostate,[58] lung and many other organs, including liver, pancreas, kidney or ovary.[41][59] In prostate cancer, the tumor-initiating cells have been identified in CD44+[60] cell subset as CD44+α2β1+,[61] TRA-1-60+CD151+CD166+ [62] or ALDH+ [63] cell populations. Putative markers for lung CSCs have been reported, including CD133+,[64] ALDH+,[65] CD44+ [66] and oncofetal protein 5T4+.[67]

13.7 Metastatic cancer stem cells

Metastasis is the major cause of tumor lethality in patients. However, not every tumor cell can metastasize. This potential depends on factors that determine growth, angiogenesis, invasion and other basic processes. In epithelial tumors, the epithelial-mesenchymal transition (EMT) is considered as a crucial event in metastasis.[68] EMT and the reverse transition from mesenchymal to an epithelial phenotype (MET) are involved in embryonic development, which involves disruption of epithelial cell homeostasis and the acquisition of a migratory mesenchymal phenotype.[69] EMT appears to be controlled by canonical pathways such as WNT and transforming growth factor β.[70] EMT's important feature is the loss of membrane E-cadherin in adherens junctions, where β-catenin may play a significant role. Translocation of β-catenin from adherens junctions to the nucleus may lead to a loss of E-cadherin, and subsequently to EMT. Nuclear β-catenin apparently can directly, transcriptionally activate EMT-associated target genes, such as the E-cadherin gene repressor SLUG (also known as SNAI2).[71] Furthermore, mechanical properties of the tumor microenvironment, or hypoxia, can influence cancer stem cell properties and metastatic behavior. [72]

Tumor cells undergoing an EMT may be precursors for metastatic cancer cells, or even metastatic CSCs.[73] In the invasive edge of pancreatic carcinoma, a subset of CD133+CXCR4+ (receptor for CXCL12 chemokine also known as a SDF1 ligand) cells was defined. These cells exhibited significantly stronger migratory activity than their counterpart CD133+CXCR4− cells, but both cell subsets showed similar tumor development capacity.[74] Moreover, inhibition of the CXCR4 receptor led to the reduced metastatic potential without altering tumorigenic capacity.[75]

In breast cancer CD44+CD24−/low cells are detectable in metastatic pleural effusions.[40] By contrast, an increased number of CD24+ cells have been identified in distant metastases in breast cancer patients.[76] It is possible that CD44+CD24−/low cells initially metastasize and in the new site they change their phenotype and undergo limited differentiation.[77] The two-phase expression pattern hypothesis proposes two forms of cancer stem cells - stationary (SCS) and mobile (MCS). SCS are embedded in tissue and persist in differentiated areas throughout tumor

progression. MCS are located at the tumor-host interface. These cells are apparently derived from SCS through the acquisition of transient EMT (Fig. 7).[78]

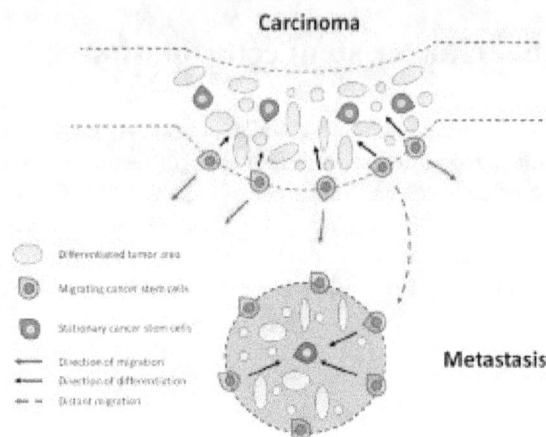

Figure 7: *The concept of migrating cancer stem cells (MSC). Stationary cancer stem cells are embedded in begin carcinomas and these cells are detectable in the differentiated central area of a tumor. The important step toward malignancy is the induction of epithelial mesenchymal transition (EMT) in the stationary cancer stem cells (SCS), which become mobile or migrating cancer stem cells. MCS cells divide asymmetrically. One daughter cell starts proliferation and differentiation. The remaining MCS migrates a short distance before undergoing new asymmetric division, or starts dissemination through blood vessela or lymphatic vessels and produces a metastasis.*

13.8 Implications for cancer treatment

CSCs have implications for cancer therapy, including disease identification, selective drug targets, prevention of metastasis and new intervention strategies.

Normal somatic stem cells are naturally resistant to chemotherapeutic agents. They produce various pumps (such as MDR) that pump out drugs and DNA repair proteins. They have a slow rate of cell turnover (chemotherapeutic agents naturally target rapidly replicating cells). CSCs that develop from normal stem cells may also produce these proteins, which could increase their resistance towards chemotherapy. The surviving CSCs then repopulate the tumor, causing a relapse.[79] Selectively targeting CSCs may allow treatment of aggressive, non-resectable tumors, as well as prevent metastasis and relapse.[79] The hypothesis suggests that upon CSC elimination, cancer could regress due to differentiation and/or cell death. What fraction of tumor cells are CSCs and therefore need to be eliminated is unclear.[80]

Studies looked for specific markers[13] and for proteomic and genomic tumor signatures that distinguish CSCs from others.[81] In 2009, scientists identified the compound salinomycin, which selectively reduces the proportion of breast CSCs in mice by more than 100-fold relative to Paclitaxel, a commonly used chemotherapeutic agent.[82] Some types of cancer cells can survive treatment with salinomycin through autophagy,[83] whereby cells use acidic organelles such as lysosomes to degrade and recycle certain types of proteins. The use of autophagy inhibitors can kill cancer stem cells that survive by autophagy.[84]

The cell surface receptor interleukin-3 receptor-alpha (CD123) is overexpressed on CD34+CD38- leukemic stem cells (LSCs) in acute myelogenous leukemia (AML) but not on normal CD34+CD38- bone marrow cells.[85] Treating AML-engrafted NOD/SCID mice with a CD123-specific monoclonal antibody impaired LSCs homing to the bone marrow and reduced overall AML cell repopulation including the proportion of LSCs in secondary mouse recipients.[86]

A 2015 study packaged nanoparticles with miR-34a and ammonium bicarbonate and delivered them to prostate CSCs in a mouse model. Then they irradiated the area with near-infrared laser light. This caused the nanoparticles to swell three times or more in size bursting the endosomes and dispersing the RNA in the cell. miR-34a can lower the levels of CD44.[87][88]

13.9   Pathways

The design of new drugs for targeting CSCs requires understanding the cellular mechanisms that regulate cell proliferation. The first advances in this area were made with hematopoietic stem cells (HSCs) and their transformed counterparts in leukemia, the disease for which the origin of CSCs is best understood. Stem cells of many organs share the same cellular pathways as leukemia-derived HSCs.

A normal stem cell may be transformed into a CSC through disregulation of the proliferation and differentiation pathways controlling it or by inducing oncoprotein activity.

13.9.1   BMI-1

The Polycomb group transcriptional repressor Bmi-1 was discovered as a common oncogene activated in lymphoma[89] and later shown to regulate HSCs.[90] The role of Bmi-1 has been illustrated in neural stem cells.[91] The pathway appears to be active in CSCs of pediatric brain tumors.[92]

13.9.2   Notch

The Notch pathway plays a role in controlling stem cell proliferation for several cell types including hematopoietic, neural and mammary[93] stem cells. Components of this pathway have been proposed to act as oncogenes in mammary[94] and other tumors.

A particular branch of the Notch signaling pathway that involves the transcription factor Hes3 regulates a number of cultured cells with cancer stem cell characteristics obtained from glioblastoma patients.[95]

13.9.3   Sonic hedgehog and Wnt

These developmental pathways are stem cell regulators.[96] Both Sonic hedgehog (SHH) and Wnt pathways are commonly hyperactivated in tumors and are necessary to sustain tumor growth. However, the Gli transcription factors that are regulated by SHH take their name from gliomas, where they are highly expressed. A degree of crosstalk exists between the two pathways and their activation commonly goes together.[97] This is common rather than universal. For instance, in colon cancer hedgehog signalling appears to antagonise Wnt.[98]

Sonic hedgehog blockers are available, such as cyclopamine. A water-soluble cyclopamine may be more effective in cancer treatment. DMAPT, a water-soluble derivative of parthenolide (induces oxidative stress, inhibits NF-κB signaling[99]) for AML (leukemia) and possibly myeloma and prostate cancer. Telomerase is a study subject in CSC physiology.[100] GRN163L (Imetelstat) was recently started in trials to target myeloma stem cells.

Wnt signaling can become independent of regular stimuli, through mutations in downstream oncogenes and tumor suppressor genes that become permanently activated even though the normal receptor has not received a signal. β-catenin binds to transcription factors such as the protein TCF4 and in combination the molecules activate the necessary genes. LF3 strongly inhibits this binding in vitro, in cell lines and reduced tumor growth in mouse models. It prevented replication and reduced their ability to migrate, all without affecting healthy cells. No cancer stem cells remained after treatment. The discovery was the product of "rational drug design", involving AlphaScreens and ELISA technologies.[101]

13.10 Cancer stem cells spheroids (3D module)

The monolayer of CSCs grown as spheroids showed better growth rate than MDA-MB 231 cells, which shows the efficacy of 3D spheroid format of CSCs. CD44 show increased expression in spheroids compared to 2D culture of MDA-MB 231. ALDH1 a key marker of breast stem cells was highly expressed in BCSCs and MDA-MB 231 grown in 3D, while they are absent in CSCs and MDA-MB 231 cells grown in 2D.

CSCs grown as spheroids showed better growth rates, which showed the efficacy of 3D spheroid format for CSCs culture. BCSC prevalence and clinical outcomes are associated and support key roles of CSCs in breast cancer metastasis and drug resistance.[102]

13.11   References

Wikipedia and research

[1] Mukherjee, Siddhartha. "The Cancer Sleeper Cell". *New York Times.* New York Times. Retrieved 15 July 2014.

[2] Heppner, GH; Miller, BE (1983).   "Tumor heterogeneity: biological implications and therapeutic consequences".   *Cancer metastasis reviews* **2** (1):   5–23. doi:10.1007/BF00046903. PMID 6616442.

[3] Reya, T; Morrison, SJ; Clarke, MF; Weissman, IL (Nov 1, 2001).   "Stem cells, cancer, and cancer stem cells". *Nature* **414** (6859): 105–11. doi:10.1038/35102167. PMID 11689955.

[4] Bonnet, D; Dick, JE (July 1997). "Human acute myeloid leukemia is organized as a hierarchy that originates from a primitive hematopoietic cell". *Nature Medicine* **3** (7): 730–7. doi:10.1038/nm0797-730. PMID 9212098.

[5] Barabé, F; Kennedy, JA; Hope, KJ; Dick, JE (Apr 27, 2007).   "Modeling the initiation and progression of human acute leukemia in mice". *Science* **316** (5824): 600–4. doi:10.1126/science.1139851. PMID 17463288.

[6] Nowell, PC (Oct 1, 1976).   "The clonal evolution of tumor cell populations".   *Science* **194** (4260): 23–8. doi:10.1126/science.959840. PMID 959840.

[7] Clark, EA; Golub, TR; Lander, ES; Hynes, RO (Aug 3, 2000).   "Genomic analysis of metastasis reveals an essential role for RhoC".   *Nature* **406** (6795): 532–5. doi:10.1038/35020106. PMID 10952316.

[8] Gupta PB, Chaffer CL, Weinberg RA (2009).   "Cancer stem cells: mirage or reality?".   *Nat Med* **15** (9): 1010–2. doi:10.1038/nm0909-1010. PMID 19734877.

[9] Hayflick L (1965).   "The Limited in Vitro Lifetime of Human Diploid Cell Strains".   *Exp Cell Res* **37**: 614–636. doi:10.1016/0014-4827(65)90211-9. PMID 14315085.

[10] Bonnet D, Dick JE (July 1997).   "Human acute myeloid leukemia is organized as a hierarchy that originates from a primitive hematopoietic cell". *Nature Medicine* **3** (7): 730–7. doi:10.1038/nm0797-730. PMID 9212098.

[11] Ignatova TN, Kukekov VG, Laywell ED, Suslov ON, Vrionis FD, Steindler DA (Sep 2002). "Human cortical glial tumors contain neural stem-like cells expressing astroglial and neuronal markers in vitro.". *Glia.* **39** (3): 193–206. doi:10.1002/glia.10094. PMID 12203386.

[12] Singh SK, Clarke ID, Terasaki M, Bonn VE, Hawkins C, Squire J, Dirks PB (September 2003). "Identification of a cancer stem cell in human brain tumors". *Cancer Research* **63** (18): 5821–8. PMID 14522905.

[13] Al-Hajj M, Wicha MS, Benito-Hernandez A, Morrison SJ, Clarke MF (April 2003). "Prospective identification of tumorigenic breast cancer cells". *Proceedings of the National Academy of Sciences of the United States of America* **100** (7): 3983–8.   doi:10.1073/pnas.0530291100.   PMC 153034. PMID 12629218.

[14] O'Brien CA, Pollett A, Gallinger S, Dick JE (January 2007). "A human colon cancer cell capable of initiating tumour growth in immunodeficient mice". *Nature* **445** (7123): 106–10. doi:10.1038/nature05372. PMID 17122772.

[15] Zhang S, Balch C, Chan MW, Lai HC, Matei D, Schilder JM, Yan PS, Huang TH, Nephew KP (June 2008). "Identification and characterization of ovarian cancer-initiating cells from primary human tumors". *Cancer Research* **68** (11): 4311–20.   doi:10.1158/0008-5472.CAN-08-0364. PMC 2553722. PMID 18519691.

[16] "Molecular phenotyping of human ovarian cancer stem cells unravels the mechanisms for repair and chemoresistance.".   *Cell Cycle* **8** (1):   158–66.   Jan 2009. doi:10.4161/cc.8.1.7533. PMID 19158483.

[17] Alvero AB, Chen R, Fu HH, Montagna M, Schwartz PE, Rutherford T, Silasi DA, Steffensen KD, Waldstrom M (Jan 2009). "Molecular phenotyping of human ovarian cancer stem cells unravels the mechanisms for repair and chemoresistance". *Cell Cycle* **8** (1): 158–66. doi:10.4161/cc.8.1.7533. PMID 19158483.

[18] Li C, Heidt DG, Dalerba P, Burant CF, Zhang L, Adsay V, Wicha M, Clarke MF, Simeone DM (February 2007). "Identification of pancreatic cancer stem cells". *Cancer Research* **67** (3): 1030–7. doi:10.1158/0008-5472.CAN-06-2030. PMID 17283135.

[19] Maitland NJ, Collins AT (June 2008). "Prostate cancer stem cells: a new target for therapy". *J. Clin. Oncol.* **26** (17): 2862–70. doi:10.1200/JCO.2007.15.1472. PMID 18539965.

[20] Lang Sh, Frame F, Collins A (January 2009). "Prostate cancer stem cells". *J. Pathol.* **217** (2): 299–306. doi:10.1002/path.2478. PMC 2673349. PMID 19040209.

[21] Schatton T, Murphy GF, Frank, NY, Yamaura K, Waaga-Gasser AM, Gasser M, Zhan Q, Jordan S, Duncan LM, Weishaupt C, Fuhlbrigge RC, Kupper TS, Sayegh MH, Frank MH (Jan 2008). "Identification of cells initiating human melanomas". *Nature* **451** (7176): 345–9. doi:10.1038/nature06489. PMC 3660705. PMID 18202660.

[22] Boiko AD, Razorenova OV, van de Rijn M, Swetter SM, Johnson DL, Ly DP, Butler PD, Yang GP, Joshua B, Kaplan MJ, Longaker MT, Weissman IL (Jul 2010). "Human melanoma-initiating cells express neural crest nerve growth factor receptor CD271". *Nature* **466** (7302): 133–7. doi:10.1038/nature09161. PMC 2898751. PMID 20596026.

[23] Schmidt P, Kopecky C, Hombach A, Zigrino P, Mauch C, Abken H. (Feb 2011). "Eradication of melanomas by targeted elimination of a minor subset of tumor cells". *PNAS* **108** (6): 2474–9. doi:10.1073/pnas.1009069108. PMC 3038763. PMID 21282657.

[24] Civenni G, Walter A, Kobert N, Mihic-Probst D, Zipser M, Belloni B, Seifert B, Moch H, Dummer R, van den Broek M, Sommer L. (Mar 2011). "Human CD271-Positive Melanoma Stem Cells Associated with Metastasis Establish Tumor Heterogeneity and Long-Term Growth". *Cancer Res.* **71** (8): 3098–109. doi:10.1158/0008-5472.CAN-10-3997. PMID 21393506.

[25] Matsui W, Huff CA, Wang Q; et al. (March 2004). "Characterization of clonogenic multiple myeloma cells". *Blood* **103** (6): 2332–6. doi:10.1182/blood-2003-09-3064. PMC 3311914. PMID 14630803.

[26] Matsui W, Wang Q, Barber JP; et al. (January 2008). "Clonogenic multiple myeloma progenitors, stem cell properties, and drug resistance". *Cancer Res.* **68** (1): 190–7. doi:10.1158/0008-5472.CAN-07-3096. PMC 2603142. PMID 18172311.

[27] Colmont CS, Benketah A, Reed SH, Hawk NV, Telford WG, Ohyama M, Udey MC, Yee CL, Vogel JC, Patel GK. (Jan 2013). "CD200-expressing human basal cell carcinoma cells initiate tumor growth" (PDF). *PNAS* **110** (4): 1434–9. doi:10.1073/pnas.1211655110. PMC 3557049. PMID 23292936.

[28] Patel GK, Yee CL, Terunuma A, Telford WG, Voong N, Yuspa SH, Vogel JC. (Feb 2012). "Identification and characterization of tumor-initiating cells in human primary cutaneous squamous cell carcinoma" (PDF). *J Invest Dermatol* **132** (2): 401–409. doi:10.1038/jid.2011.317. PMC 3258300. PMID 22011906.

[29] Preziosi, Luigi (2003). *Cancer Modelling and Simulation.* Boca Raton: CRC Press. ISBN 1-58488-361-8.

[30] Ganguly R, Puri IK (February 2006). "Mathematical model for the cancer stem cell hypothesis". *Cell proliferation* **39** (1): 3–14. doi:10.1111/j.1365-2184.2006.00369.x. PMID 16426418.

[31] Ganguly R, Puri IK (June 2007). "Mathematical model for chemotherapeutic drug efficacy in arresting tumour growth based on the cancer stem cell hypothesis". *Cell proliferation* **40** (3): 338–354. doi:10.1111/j.1365-2184.2007.00434.x. PMID 17531079.

[32] Wang Y, Yang J, Zheng H, Tomasek GJ, Zhang P, McKeever PE, Lee EY, Zhu Y (June 2009). "Expression of mutant p53 proteins implicates a lineage relationship between neural stem cells and malignant astrocytic glioma in a murine model". *Cancer Cell* **15** (6): 514–26. doi:10.1016/j.ccr.2009.04.001. PMC 2721466. PMID 19477430.

[33] López-Lázaro, Miguel (2015-01-01). "The migration ability of stem cells can explain the existence of cancer of unknown primary site. Rethinking metastasis". *Oncoscience* **2** (5): 467–475. doi:10.18632/oncoscience.159. ISSN 2331-4737. PMC 4468332. PMID 26097879.

[34] López-Lázaro, Miguel (2015-08-18). "Stem cell division theory of cancer". *Cell Cycle (Georgetown, Tex.)* **14** (16): 2547–2548. doi:10.1080/15384101.2015.1062330. ISSN 1551-4005. PMID 26090957.

[35] Clarke MF, Dick JE, Dirks PB, Eaves CJ, Jamieson CH, Jones DL, Visvader J, Weissman IL, Wahl GM (October 2006). "Cancer stem cells--perspectives on current status and future directions: AACR Workshop on cancer stem cells". *Cancer Research* **66** (19): 9339–44. doi:10.1158/0008-5472.CAN-06-3126. PMID 16990346.

[36] Golebiewska, A; Brons, NH; Bjerkvig, R; Niclou, SP (Feb 4, 2011). "Critical appraisal of the side population assay in stem cell and cancer stem cell research". *Cell stem cell* **8** (2): 136–47. doi:10.1016/j.stem.2011.01.007. PMID 21295271.

[37] Scharenberg, CW; Harkey, MA; Torok-Storb, B (Jan 15, 2002). "The ABCG2 transporter is an efficient Hoechst 33342 efflux pump and is preferentially expressed by immature human hematopoietic progenitors". *Blood* **99** (2): 507–12. doi:10.1182/blood.V99.2.507. PMID 11781231.

[38] Pastrana, E; Silva-Vargas, V; Doetsch, F (May 6, 2011). "Eyes wide open: a critical review of sphere-formation as an assay for stem cells". *Cell stem cell* **8** (5): 486–98. doi:10.1016/j.stem.2011.04.007. PMC 3633588. PMID 21549325.

[39] Nicolis, SK (February 2007). "Cancer stem cells and "stemness" genes in neuro-oncology". *Neurobiology of disease* **25** (2): 217–29. doi:10.1016/j.nbd.2006.08.022. PMID 17141509.

[40] Al-Hajj, M; Wicha, MS; Benito-Hernandez, A; Morrison, SJ; Clarke, MF (Apr 1, 2003). "Prospective identification

of tumorigenic breast cancer cells". *Proceedings of the National Academy of Sciences of the United States of America* **100** (7): 3983–8. doi:10.1073/pnas.0530291100. PMC 153034. PMID 12629218.

[41] Meng E, Mitra A, Tripathi K, Finan MA, Scalici J, McClellan S, Madeira da Silva L, Reed E, Shevde LA, Palle K, Rocconi RP (Sep 12, 2014). "ALDH1A1 Maintains Ovarian Cancer Stem Cell-Like Properties by Altered Regulation of Cell Cycle Checkpoint and DNA Repair Network Signaling.". *PLOS ONE* **9** (9): e107142. doi:10.1371/journal.pone.0107142. PMID 25216266.

[42] Visvader, JE; Lindeman, GJ (October 2008). "Cancer stem cells in solid tumours: accumulating evidence and unresolved questions". *Nature Reviews Cancer* **8** (10): 755–68. doi:10.1038/nrc2499. PMID 18784658.

[43] Hirschmann-Jax, C; Foster, AE; Wulf, GG; Nuchtern, JG; Jax, TW; Gobel, U; Goodell, MA; Brenner, MK (Sep 28, 2004). "A distinct "side population" of cells with high drug efflux capacity in human tumor cells". *Proceedings of the National Academy of Sciences of the United States of America* **101** (39): 14228–33. doi:10.1073/pnas.0400067101. PMC 521140. PMID 15381773.

[44] Ginestier, C; Hur, MH; Charafe-Jauffret, E; Monville, F; Dutcher, J; Brown, M; Jacquemier, J; Viens, P; Kleer, CG; Liu, S; Schott, A; Hayes, D; Birnbaum, D; Wicha, MS; Dontu, G (November 2007). "ALDH1 is a marker of normal and malignant human mammary stem cells and a predictor of poor clinical outcome". *Cell stem cell* **1** (5): 555–67. doi:10.1016/j.stem.2007.08.014. PMC 2423808. PMID 18371393.

[45] Pece, S; Tosoni, D; Confalonieri, S; Mazzarol, G; Vecchi, M; Ronzoni, S; Bernard, L; Viale, G; Pelicci, PG; Di Fiore, PP (Jan 8, 2010). "Biological and molecular heterogeneity of breast cancers correlates with their cancer stem cell content". *Cell* **140** (1): 62–73. doi:10.1016/j.cell.2009.12.007. PMID 20074520.

[46] Deng, S; Yang, X; Lassus, H; Liang, S; Kaur, S; Ye, Q; Li, C; Wang, LP; Roby, KF; Orsulic, S; Connolly, DC; Zhang, Y; Montone, K; Bützow, R; Coukos, G; Zhang, L (Apr 21, 2010). Cao, Yihai, ed. "Distinct expression levels and patterns of stem cell marker, aldehyde dehydrogenase isoform 1 (ALDH1), in human epithelial cancers". *PLOS ONE* **5** (4): e10277. doi:10.1371/journal.pone.0010277. PMC 2858084. PMID 20422001.

[47] Meyer, MJ; Fleming, JM; Lin, AF; Hussnain, SA; Ginsburg, E; Vonderhaar, BK (Jun 1, 2010). "CD44posCD49fhiCD133/2hi defines xenograft-initiating cells in estrogen receptor-negative breast cancer". *Cancer Research* **70** (11): 4624–33. doi:10.1158/0008-5472.CAN-09-3619. PMID 20484027.

[48] Singh, SK; Hawkins, C; Clarke, ID; Squire, JA; Bayani, J; Hide, T; Henkelman, RM; Cusimano, MD; Dirks, PB (Nov 18, 2004). "Identification of human brain tumour initiating cells". *Nature* **432** (7015): 396–401. doi:10.1038/nature03128. PMID 15549107.

[49] Son, MJ; Woolard, K; Nam, DH; Lee, J; Fine, HA (May 8, 2009). "SSEA-1 is an enrichment marker for tumor-initiating cells in human glioblastoma". *Cell stem cell* **4** (5): 440–52. doi:10.1016/j.stem.2009.03.003. PMID 19427293.

[50] Mazzoleni, S; Politi, LS; Pala, M; Cominelli, M; Franzin, A; Sergi Sergi, L; Falini, A; De Palma, M; Bulfone, A; Poliani, PL; Galli, R (Oct 1, 2010). "Epidermal growth factor receptor expression identifies functionally and molecularly distinct tumor-initiating cells in human glioblastoma multiforme and is required for gliomagenesis". *Cancer Research* **70** (19): 7500–13. doi:10.1158/0008-5472.CAN-10-2353. PMID 20858720.

[51] Anido, J; Sáez-Borderías, A; Gonzàlez-Juncà, A; Rodón, L; Folch, G; Carmona, MA; Prieto-Sánchez, RM; Barba, I; Martínez-Sáez, E; Prudkin, L; Cuartas, I; Raventós, C; Martínez-Ricarte, F; Poca, MA; García-Dorado, D; Lahn, MM; Yingling, JM; Rodón, J; Sahuquillo, J; Baselga, J; Seoane, J (Dec 14, 2010). "TGF-β Receptor Inhibitors Target the CD44(high)/Id1(high) Glioma-Initiating Cell Population in Human Glioblastoma". *Cancer Cell* **18** (6): 655–68. doi:10.1016/j.ccr.2010.10.023. PMID 21156287.

[52] O'Brien, CA; Pollett, A; Gallinger, S; Dick, JE (Jan 4, 2007). "A human colon cancer cell capable of initiating tumour growth in immunodeficient mice". *Nature* **445** (7123): 106–10. doi:10.1038/nature05372. PMID 17122772.

[53] Dalerba, P; Dylla, SJ; Park, IK; Liu, R; Wang, X; Cho, RW; Hoey, T; Gurney, A; Huang, EH; Simeone, DM; Shelton, AA; Parmiani, G; Castelli, C; Clarke, MF (Jun 12, 2007). "Phenotypic characterization of human colorectal cancer stem cells". *Proceedings of the National Academy of Sciences of the United States of America* **104** (24): 10158–63. doi:10.1073/pnas.0703478104. PMC 1891215. PMID 17548814.

[54] Wilson, BJ; Schatton, T; Zhan, Q; Gasser, M; Ma, J; Saab, KR; Schanche, R; Waaga-Gasser, AM; Gold, JS; Huang, Q; Murphy, GF; Frank, MH; Frank, NY (Aug 1, 2011). "ABCB5 identifies a therapy-refractory tumor cell population in colorectal cancer patients". *Cancer Research* **71** (15): 5307–16. doi:10.1158/0008-5472.CAN-11-0221. PMC 3395026. PMID 21652540.

[55] Odoux, C; Fohrer, H; Hoppo, T; Guzik, L; Stolz, DB; Lewis, DW; Gollin, SM; Gamblin, TC; Geller, DA; Lagasse, E (Sep 1, 2008). "A stochastic model for cancer stem cell origin in metastatic colon cancer". *Cancer Research* **68** (17): 6932–41. doi:10.1158/0008-5472.CAN-07-5779. PMC 2562348. PMID 18757407.

[56] Huang, EH; Hynes, MJ; Zhang, T; Ginestier, C; Dontu, G; Appelman, H; Fields, JZ; Wicha, MS; Boman, BM (Apr 15, 2009). "Aldehyde dehydrogenase 1 is a marker for normal and malignant human colonic stem cells (SC) and tracks

SC overpopulation during colon tumorigenesis". *Cancer Research* **69** (8): 3382–9. doi:10.1158/0008-5472.CAN-08-4418. PMC 2789401. PMID 19336570.

[57] Kemper, K; Sprick, MR; de Bree, M; Scopelliti, A; Vermeulen, L; Hoek, M; Zeilstra, J; Pals, ST; Mehmet, H; Stassi, G; Medema, JP (Jan 15, 2010). "The AC133 epitope, but not the CD133 protein, is lost upon cancer stem cell differentiation". *Cancer Research* **70** (2): 719–29. doi:10.1158/0008-5472.CAN-09-1820. PMID 20068153.

[58] Liu, C; Kelnar, K; Liu, B; Chen, X; Calhoun-Davis, T; Li, H; Patrawala, L; Yan, H; Jeter, C; Honorio, S; Wiggins, JF; Bader, AG; Fagin, R; Brown, D; Tang, DG (February 2011). "The microRNA miR-34a inhibits prostate cancer stem cells and metastasis by directly repressing CD44". *Nature Medicine* **17** (2): 211–5. doi:10.1038/nm.2284. PMC 3076220. PMID 21240262.

[59] Ho, MM; Ng, AV; Lam, S; Hung, JY (May 15, 2007). "Side population in human lung cancer cell lines and tumors is enriched with stem-like cancer cells". *Cancer Research* **67** (10): 4827–33. doi:10.1158/0008-5472.CAN-06-3557. PMID 17510412.

[60] Patrawala, L; Calhoun, T; Schneider-Broussard, R; Li, H; Bhatia, B; Tang, S; Reilly, JG; Chandra, D; Zhou, J; Claypool, K; Coghlan, L; Tang, DG (Mar 16, 2006). "Highly purified CD44+ prostate cancer cells from xenograft human tumors are enriched in tumorigenic and metastatic progenitor cells". *Oncogene* **25** (12): 1696–708. doi:10.1038/sj.onc.1209327. PMID 16449977.

[61] Dubrovska, A; Kim, S; Salamone, RJ; Walker, JR; Maira, SM; García-Echeverría, C; Schultz, PG; Reddy, VA (Jan 6, 2009). "The role of PTEN/Akt/PI3K signaling in the maintenance and viability of prostate cancer stem-like cell populations". *Proceedings of the National Academy of Sciences of the United States of America* **106** (1): 268–73. doi:10.1073/pnas.0810956106. PMC 2629188. PMID 19116269.

[62] Rajasekhar, VK; Studer, L; Gerald, W; Socci, ND; Scher, HI (Jan 18, 2011). "Tumour-initiating stem-like cells in human prostate cancer exhibit increased NF-κB signalling". *Nature Communications* **2** (1): 162–. doi:10.1038/ncomms1159. PMC 3105310. PMID 21245843.

[63] Li, T; Su, Y; Mei, Y; Leng, Q; Leng, B; Liu, Z; Stass, SA; Jiang, F (February 2010). "ALDH1A1 is a marker for malignant prostate stem cells and predictor of prostate cancer patients' outcome". *Laboratory Investigation* **90** (2): 234–44. doi:10.1038/labinvest.2009.127. PMC 3552330. PMID 20010854.

[64] Eramo, A; Lotti, F; Sette, G; Pilozzi, E; Biffoni, M; Di Virgilio, A; Conticello, C; Ruco, L; Peschle, C; De Maria, R (March 2008). "Identification and expansion of the tumorigenic lung cancer stem cell population". *Cell death and differentiation* **15** (3): 504–14. doi:10.1038/sj.cdd.4402283. PMID 18049477.

[65] Sullivan, JP; Spinola, M; Dodge, M; Raso, MG; Behrens, C; Gao, B; Schuster, K; Shao, C; Larsen, JE; Sullivan, LA; Honorio, S; Xie, Y; Scaglioni, PP; DiMaio, JM; Gazdar, AF; Shay, JW; Wistuba, II; Minna, JD (Dec 1, 2010). "Aldehyde dehydrogenase activity selects for lung adenocarcinoma stem cells dependent on notch signaling". *Cancer Research* **70** (23): 9937–48. doi:10.1158/0008-5472.CAN-10-0881. PMC 3058307. PMID 21118965.

[66] Leung, EL; Fiscus, RR; Tung, JW; Tin, VP; Cheng, LC; Sihoe, AD; Fink, LM; Ma, Y; Wong, MP (Nov 19, 2010). Jin, Dong-Yan, ed. "Non-small cell lung cancer cells expressing CD44 are enriched for stem cell-like properties". *PLOS ONE* **5** (11): e14062. doi:10.1371/journal.pone.0014062. PMC 2988826. PMID 21124918.

[67] Damelin, M; Geles, KG; Follettie, MT; Yuan, P; Baxter, M; Golas, J; DiJoseph, JF; Karnoub, M; Huang, S; Diesl, V; Behrens, C; Choe, SE; Rios, C; Gruzas, J; Sridharan, L; Dougher, M; Kunz, A; Hamann, PR; Evans, D; Armellino, D; Khandke, K; Marquette, K; Tchistiakova, L; Boghaert, ER; Abraham, RT; Wistuba, II; Zhou, BB (Jun 15, 2011). "Delineation of a cellular hierarchy in lung cancer reveals an oncofetal antigen expressed on tumor-initiating cells". *Cancer Research* **71** (12): 4236–46. doi:10.1158/0008-5472.CAN-10-3919. PMID 21540235.

[68] Thiery, JP (June 2002). "Epithelial-mesenchymal transitions in tumour progression". *Nature Reviews Cancer* **2** (6): 442–54. doi:10.1038/nrc822. PMID 12189386.

[69] Angerer, LM; Angerer, RC (June 1999). "Regulative development of the sea urchin embryo: signalling cascades and morphogen gradients". *Seminars in cell & developmental biology* **10** (3): 327–34. doi:10.1006/scdb.1999.0292. PMID 10441547.

[70] Mani, SA; Yang, J; Brooks, M; Schwaninger, G; Zhou, A; Miura, N; Kutok, JL; Hartwell, K; Richardson, AL; Weinberg, RA (Jun 12, 2007). "Mesenchyme Forkhead 1 (FOXC2) plays a key role in metastasis and is associated with aggressive basal-like breast cancers". *Proceedings of the National Academy of Sciences of the United States of America* **104** (24): 10069–74. doi:10.1073/pnas.0703900104. PMC 1891217. PMID 17537911.

[71] Conacci-Sorrell, M; Simcha, I; Ben-Yedidia, T; Blechman, J; Savagner, P; Ben-Ze'ev, A (Nov 24, 2003). "Autoregulation of E-cadherin expression by cadherin-cadherin interactions: the roles of beta-catenin signaling, Slug, and MAPK". *The Journal of Cell Biology* **163** (4): 847–57. doi:10.1083/jcb.200308162. PMC 2173691. PMID 14623871.

[72] Spill, F.; Reynolds, D.S.; Kamm, R.D.; Zaman, M.H. "Impact of the physical microenvironment on tumor progression and metastasis". Current Opinion in Biotechnology. pp. 41–48. doi:10.1016/j.copbio.2016.02.007.

[73] Kaplan, RN; Riba, RD; Zacharoulis, S; Bramley, AH; Vincent, L; Costa, C; MacDonald, DD; Jin, DK; Shido, K;

Kerns, SA; Zhu, Z; Hicklin, D; Wu, Y; Port, JL; Altorki, N; Port, ER; Ruggero, D; Shmelkov, SV; Jensen, KK; Rafii, S; Lyden, D (Dec 8, 2005). "VEGFR1-positive haematopoietic bone marrow progenitors initiate the pre-metastatic niche". *Nature* **438** (7069): 820–7. doi:10.1038/nature04186. PMC 2945882. PMID 16341007.

[74] Hermann, PC; Huber, SL; Herrler, T; Aicher, A; Ellwart, JW; Guba, M; Bruns, CJ; Heeschen, C (Sep 13, 2007). "Distinct populations of cancer stem cells determine tumor growth and metastatic activity in human pancreatic cancer". *Cell stem cell* **1** (3): 313–23. doi:10.1016/j.stem.2007.06.002. PMID 18371365.

[75] Yang, ZF; Ho, DW; Ng, MN; Lau, CK; Yu, WC; Ngai, P; Chu, PW; Lam, CT; Poon, RT; Fan, ST (February 2008). "Significance of CD90+ cancer stem cells in human liver cancer". *Cancer Cell* **13** (2): 153–66. doi:10.1016/j.ccr.2008.01.013. PMID 18242515.

[76] Shipitsin, M; Campbell, LL; Argani, P; Weremowicz, S; Bloushtain-Qimron, N; Yao, J; Nikolskaya, T; Serebryiskaya, T; Beroukhim, R; Hu, M; Halushka, MK; Sukumar, S; Parker, LM; Anderson, KS; Harris, LN; Garber, JE; Richardson, AL; Schnitt, SJ; Nikolsky, Y; Gelman, RS; Polyak, K (March 2007). "Molecular definition of breast tumor heterogeneity". *Cancer Cell* **11** (3): 259–73. doi:10.1016/j.ccr.2007.01.013. PMID 17349583.

[77] Shmelkov, SV; Butler, JM; Hooper, AT; Hormigo, A; Kushner, J; Milde, T; St Clair, R; Baljevic, M; White, I; Jin, DK; Chadburn, A; Murphy, AJ; Valenzuela, DM; Gale, NW; Thurston, G; Yancopoulos, GD; D'Angelica, M; Kemeny, N; Lyden, D; Rafii, S (June 2008). "CD133 expression is not restricted to stem cells, and both CD133+ and CD133- metastatic colon cancer cells initiate tumors". *The Journal of Clinical Investigation* **118** (6): 2111–20. doi:10.1172/JCI34401. PMC 2391278. PMID 18497886.

[78] Brabletz, T; Jung, A; Spaderna, S; Hlubek, F; Kirchner, T (September 2005). "Opinion: migrating cancer stem cells - an integrated concept of malignant tumour progression". *Nature Reviews Cancer* **5** (9): 744–9. doi:10.1038/nrc1694. PMID 16148886.

[79] Mraz, M.; Zent, C. S.; Church, A. K.; Jelinek, D. F.; Wu, X.; Pospisilova, S.; Ansell, S. M.; Novak, A. J.; Kay, N. E.; Witzig, T. E.; Nowakowski, G. S. (2011). "Bone marrow stromal cells protect lymphoma B-cells from rituximab-induced apoptosis and targeting integrin α−4−β−1 (VLA-4) with natalizumab can overcome this resistance". *British Journal of Haematology* **155** (1): 53–64. doi:10.1111/j.1365-2141.2011.08794.x. PMID 21749361.

[80] Dirks, P (Jul 1, 2010). "Cancer stem cells: Invitation to a second round". *Nature* **466** (7302): 40–1. doi:10.1038/466040a. PMID 20596007.

[81] "Insights on neoplastic stem cells from gel-based proteomics of childhood germ cell tumors". PMID 21793190.

[82] Gupta, PB; Onder, TT; Jiang, G; Tao, K; Kuperwasser, C; Weinberg, RA; Lander, ES (Aug 21, 2009). "Identification of selective inhibitors of cancer stem cells by high-throughput screening". *Cell* **138** (4): 645–59. doi:10.1016/j.cell.2009.06.034. PMID 19682730.

[83] Jangamreddy JR1, Ghavami S, Grabarek J, Kratz G, Wiechec E, Fredriksson BA, Rao Pariti RK, Cieślar-Pobuda A, Panigrahi S, Los MJ (2013). "Salinomycin induces activation of autophagy, mitophagy and affects mitochondrial polarity: differences between primary and cancer cells.". *Biochim Biophys Acta* **1833** (9): 2057–69. doi:10.1016/j.bbamcr.2013.04.011. PMID 23639289.

[84] Vlahopoulos S, Critselis E, Voutsas IF, Perez SA, Moschovi M, Baxevanis CN, Chrousos GP (2014). "New use for old drugs? Prospective targets of chloroquines in cancer therapy". *Curr Drug Targets*. **15** (9): 843–51. doi:10.2174/1389450115666140714121514. PMID 25023646.

[85] Jordan, C. T.; Upchurch, D.; Szilvassy, S. J.; Guzman, M. L.; Howard, D. S.; Pettigrew, A. L.; Meyerrose, T.; Rossi, R.; Grimes, B. (2000-10-01). "The interleukin-3 receptor alpha chain is a unique marker for human acute myelogenous leukemia stem cells". *Leukemia* **14** (10): 1777–1784. doi:10.1038/sj.leu.2401903. ISSN 0887-6924. PMID 11021753.

[86] Jin, Liqing; Lee, Erwin M.; Ramshaw, Hayley S.; Busfield, Samantha J.; Peoppl, Armando G.; Wilkinson, Lucy; Guthridge, Mark A.; Thomas, Daniel; Barry, Emma F. (2009-07-02). "Monoclonal antibody-mediated targeting of CD123, IL-3 receptor alpha chain, eliminates human acute myeloid leukemic stem cells". *Cell Stem Cell* **5** (1): 31–42. doi:10.1016/j.stem.2009.04.018. ISSN 1875-9777. PMID 19570512.

[87] http://helldesign.net (2015-12-07). "'Nanobombs' that blow up cancer cells | KurzweilAI". *www.kurzweilai.net*. Retrieved 2016-02-20.

[88] Wang, Hai; Agarwal, Pranay; Zhao, Shuting; Yu, Jianhua; Lu, Xiongbin; He, Xiaoming (2016-01-01). "A Near-Infrared Laser-Activated "Nanobomb" for Breaking the Barriers to MicroRNA Delivery". *Advanced Materials* **28** (2): 347–355. doi:10.1002/adma.201504263. ISSN 1521-4095.

[89] Haupt Y, Bath ML, Harris AW, Adams JM (November 1993). "bmi-1 transgene induces lymphomas and collaborates with myc in tumorigenesis". *Oncogene* **8** (11): 3161–4. PMID 8414519.

[90] Park IK, Qian D, Kiel M, Becker MW, Pihalja M, Weissman IL, Morrison SJ, Clarke MF (May 2003). "Bmi-1 is required for maintenance of adult self-renewing haematopoietic stem cells". *Nature* **423** (6937): 302–5. doi:10.1038/nature01587. PMID 12714971.

[91] Molofsky AV, Pardal R, Iwashita T, Park IK, Clarke MF, Morrison SJ (October 2003). "Bmi-1 dependence distinguishes neural stem cell self-renewal from progenitor proliferation". *Nature* **425** (6961): 962–7. doi:10.1038/nature02060. PMC 2614897. PMID 14574365.

[92] Hemmati HD, Nakano I, Lazareff JA, Masterman-Smith M, Geschwind DH, Bronner-Fraser M, Kornblum HI (December 2003). "Cancerous stem cells can arise from pediatric brain tumors". *Proceedings of the National Academy of Sciences of the United States of America* **100** (25): 15178–83. doi:10.1073/pnas.2036535100. PMC 299944. PMID 14645703.

[93] Dontu G, Jackson KW, McNicholas E, Kawamura MJ, Abdallah WM, Wicha MS (2004). "Role of Notch signaling in cell-fate determination of human mammary stem/progenitor cells". *Breast cancer research : BCR* **6** (6): R605–15. doi:10.1186/bcr920. PMC 1064073. PMID 15535842.

[94] Diévart A, Beaulieu N, Jolicoeur P (October 1999). "Involvement of Notch1 in the development of mouse mammary tumors". *Oncogene* **18** (44): 5973–81. doi:10.1038/sj.onc.1202991. PMID 10557086.

[95] Park DM, Jung J, Masjkur J; et al. (2013). "Hes3 regulates cell number in cultures from glioblastoma multiforme with stem cell characteristics". *Sci Rep* **3**: 1095. doi:10.1038/srep01095. PMC 3566603. PMID 23393614.

[96] Beachy PA, Karhadkar SS, Berman DM (November 2004). "Tissue repair and stem cell renewal in carcinogenesis". *Nature* **432** (7015): 324–31. doi:10.1038/nature03100. PMID 15549094.

[97] Zhou BP, Hung MC (June 2005). "Wnt, hedgehog and snail: sister pathways that control by GSK-3beta and beta-Trcp in the regulation of metastasis". *Cell cycle (Georgetown, Tex.)* **4** (6): 772–6. doi:10.4161/cc.4.6.1744. PMID 15917668.

[98] Akiyoshi T, Nakamura M, Koga K, Nakashima H, Yao T, Tsuneyoshi M, Tanaka M, Katano M (July 2006). "Gli1, downregulated in colorectal cancers, inhibits proliferation of colon cancer cells involving Wnt signalling activation". *Gut* **55** (7): 991–9. doi:10.1136/gut.2005.080333. PMC 1856354. PMID 16299030.

[99] She M, Chen X (2009). "Targeting signal pathways active in cancer stem cells to overcome drug resistance". *Chin J Lung Cancer* **12** (1): 3–7. doi:10.3779/j.issn.1009-3419.2009.01.001. PMID 20712949.

[100] Bollmann FM (August 2008). "The many faces of telomerase: emerging extratelomeric effects". *BioEssays* **30** (8): 728–32. doi:10.1002/bies.20793. PMID 18623070.

[101] Hodge, Russ (2016-01-25). "Hacking the programs of cancer stem cells". *medicalxpress.com*. Medical Express. Retrieved 2016-02-12.

[102] Abboodi MA. Isolation, identification, and spheroids formation of breast cancer stem cells, therapeutics implications. Clin Cancer Investig J 2014;3:322-5

13.12 Further reading

- Polyak K, Weinberg RA (April 2009). "Transitions between epithelial and mesenchymal states: acquisition of malignant and stem cell traits". *Nature Reviews Cancer* **9** (4): 265–73. doi:10.1038/nrc2620. PMID 19262571.

- Sánchez-García I, Vicente-Dueñas C, Cobaleda C (December 2007). "The theoretical basis of cancer-stem-cell-based therapeutics of cancer: can it be put into practice?". *BioEssays* **29** (12): 1269–80. doi:10.1002/bies.20679. PMID 18022789.

- Gao JX (2008). "Cancer stem cells: the lessons from pre-cancerous stem cells". *Journal of Cellular and Molecular Medicine* **12** (1): 67–96. doi:10.1111/j.1582-4934.2007.00170.x. PMID 18053092.

- Yanyan Lia, Tao Zhang (May 2014). "Targeting cancer stem cells by curcumin and clinical applications". *Cancer Letters* **346** (2): 197–205. doi:10.1016/j.canlet.2014.01.012.

13.13 External links

- European Cancer Stem Cell Research Institute A new Institute dedicated to research into cancer stem cells and related work.

- Cancer Stem Cell News A blog of news items related to cancer stem cells, with an emphasis on recent research and articles that are openly accessible

- Exploring the role of cancer stem cells in radioresistance Abstract of a review by Michael Baumann, Mechthild Krause, Richard Hill, "Nature Reviews Cancer " 2008(Jul): 8(7) 545-54

- "A Tumor's Lifeblood", Jessica Gorman, *CR* magazine, Summer 2006

- "Cancer Stem Cell Scientific Literature Review", *UMDNJ Stem Cell Research and Regenerative Medicine*, June 17, 2006

- "Stem cells may cause some forms of bone cancer", *News-Medical.Net*, December 7, 2005

- "The Bad Seed: Rare stem cells appear to drive cancers", *Science News Online*, March 20, 2004

- "The Real Problem in Breast Tumors: Cancer Stem Cells", *Genome News Network*, March 7, 2003

- Differentiation Therapy - A Different Approach to Treating Tumors (from Beaker Blog)

- Characteristics of Cancer Cells Cancer Inform Blog

- Cancer stem cells may be cause of brain tumors (research of John A. Boockvar)

Chapter 14

Induced pluripotent stem cell

Induced pluripotent stem cells (also known as **iPS** cells or **iPSCs**) are a type of pluripotent stem cell that can be generated directly from adult cells. The iPSC technology was pioneered by Shinya Yamanaka's lab in Kyoto, Japan, who showed in 2006 that the introduction of four specific genes encoding transcription factors could convert adult cells into pluripotent stem cells.[1] He was awarded the 2012 Nobel Prize along with Sir John Gurdon "for the discovery that mature cells can be reprogrammed to become pluripotent." [2]

Pluripotent stem cells hold great promise in the field of regenerative medicine. Because they can propagate indefinitely, as well as give rise to every other cell type in the body (such as neurons, heart, pancreatic, and liver cells), they represent a single source of cells that could be used to replace those lost to damage or disease.

The most well-known type of pluripotent stem cell is the embryonic stem cell. However, since the generation of embryonic stem cells involves destruction (or at least manipulation) [3] of the pre-implantation stage embryo (a spherical structure termed a "blastocyst"), there has been much controversy surrounding their use. Further, because embryonic stem cells can only be derived from embryos, it has so far not been feasible to create patient-matched embryonic stem cell lines.

Since iPSCs can be derived directly from adult tissues, they not only bypass the need for embryos, but can be made in a patient-matched manner, which means that each individual could have their own pluripotent stem cell line. These unlimited supplies of autologous cells could be used to generate transplants without the risk of immune rejection. While the iPSC technology has not yet advanced to a stage where therapeutic transplants have been deemed safe, iPSCs are readily being used in personalized drug discovery efforts and understanding the patient-specific basis of disease.

Depending on the methods used, reprogramming of adult cells to obtain iPSCs may pose significant risks that could limit their use in humans. For example, if viruses are used to genomically alter the cells, the expression of oncogenes (cancer-causing genes) may potentially be triggered. In February 2008, scientists announced the discovery of a technique that could remove oncogenes after the induction of pluripotency, thereby increasing the potential use of iPS cells in human diseases.[4] In April 2009, it was demonstrated that generation of iPS cells is possible without any genetic alteration of the adult cell: a repeated treatment of the cells with certain proteins channeled into the cells via poly-arginine anchors was sufficient to induce pluripotency.[5] The acronym given for those iPSCs is **piP-SCs** (protein-induced pluripotent stem cells).

14.1 Production of iPSCs

See also: Reprogramming

iPSCs are typically derived by introducing a specific set of

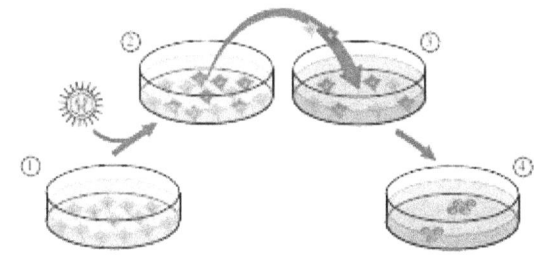

A scheme of the generation of induced pluripotent stem (IPS) cells. (1)Isolate and culture donor cells. (2)Transduce stem cell-associated genes into the cells by viral vectors. Red cells indicate the cells expressing the exogenous genes. (3)Harvest and culture the cells according to ES cell culture, using mitotically inactivated feeder cells (lightgray). (4)A small subset of the transfected cells become iPS cells and generate ES-like colonies.

pluripotency-associated genes, or "reprogramming factors", into a given cell type. The original set of reprogramming factors (also dubbed Yamanaka factors) are the genes Oct4 (Pou5f1), Sox2, cMyc, and Klf4. While this combination is most conventional in producing iPSCs, each of the factors can be functionally replaced by related transcription factors, miRNAs, small molecules, or even non-related genes such as lineage specifiers.

iPSC derivation is typically a slow and inefficient process, taking 1–2 weeks for mouse cells and 3–4 weeks for human cells, with efficiencies around 0.01%–0.1%. However, considerable advances have been made in improving the efficiency and the time it takes to obtain iPSCs. Upon introduction of reprogramming factors, cells begin to form colonies that resemble pluripotent stem cells, which can be isolated based on their morphology, conditions that select for their growth, or through expression of surface markers or reporter genes.

14.1.1   First generation (mouse)

Induced pluripotent stem cells were first generated by Shinya Yamanaka's team at Kyoto University, Japan, in 2006.[1] Their hypothesis was that genes important to embryonic stem cell function might be able to induce an embryonic state in adult cells. They began by choosing twenty-four genes that were previously identified as important in embryonic stem cells, and used retroviruses to deliver these genes to fibroblasts from mice. The mouse fibroblasts were engineered so that any cells that reactivated the ESC-specific gene, Fbx15, could be isolated using antibiotic selection.

Upon delivery of all twenty-four factors, colonies emerged that had reactivated the Fbx15 reporter, resembled ESCs, and could propagate indefinitely. They then narrowed their candidates by removing one factor at a time from the pool of twenty-four. By this process, they identified four factors, Oct4, Sox2, cMyc, and Klf4, which as a group were both necessary and sufficient to obtain ESC-like colonies under selection for reactivation of Fbx15.

Similar to ESCs, these first-generation iPSCs showed unlimited self-renewal and demonstrated pluripotency by contributing to lineages from all three germ layers in the context of embryoid bodies, teratomas, fetal chimeras. However, the molecular makeup of these cells, including gene expression and epigenetic marks, was somewhere between that of a fibroblast and an ESC, and the cells also failed to produce viable chimeras when injected into developing embryos.

14.1.2   Second generation (mouse)

In June 2007, the same group published a breakthrough study along with two other independent research groups from Harvard, MIT, and the University of California, Los Angeles, showing successful reprogramming of mouse fibroblasts into iPS cells. Unlike the first generation of iPS cells, these cells could produce viable chimeric mice and could contribute to the germline, the 'gold standard' for pluripotent stem cells. These cells were derived from mouse fibroblasts by retroviral-mediated expression of the same four transcription factors (Oct4, Sox2, cMyc, Klf4), but the researchers used a different marker to select for pluripotent cells. Instead of Fbx15, they used Nanog, a gene that is functionally important in ESCs. By using this different strategy, the researchers were able to create iPS cells that were more similar to ESCs than the first generation of iPS cells, and independently proved that it was possible to create iPS cells that are functionally identical to ESCs.[6][7][8][9]

Unfortunately, two of the four genes used (namely, c-Myc and KLF4) are oncogenic, and 20% of the chimeric mice developed cancer. In a later study, Yamanaka reported that one can create iPSCs even without c-Myc. The process takes longer and is not as efficient, but the resulting chimeras didn't develop cancer.[10]

Induced pluripotent cells have been made from adult stomach, liver, skin cells, blood cells, prostate cells and urinary tract cells.[11]

14.1.3   Human   induced   pluripotent   stem cells

Generation from human fibroblasts

In November 2007, a milestone was achieved[12][13] by creating iPSCs from adult human cells; two independent research teams' studies were released – one in *Science* by James Thomson at University of Wisconsin–Madison[14] and another in *Cell* by Shinya Yamanaka and colleagues at Kyoto University, Japan.[15] With the same principle used earlier in mouse models, Yamanaka had successfully transformed human fibroblasts into pluripotent stem cells using the same four pivotal genes: Oct3/4, Sox2, Klf4, and c-Myc with a retroviral system. Thomson and colleagues used OCT4, SOX2, NANOG, and a different gene LIN28 using a lentiviral system.

Generation from human renal epithelial cells in urine

On 8 November 2012, researchers from Austria, Hong Kong and China presented a protocol for generating human iPSCs from exfoliated renal epithelial cells present in urine on Nature Protocols.[16] This method of acquiring donor cells is comparatively less invasive and simple. The team reported the induction procedure to take less time, around 2 weeks for the urinary cell culture and 3 to 4 weeks for the reprogramming; and higher yield, up to 4% using retroviral delivery of exogenous factors. Urinary iPSCs (UiPSCs) were found to show good differentiation potential, and thus represent an alternative choice for producing pluripotent cells from normal individuals or patients with genetic diseases, including those affecting the kidney.[16]

14.1.4 Challenges in reprogramming cells to pluripotency

Although the methods pioneered by Yamanaka and others have demonstrated that adult cells can be reprogrammed to iPS cells, there are still challenges associated with this technology:

1. Low efficiency: in general, the conversion to iPS cells has been incredibly low. For example, the rate at which somatic cells were reprogrammed into iPS cells in Yamanaka's original mouse study was 0.01–0.1%.[11] The low efficiency rate may reflect the need for precise timing, balance, and absolute levels of expression of the reprogramming genes. It may also suggest a need for rare genetic and/or epigenetic changes in the original somatic cell population or in the prolonged culture. However, recently a path was found for efficient reprogramming which required downregulation of the nucleosome remodeling and deacetylation (NuRD) complex. Overexpression of Mbd3, a subunit of NuRD, inhibits induction of iPSCs. Depletion of Mbd3, on the other hand, improves reprogramming efficiency,[17] that results in deterministic and synchronized iPS cell reprogramming (near 100% efficiency within seven days from mouse and human cells).[18]

2. Genomic Insertion: genomic integration of the transcription factors limits the utility of the transcription factor approach because of the risk of mutations being inserted into the target cell's genome.[19] A common strategy for avoiding genomic insertion has been to use a different vector for input. Plasmids, adenoviruses, and transposon vectors have all been explored, but these often come with the tradeoff of lower throughput.[20][21][22]

3. Tumors: another main challenge was mentioned above – some of the reprogramming factors are oncogenes that bring on a potential tumor risk. Inactivation or deletion of the tumor suppressor p53, which is the master regulator of cancer, significantly increases reprogramming efficiency.[23] Thus there seems to be a tradeoff between reprogramming efficiency and tumor generation.

4. Incomplete reprogramming: reprogramming also faces the challenge of completeness. This is particularly challenging because the genome-wide epigenetic code must be reformatted to that of the target cell

type in order to fully reprogram a cell. However, three separate groups were able to find mouse embryonic fibroblast (MEF)-derived iPS cells that could be injected into tetraploid blastocysts and resulted in the live birth of mice derived entirely from iPS cells, thus ending the debate over the equivalence of embryonic stem cells (ESCs) and iPS with regard to pluripotency.[24]

The table at right summarizes the key strategies and techniques used to develop iPS cells over the past half-decade. Rows of similar colors represents studies that used similar strategies for reprogramming.

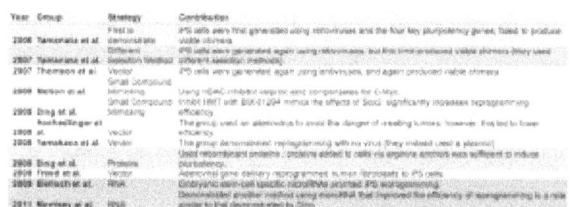

This timeline summarizes the key strategies and techniques used to develop iPS cells over the past half-decade. Rows of similar colors represents studies that used similar strategies for reprogramming.

14.1.5 Alternative approaches

Mimicking transcription factors with chemicals

One of the main strategies for avoiding problems (1) and (2) has been to use small compounds that can mimic the effects of transcription factors. These molecule compounds can compensate for a reprogramming factor that does not effectively target the genome or fails at reprogramming for another reason; thus they raise reprogramming efficiency. They also avoid the problem of genomic integration, which in some cases contributes to tumor genesis. Key studies using such strategy were conducted in 2008. Melton et al. studied the effects of histone deacetylase (HDAC) inhibitor valproic acid. They found that it increased reprogramming efficiency 100-fold (compared to Yamanaka's traditional transcription factor method).[25] The researchers proposed that this compound was mimicking the signaling that is usually caused by the transcription factor c-Myc. A similar type of compensation mechanism was proposed to mimic the effects of Sox2. In 2008, Ding et al. used the inhibition of histone methyl transferase (HMT) with BIX-01294 in combination with the activation of calcium channels in the plasma membrane in order to increase reprogramming efficiency.[26] Deng et al. of Beijing University reported on July 2013 that induced pluripotent stem cells can be created without any genetic modification. They used a cocktail of seven small-molecule compounds including DZNep to induce the mouse somatic cells into stem cells which they called CiPS cells with the efficiency – at 0.2% – comparable to those using standard iPSC production techniques. The CiPS cells were introduced into developing mouse embryos and were found to contribute to all major cells types, proving its pluripotency.[27][28]

Ding et al. demonstrated an alternative to transcription factor reprogramming through the use of drug-like chemicals. By studying the MET (mesenchymal-epithelial transition) process in which fibroblasts are pushed to a stem-cell like state, Ding's group identified two chemicals – ALK5 inhibitor SB431412 and MEK (mitogen-activated protein kinase) inhibitor PD0325901 – which was found to increase the efficiency of the classical genetic method by 100 fold. Adding a third compound known to be involved in the cell survival pathway, Thiazovivin further increases the efficiency by 200 fold. Using the combination of these three compounds also decreased the reprogramming process of the human fibroblasts from four weeks to two weeks. [29][30]

Alternate vectors

Another key strategy for avoiding problems such as tumor genesis and low throughput has been to use alternate forms of vectors: adenovirus, plasmids, and naked DNA and/or protein compounds.

In 2008, Hochedlinger et al. used an adenovirus to transport the requisite four transcription factors into the DNA of skin and liver cells of mice, resulting in cells identical to ESCs. The adenovirus is unique from other vectors like viruses and retroviruses because it does not incorporate any of its own genes into the targeted host and avoids the potential for insertional mutagenesis.[31] In 2009, Freed et al. demonstrated successful reprogramming of human fibroblasts to iPS cells.[32] Another advantage of using adenoviruses is that they only need to present for a brief amount of time in order for effective reprogramming to take place.

Also in 2008, Yamanaka et al. found that they could transfer the four necessary genes with a plasmid.[33] The Yamanaka group successfully reprogrammed mouse cells by transfection with two plasmid constructs carrying the reprogramming factors; the first plasmid expressed c-Myc, while the second expressed the other three factors (Oct4, Klf4, and Sox2). Although the plasmid methods avoid viruses, they still require cancer-promoting genes to accomplish reprogramming. The other main issue with these methods is that they tend to be much less efficient compared to retroviral methods. Furthermore, transfected plasmids have been shown to integrate into the host genome and therefore they still pose the risk of insertional mutagenesis. Because non-retroviral approaches have demonstrated such low ef-

ficiency levels, researchers have attempted to effectively rescue the technique with what is known as the piggyBac transposon system. The lifecycle of this system is shown below. Several studies have demonstrated that this system can effectively deliver the key reprogramming factors without leaving any footprint mutations in the host cell genome. As demonstrated in the figure, the piggyBac transposon system involves the re-excision of exogenous genes, which eliminates issues like insertional mutagenesis

Lifecycle of the Piggybac Transposon System

Stimulus-Triggered Acquisition of Pluripotency Cell

Main article: Stimulus-triggered acquisition of pluripotency

In January 2014, two articles were published claiming that a type of pluripotent stem cell can be generated by subjecting the cells to certain types of stress (bacterial toxin, a low pH of 5.7, or physical squeezing); the resulting cells were called STAP cells, for stimulus-triggered acquisition of pluripotency.[34]

In light of difficulties that other labs had replicating the results of the surprising study, in March 2014, one of the co-authors has called for the articles to be retracted.[35] On 4 June 2014, the lead author, Obokata agreed to retract both the papers [36] after she was found to have committed 'research misconduct' as concluded in an investigation by RIKEN on 1 April 2014.[37]

RNA molecules

Studies by Blelloch et al. in 2009 demonstrated that expression of ES cell-specific microRNA molecules (such as miR-291, miR-294 and miR-295) enhances the efficiency of induced pluripotency by acting downstream of c-Myc .[38] More recently (in April 2011), Morrisey et al. demonstrated another method using microRNA that improved the efficiency of reprogramming to a rate similar to that demonstrated by Ding. MicroRNAs are short RNA molecules that bind to complementary sequences on messenger RNA and block expression of a gene. Morrisey's team worked on microRNAs in lung development, and hypothesized that their microRNAs perhaps blocked expression of repressors of Yamanaka's four transcription factors. Possible mechanisms by which microRNAs can induce reprogramming even in the absence of added exogenous transcription factors, and how variations in microRNA expression of iPS cells can predict their differentiation potential discussed by Xichen Bao et al.[39]

14.1.6   Genes of induction

The generation of iPS cells is crucially dependent on the genes used for the induction.

Oct-3/4 and certain members of the Sox gene family (Sox1, Sox2, Sox3, and Sox15) have been identified as crucial transcriptional regulators involved in the induction process whose absence makes induction impossible. Additional genes, however, including certain members of the Klf family (Klf1, Klf2, Klf4, and Klf5), the Myc family (c-myc, L-myc, and N-myc), Nanog, and LIN28, have been identified to increase the induction efficiency.

- **Oct-3/4** (Pou5f1) Oct-3/4 is one of the family of octamer ("Oct") transcription factors, and plays a crucial role in maintaining pluripotency. The absence of Oct-3/4 in Oct-3/4+ cells, such as blastomeres and embryonic stem cells, leads to spontaneous trophoblast differentiation, and presence of Oct-3/4 thus gives rise to the pluripotency and differentiation potential of embryonic stem cells. Various other genes in the "Oct" family, including Oct-3/4's close relatives, Oct1 and Oct6, fail to elicit induction, thus demonstrating the exclusivity of Oct-3/4 to the induction process.

- **Sox family**: The Sox family of genes is associated with maintaining pluripotency similar to Oct-3/4, although it is associated with multipotent and unipotent stem cells in contrast with Oct-3/4, which is exclusively expressed in pluripotent stem cells. While Sox2 was the initial gene used for induction by Yamanaka et al., Jaenisch et al., and Thomson et al., other genes in the Sox family have been found to work as well in the induction process. Sox1 yields iPS cells with a similar efficiency as Sox2, and genes Sox3, Sox15, and Sox18 also generate iPS cells, although with decreased efficiency.

- **Klf family**: Klf4 of the Klf family of genes was initially identified by Yamanaka et al. and confirmed by Jaenisch et al. as a factor for the generation of mouse iPS cells and was demonstrated by Yamanaka et al. as a factor for generation of human iPS cells. However,

Thomson et al. reported that Klf4 was unnecessary for generation of human iPS cells and in fact failed to generate human iPS cells. Klf2 and Klf4 were found to be factors capable of generating iPS cells, and related genes Klf1 and Klf5 did as well, although with reduced efficiency.

- **Myc family**: The Myc family of genes are proto-oncogenes implicated in cancer. Yamanaka et al. and Jaenisch et al. demonstrated that c-myc is a factor implicated in the generation of mouse iPS cells and Yamanaka et al. demonstrated it was a factor implicated in the generation of human iPS cells. However, Thomson et al., Yamanaka et al. Usage of the "myc" family of genes in induction of iPS cells is troubling for the eventuality of iPS cells as clinical therapies, as 25% of mice transplanted with c-myc-induced iPS cells developed lethal teratomas. N-myc and L-myc have been identified to induce instead of c-myc with similar efficiency.

- **Nanog**: In embryonic stem cells, Nanog, along with Oct-3/4 and Sox2, is necessary in promoting pluripotency. Therefore, it was surprising when Yamanaka et al. reported that Nanog was unnecessary for induction although Thomson et al. has reported it is possible to generate iPS cells with Nanog as one of the factors.

- **LIN28**: LIN28 is an mRNA binding protein[40] expressed in embryonic stem cells and embryonic carcinoma cells associated with differentiation and proliferation. Thomson et al. demonstrated it is a factor in iPSC generation, although it is unnecessary.

- **Glis1**: Glis1 is transcription factor that can be used with Oct-3/4, Sox2 and Klf4 to induce pluripotency. It poses numerous advantages when used instead of C-myc.[41]

14.2   Identity

Induced pluripotent stem cells are similar to natural pluripotent stem cells, such as embryonic stem (ES) cells, in many aspects, such as the expression of certain stem cell genes and proteins, chromatin methylation patterns, doubling time, embryoid body formation, teratoma formation, viable chimera formation, and potency and differentiability, but the full extent of their relation to natural pluripotent stem cells is still being assessed.[42]

Gene expression and genome-wide H3K4me3 and H3K27me3 were found to be extremely similar between ES and iPS cells.[43] The generated iPSCs were remarkably similar to naturally isolated pluripotent stem cells (such as mouse and human embryonic stem cells, mESCs and hESCs, respectively) in the following respects, thus confirming the identity, authenticity, and pluripotency of iPSCs to naturally isolated pluripotent stem cells:

- Cellular biological properties

  - Morphology: iPSCs were morphologically similar to ESCs. Each cell had round shape, large nucleolus and scant cytoplasm. Colonies of iPSCs were also similar to that of ESCs. Human iPSCs formed sharp-edged, flat, tightly packed colonies similar to hESCs and mouse iPSCs formed the colonies similar to mESCs, less flat and more aggregated colonies than that of hESCs.

  - Growth properties: Doubling time and mitotic activity are cornerstones of ESCs, as stem cells must self-renew as part of their definition. iPSCs were mitotically active, actively self-renewing, proliferating, and dividing at a rate equal to ESCs.

  - Stem cell markers: iPSCs expressed cell surface antigenic markers expressed on ESCs. Human iPSCs expressed the markers specific to hESC, including SSEA-3, SSEA-4, TRA-1-60, TRA-1-81, TRA-2-49/6E, and Nanog. Mouse iPSCs expressed SSEA-1 but not SSEA-3 nor SSEA-4, similarly to mESCs.

  - Stem Cell Genes: iPSCs expressed genes expressed in undifferentiated ESCs, including Oct-3/4, Sox2, Nanog, GDF3, REX1, FGF4, ESG1, DPPA2, DPPA4, and hTERT.

  - Telomerase activity: Telomerases are necessary to sustain cell division unrestricted by the Hayflick limit of ~50 cell divisions. hESCs express high telomerase activity to sustain self-renewal and proliferation, and iPSCs also demonstrate high telomerase activity and express hTERT (human telomerase reverse transcriptase), a necessary component in the telomerase protein complex.

- Pluripotency: iPSCs were capable of differentiation in a fashion similar to ESCs into fully differentiated tissues.

  - Neural differentiation: iPSCs were differentiated into neurons, expressing βIII-tubulin, tyrosine hydroxylase, AADC, DAT, ChAT, LMX1B, and MAP2. The presence of catecholamine-associated enzymes may indicate that iPSCs, like hESCs, may be differentiable into dopaminergic neurons. Stem cell-associated genes were downregulated after differentiation.

  - Cardiac differentiation: iPSCs were differentiated into cardiomyocytes that spontaneously began beating. Cardiomyocytes expressed TnTc, MEF2C, MYL2A, MYHCβ, and NKX2.5. Stem cell-associated genes were downregulated after differentiation.

  - Teratoma formation: iPSCs injected into immunodeficient mice spontaneously formed teratomas after nine weeks. Teratomas are tumors of multiple lineages containing tissue derived from the three germ layers endoderm, mesoderm and ectoderm; this is unlike other tumors, which typically are of only one cell type. Teratoma formation is a landmark test for pluripotency.

  - Embryoid body: hESCs in culture spontaneously form ball-like embryo-like structures termed "embryoid bodies", which consist of a core of mitotically active and differentiating hESCs and a periphery of fully differentiated cells from all three germ layers. iPSCs also form embryoid bodies and have peripheral differentiated cells.

  - Chimeric mice: hESCs naturally reside within the inner cell mass (embryoblast) of blastocysts, and in the embryoblast, differentiate into the embryo while the blastocyst's shell (trophoblast) differentiates into extraembryonic tissues. The hollow trophoblast is unable to form a living embryo, and thus it is necessary for the embryonic stem cells within the embryoblast to differentiate and form the embryo. iPSCs were injected by micropipette into a trophoblast, and the blastocyst was transferred to recipient females. Chimeric living mouse pups were created: mice with iPSC derivatives incorporated all across their bodies with 10%–90% chimerism.

  - Tetraploid complementation: iPS cells from mouse fetal fibroblasts injected into tetraploid blastocysts (which themselves can only form extra-embryonic tissues) can form whole, non-chimeric, fertile mice, although with low success rate.[44][45][46]

- Epigenetic reprogramming

  - Promoter demethylation: Methylation is the transfer of a methyl group to a DNA base, typically the transfer of a methyl group to a cytosine molecule in a CpG site (adjacent cytosine/guanine sequence). Widespread methylation of a gene interferes with expression by preventing the activity of expression proteins, or by recruiting enzymes that interfere with expression. Thus, methylation of a gene effectively silences it by preventing transcription. Promoters of pluripotency-associated genes, including Oct-3/4, Rex1, and Nanog, were demethylated in iPSCs, demonstrating their promoter activity and the active promotion and expression of pluripotency-associated genes in iPSCs.

- DNA methylation globally: Human iPS cells are highly similar to ES cells in their patterns of which cytosines are methylated, more than to any other cell type. However, on the order of a thousand sites show differences in several iPS cell lines. Half of these resemble the somatic cell line the iPS cells were derived from, the rest are iPSC-specific. Tens of regions which are megabases in size have also been found where iPS cells are not reprogrammed to the ES cell state.[47]

- Histone demethylation: Histones are compacting proteins that are structurally localized to DNA sequences that can affect their activity through various chromatin-related modifications. H3 histones associated with Oct-3/4, Sox2, and Nanog were demethylated, indicating the expression of Oct-3/4, Sox2, and Nanog.

14.3 Safety for regenerative medicine

- The major concern with the potential clinical application of iPSCs is their propensity to form tumors.[48] Much the same as ESC, iPSCs readily form teratoma when injected into immunodeficient mice. Teratoma formation is considered a major obstacle to stem-cell based regenerative medicine by the FDA.

- A more recent study on motor functional recovery after spinal cord injuries in mice showed that after human-induced pluripotent stem cells were transplanted into the mice, the cells differentiated into three neural lineages in the spinal cord. The cells stimulated regrowth of the damaged spinal cord, maintained myelination, and formed synapses. These positive outcomes were observed for over 112 days after the spinal cord injury, without tumor formation.[49] Nevertheless, a follow-up study by the same group showed distinct clones of human-induced pluripotent stem cells eventually formed tumors.[50]

- Since iPSCs can only be produced with high efficiency at this time using modifications, they are generally predicted to be less safe and more tumorigenic than hESC. All the genes that have been shown to promote iPSC formation have also been linked to cancer

in one way or another. Some of the genes are known oncogenes, including the members of the Myc family. While omitting Myc allows for IPSC formation, the efficiency is reduced up to 100 fold.

- A non-genetic method of producing iPSCs has been demonstrated using recombinant proteins, but its efficiency was quite low.[5] However, refinements to this methodology yielding higher efficiency may lead to production of safer iPSCs. Other approaches such as using adenovirus or plasmids are generally thought to be safer than retroviral methods.

- An important area for future studies in the iPSC field is directly testing iPSC tumorigenicity using methods that mimic the approaches that would be used for regenerative medicine therapies. Such studies are crucial since iPSCs not only form teratoma, but also mice derived from iPSCs have a high incidence of death from malignant cancer.[51] A 2010 paper was published in the journal Stem Cells indicating that iPS cells are far more tumorigenic than ESC, supporting the notion that iPS cell safety is a serious concern.[52]

- Concern regarding the immunogenicity of IPS cells arose in 2011 when Zhou et al. performed a study involving a teratomaformation assay and demonstrated that IPS cells produced an immune response strong enough to cause rejection of the cells. When a similar procedure was performed on genetically equivalent ES cells however, Zhou et al. found teratomas, which indicated that the cells were tolerated by the immune system.[53] In 2013, Araki et al. attempted to reproduce the conclusion obtained by Zhou et al. using a different procedure. They took cells from a chimera that had been grown from IPSC clones and a mouse embryo, this tissue was then transplanted into syngenic mice. They conducted a similar trial using ES cells instead of IPSC clone and compared the results. Findings indicate that there was no significant difference in the immunogenic response produced by the IPS cells and the ES cells. Furthermore, Araki et al. reported little or no immunogenic response for both cell lines.[54] Thus, Araki et al. was unable to come to the same conclusion as Zhou et al.

Recent achievements and future tasks for safe iPSC-based cell therapy are collected in the review of Okano et al.[55]

14.4 An open future

The task of producing iPS cells continues to be challenging due to the six problems mentioned above. A key tradeoff to overcome is that between efficiency and genomic integration. Most methods that do not rely on the integration of transgenes are inefficient, while those that do rely on the integration of transgenes face the problems of incomplete reprogramming and tumor genesis, although a vast number of techniques and methods have been attempted. Another large set of strategies is to perform a proteomic characterization of iPS cells. The Wu group at Stanford University has made significant progress with this strategy.[56] Further studies and new strategies should generate optimal solutions to the five main challenges. One approach might attempt to combine the positive attributes of these strategies into an ultimately effective technique for reprogramming cells to iPS cells.

Another approach is the use of iPS cells derived from patients to identify therapeutic drugs able to rescue a phenotype. For instance, iPS cell lines derived from patients affected by ectodermal dysplasia syndrome (EEC), in which the p63 gene is mutated, display abnormal epithelial commitment that could be partially rescued by a small compound[57]

14.5 Medical research

14.5.1 Disease modeling and drug development

An attractive feature of human iPS cells is the ability to derive them from adult patients to study the cellular basis of human disease. Since iPS cells are self-renewing and pluripotent, they represent a theoretically unlimited source of patient-derived cells which can be turned into any type of cell in the body. This is particularly important because many other types of human cells derived from patients tend to stop growing after a few passages in laboratory culture. iPS cells have been generated for a wide variety of human genetic diseases, including common disorders such as Down syndrome and polycystic kidney disease.[58][59] In many instances, the patient-derived iPS cells exhibit cellular defects not observed in iPS cells from healthy patients, providing insight into the pathophysiology of the disease.[60] An international collaborated project, StemBANCC, was formed in 2012 to build a collection of iPS cell lines for drug screening for a variety of disease. Managed by the University of Oxford, the effort pooled funds and resources from 10 pharmaceutical companies and 23 universities. The goal is to generate a library of 1,500 iPS cell lines which will be used in early drug testing by providing a simulated human disease environment.[61] Furthermore, combining hiPSC technology and genetically-encoded voltage and calcium indicators provided a large-scale and high-throughput platform for cardiovascular drug safety screening.[62]

14.5.2 Organ synthesis

A proof-of-concept of using induced pluripotent stem cells (iPSCs) to generate human organ for transplantation was reported by researchers from Japan. Human 'liver buds' (iPSC-LBs) were grown from a mixture of three different kinds of stem cells: hepatocytes (for liver function) coaxed from iPSCs; endothelial stem cells (to form lining of blood vessels) from umbilical cord blood; and mesenchymal stem cells (to form connective tissue). This new approach allows different cell types to self-organize into a complex organ, mimicking the process in fetal development. After growing *in vitro* for a few days, the liver buds were transplanted into mice where the 'liver' quickly connected with the host blood vessels and continued to grow. Most importantly, it performed regular liver functions including metabolizing drugs and producing liver-specific proteins. Further studies will monitor the longevity of the transplanted organ in the host body (ability to integrate or avoid rejection) and whether it will transform into tumors.[63][64] Using this method, cells from one mouse could be used to test 1,000 drug compounds to treat liver disease, and reduce animal use by up to 50,000.[65]

14.5.3 Tissue repair

Embryonic cord-blood cells were induced into pluripotent stem cells using plasmid DNA. Using cell surface endothelial/pericytic markers CD31 and CD146, researchers identified 'vascular progenitor', the high-quality, multipotent vascular stem cells. After the iPS cells were injected directly into the vitreous of the damaged retina of mice, the stem cells engrafted into the retina, grew and repaired the vascular vessels.[66][67]

In a study conducted in China in 2013, Superparamagnetic iron oxide (SPIO) particles were used to label iPSCs-derived NSCs in vitro. Labeled NSCs were implanted into TBI rats and SCI monkeys 1 week after injury, and then imaged using gradient reflection echo (GRE) sequence by 3.0T magnetic resonance imaging (MRI) scanner. MRI analysis was performed at 1, 7, 14, 21, and 30 days, respectively, following cell transplantation. Pronounced hypointense signals were initially detected at the cell injection sites in rats and monkeys and were later found to extend progressively to the lesion regions, demonstrating that iPSCs-derived NSCs could migrate to the lesion area from the primary sites. The therapeutic efficacy of iPSCs-derived NSCs was examined concomitantly through functional recovery tests of the animals. In this study, we tracked iPSCs-derived NSCs migration in the CNS of TBI rats and SCI monkeys in vivo for the first time. Functional recovery tests showed obvious motor function improvement in transplanted animals. These data provide the necessary foundation for future clinical application of iPSCs for CNS injury.[68]

14.5.4 Red blood cells

In 2014, type O red blood cells were synthesized at the Scottish National Blood Transfusion Service from iPSC. The cells were induced to become a mesoderm and then blood cells and then red blood cells. The final step was to make them eject their nuclei and mature properly. Type O can be transfused into all patients. Each pint of blood contains about two trillion red blood cells, while some 107 million blood donations are collected globally every year. Human transfusions were not expected to begin until 2016.[69]

14.6 Clinical trial

The first human clinical trial using autologous iPSCs is approved by the Japan Ministry Health and will be conducted in 2014 in Kobe. iPSCs derived from skin cells from six patients suffering from wet age-related macular degeneration will be reprogrammed to differentiate into retinal pigment epithelial (RPE) cells. The cell sheet will be transplanted into the affected retina where the degenerated RPE tissue has been excised. Safety and vision restoration monitoring is expected to last one to three years.[70][71] The benefits of using autologous iPSCs are that there is theoretically no risk of rejection and it eliminates the need to use embryonic stem cells.[71]

14.7 See also

- Induced stem cells

- Stem cell treatments

- Stimulus-triggered acquisition of pluripotency cell, a type of pluripotent stem cell that can be generated by immersing cells in an acid

- Induced pluripotent stem cells vs embryonic stem cells lines obtained by SCNT (discussion)

14.8 References

[1] Takahashi, K; Yamanaka, S (2006). "Induction of pluripotent stem cells from mouse embryonic and adult fibroblast cultures by defined factors". *Cell* **126** (4): 663–76. doi:10.1016/j.cell.2006.07.024. PMID 16904174.

[2] "The Nobel Prize in Physiology or Medicine – 2012 Press Release". Nobel Media AB. 8 October 2012.

[3] Klimanskaya; et al. (2006). "Human embryonic stem cell lines derived from single blastomeres". *Nature* (Nature) **444** (7118): 484–485. doi:10.1038/nature05142. PMID 16929302.

[4] Kaplan, Karen (2009-03-06). "Cancer threat removed from stem cells, scientists say". *Los Angeles Times*.

[5] Zhou H, Wu S, Joo JY, et al. (May 2009). "Generation of Induced Pluripotent Stem Cells Using Recombinant Proteins". *Cell Stem Cell* **4** (5): 381–4. doi:10.1016/j.stem.2009.04.005. PMID 19398399. Retrieved April 23, 2009.

[6] Okita, K; Ichisaka, T; Yamanaka, S (2007). "Generation of germline-competent induced pluripotent stem cells". *Nature* **448** (7151): 313–7. doi:10.1038/nature05934. PMID 17554338.

[7] Wernig, M; Meissner, A; Foreman, R; Brambrink, T; Ku, M; Hochedlinger, K; Bernstein, BE; Jaenisch, R (2007). "In vitro reprogramming of fibroblasts into a pluripotent ES-cell-like state". *Nature* **448** (7151): 318–24. doi:10.1038/nature05944. PMID 17554336.

[8] Maherali N, et al. (2007). "Directly reprogrammed fibroblasts show global epigenetic remodeling and widespread tissue contribution". *Cell Stem Cell* **1** (1): 55–70. doi:10.1016/j.stem.2007.05.014. PMID 18371336.

[9] Generations of iPSCs and related references

[10] Swaminathan, Nikhil (2007-11-30). "Stem Cells – This Time Without the Cancer". *Scientific American News*. Retrieved 2007-12-11.

[11] Moad, Mohammad; Pal, Deepali; Hepburn, Anastasia C; Williamson, Stuart C; Wilson, Laura; Lako, Majlinda; Armstrong, Lyle; Hayward, Simon W; Franco, Omar E; Cates, Justin M; Fordham, Sarah E; Przyborski, Stefan; Carr-Wilkinson, Jane; Robson, Craig N; Heer, Rakesh (2013). "A Novel Model of Urinary Tract Differentiation, Tissue Regeneration, and Disease: Reprogramming Human Prostate and Bladder Cells into Induced Pluripotent Stem Cells.". *European Urology* **64** (5): 753–761. doi:10.1016/j.eururo.2013.03.054. PMID 23582880. Cite uses deprecated parameter |coauthors= (help)

[12] Baker, Monya (2007-12-06). "Adult cells reprogrammed to pluripotency, without tumors". *Nature Reports Stem Cells*. doi:10.1038/stemcells.2007.124. Retrieved 2007-12-11.

[13] Kolata, Gina (2007-11-21). "Scientists Bypass Need for Embryo to Get Stem Cells". *The New York Times*. ISSN 0362-4331. Retrieved 2007-12-11.

[14] Yu J, Vodyanik MA, et al. (2007). "Induced Pluripotent Stem Cell Lines Derived from Human Somatic Cells". *Science* **318** (5858): 1917–1920. doi:10.1126/science.1151526. PMID 18029452.

[15] Takahashi K, et al. (2007). "Induction of Pluripotent Stem Cells from Adult Human Fibroblasts by Defined Factors". *Cell* **131** (5): 861–872. doi:10.1016/j.cell.2007.11.019. PMID 18035408.

[16] Zhou, Ting; Benda, Christina; Dunzinger, Sarah; Huang, Yinghua; Ho, Jenny Cy; Yang, Jiayin; Wang, Yu; Zhang, Ya; Zhuang, Qiang; Li, Yanhua; Bao, Xichen; Tse, Hung-Fat; Grillari, Johannes; Grillari-Voglauer, Regina; Pei, Duanqing; Esteban, Miguel A (7 November 2012). "Generation of human induced pluripotent stem cells from urine samples". *Nature Protocols* **7** (12): 2080–2089. doi:10.1038/nprot.2012.115. PMID 23138349. Retrieved 1 December 2012. Cite uses deprecated parameter |coauthors= (help)

[17] Luo M, et al. "& Tao, W., Lu, Z., Grummt, I. (2013). NuRD Blocks Reprogramming of Mouse Somatic Cells into Pluripotent Stem Cells". *STEM CELLS* **31**: 1278–1286. doi:10.1002/stem.1374.

[18] Yoach Rais, Asaf Zviran, Shay Geula, et al. (2013) Deterministic direct reprogramming of somatic cells to pluripotency. Nature. doi:10.1038/nature12587

[19] Selvaraj V, Plane JM, Williams AJ, Deng W (April 2010). "Switching cell fate: the remarkable rise of induced pluripotent stem cells and lineage reprogramming technologies". *Trends in Biotechnology* **28** (4): 214–23. doi:10.1016/j.tibtech.2010.01.002. PMC 2843790. PMID 20149468.

[20] Okita, K; Nakagawa, M.; Hyenjong, H.; Ichisaka, T.; Yamanaka, S. (2008). "Generation of mouse induced pluripotent stem cells without viral vectors". *Science* **322** (5903): 949–953. doi:10.1126/science.1164270. PMID 18845712.

[21] Stadtfeld, M; Nakagawa, M; Hyenjong, H; Ichisaka, T; Yamanaka, S (2008). "Induced pluripotent stem cells generated without viral integration". *Science* **322** (5903): 949–53. doi:10.1126/science.1164270. PMID 18845712.

[22] Woltjen, K; Michael, IP; Mohseni, P; Desai, R; Mileikovsky, M; Hämäläinen, R; Cowling, R; Wang, W; Liu, P (2009). "piggyBac transposition reprograms fibroblasts to induced pluripotent stem cells". *Nature* **458** (7239): 766–770. doi:10.1038/nature07863. PMID 19252478.

[23] RM, Mario; Strati, Katerina; Li, Han; Murga, Matilde; Blanco, Raquel; Ortega, Sagrario; Fernandez-Capetillo, Oscar; Serrano, Manuel; Blasco, Maria A. (2009). "A p53-mediated DNA damage response limits reprogramming to ensure iPS cell genomic integrity". *Nature* **460** (7259): 1149–1153. doi:10.1038/nature08287. PMID 19668189.

[24] Zhao, XY; Li, Wei; Lv, Zhuo; Liu, Lei; Tong, Man; Hai, Tang; Hao, Jie; Guo, Chang-Long; Ma, Qingwen (2009). "iPS cells produce viable mice through tetraploid complementation". *Nature* **461** (7260): 86–90. doi:10.1038/nature08267. PMID 19672241.

[25] Huangfu D, Maehr R, Guo W, et al. (2008). "Induction of pluripotent stem cells by defined factors is greatly improved by small-molecule compounds". *Nat Biotechnol* **26** (7): 795–7. doi:10.1038/nbt1418. PMID 18568017.

[26] Desponts, Shi; Desponts, Caroline; Do, Jeong Tae; Hahm, Heung Sik; Schöler, Hans R.; Ding, Sheng (November 2008). "Induction of pluripotent stem cells from mouse embryonic fibroblasts by Oct4 and Klf4 with small-molecule compounds". *Cell Stem Cell* **3** (5): 568–74. doi:10.1016/j.stem.2008.10.004. PMID 18983970.

[27] Cyranoski, David (18 July 2013). "Stem cells reprogrammed using chemicals alone". Nature. doi:10.1038/nature.2013.13416. Retrieved 22 July 2013.

[28] Deng, Hongkui; Hou, Pingping; Li, Yanqin; Zhang, Xu; Liu, Chun; Guan, Jingyang; Li, Honggang; Zhao, Ting; Ye, Junqing (18 July 2013). "Pluripotent Stem Cells Induced from Mouse Somatic Cells by Small-Molecule Compounds". *Science* (sciencemag.org) **341**: 651–654. doi:10.1126/science.1239278. PMID 23868920.

[29] "Major Step In Making Better Stem Cells From Adult Tissue". *Science Daily*. 19 October 2009. Retrieved 30 September 2013.

[30] Lin, Tongxiang; Ambasudhan, Rajesh; Yuan, Xu; Li, Wenlin; Hilcove, Simon; Abujarour, Ramzey; Lin, Xiangyi; Hahm, Heung Sol; Hao, Ergeng; Hayek, Alberto; Ding, Sheng (2009). "A chemical platform for improved induction of human iPSCs". *Nature Methods* (Nature) **6**: 805–808. doi:10.1038/nmeth.1393.

[31] Desponts, Shi; Desponts, Caroline; Do, Jeong Tae; Hahm, Heung Sik; Schöler, Hans R.; Ding, Sheng (2008). "Induction of pluripotent stem cells from mouse embryonic fibroblasts by Oct4 and Klf4 with small-molecule compounds". *Cell Stem Cell* **3** (5): 568–74. doi:10.1016/j.stem.2008.10.004. PMID 18983970.

[32] Zhou, Wi; Freed, Curt R. (2009). "Adenoviral gene delivery can reprogram human fibroblasts to induced pluripotent stem cells". *Stem Cells* **27** (11): 2667–74. doi:10.1002/stem.201. PMID 19697349.

[33] Yamanaka, K.; Nakagawa, M.; Hyenjong, H.; Ichisaka, T.; Yamanaka, S. (2008). "Generation of Mouse Induced Pluripotent Stem Cells Without Viral Vectors". *Science* **322** (5903): 949–53. doi:10.1126/science.1164270. PMID 18845712.

[34] David Cyranoski for Nature News. January 29, 2014 Acid bath offers easy path to stem cells

[35] Tracy Vence for the Scientist. March 11, 2014 Call for STAP Retractions

[36] Elaine Lies (4 June 2014). "Japan researcher agrees to withdraw disputed stem cell paper". *Reuters*. Retrieved 4 June 2014.

[37] Press Release (1 April 2014). "Report on STAP Cell Research Paper Investigation". RIKEN. Retrieved 2 June 2014.

[38] Judson, RL. "Embryonic stem cell-specific microRNAs promote induced pluripotency". *Source the Eli and Edythe Broad Center of Regeneration Medicine and Stem Cell Research, University of California, San Francisco, San Francisco, California, USA.*

[39] Bao Xichen, Zhu Xihua, Liao Baojian; et al. (Apr 2013). "& Miguel A Esteban (2013) MicroRNAs in somatic cell reprogramming". *Current Opinion in Cell Biology* 25 (2): 208–214. doi:10.1016/j.ceb.2012.12.004. PMID 23332905.

[40] Ali, P. S.; Ghoshdastider, U; Hoffmann, J; Brutschy, B; Filipek, S (2012). "Recognition of the let-7g miRNA precursor by human Lin28B". *FEBS Letters* 586 (22): 3986–90. doi:10.1016/j.febslet.2012.09.034. PMID 23063642.

[41] Maekawa M, et al. (2011). "Direct reprogramming of somatic cells is promoted by maternal transcription factor Glis1". *Nature* 474 (7350): 225–229. doi:10.1038/nature10106. PMID 21654807.

[42] Takahashi K, Yamanaka S (August 2006). "Induction of pluripotent stem cells from mouse embryonic and adult fibroblast cultures by defined factors". *Cell* 126 (4): 663–76. doi:10.1016/j.cell.2006.07.024. PMID 16904174.

[43] Guenther, M.G.; Frmapton, G.M.; Soldner, F.; Hockemeyer, D.; Mitalipova, M.; Jaenisch, R.; Young, R.A. (2010). "Chromatin structure and gene expression programs of human embryonic and induced pluripotent stem cells.". *Cell Stem Cell* 7 (?): 249–257. doi:10.1016/j.stem.2010.06.015. PMC 3010384. PMID 20682450.

[44] Zhao, Xiao-Yang; Li, Wei; Lv, Zhuo; Liu, Lei; Tong, Man; Hai, Tang; Hao, Jie; Guo, Chang-Long; Ma, Qingwen (2009). "iPS cells produce viable mice through tetraploid complementation". *Nature* 461 (7260): 86–90. doi:10.1038/nature08267. PMID 19672241.

[45] Kang, Lan; Wang, Jianle; Zhang, Yu; Kou, Zhaohui; Gao, Shaorong (2009). "iPS Cells Can Support Full-Term Development of Tetraploid Blastocyst-Complemented Embryos". *Cell Stem Cell* 5 (2): 135–138. doi:10.1016/j.stem.2009.07.001. PMID 19631602.

[46] Michael J. Boland; et al. (2009). "Adult mice generated from induced pluripotent stem cells". *Nature* 461 (7260): 91–4. doi:10.1038/nature08310. PMID 19672243.

[47] Lister R, Pelizzola M, Kida YS, Hawkins D, Nery JR, et al. (2011). "Hotspots of aberrant epigenomic reprogramming in human induced pluripotent stem cells". *Nature* 471 (7336): 68–73. doi:10.1038/nature09798. PMC 3100360. PMID 21289626.

[48] Knoepfler, Paul S. (2009). "Deconstructing Stem Cell Tumorigenicity: A Roadmap to Safe Regenerative Medicine".

[49] Satoshi, Nori (2011). "Grafted Human-induced Pluripotent Stem-Cell–Derived Neurospheres Promote Motor Functional Recovery After Spinal Cord Injury In Mice". *PNAS* 108 (40): 16825–16830. doi:10.1073/pnas.1108077108. PMC 3189018. PMID 21949375.

[50] Nori, Satoshi; Okada, Yohei; Nishimura, Soraya; Sasaki, Takashi; Itakura, Go; Kobayashi, Yoshiomi; Renault-Mihara, Francois; Shimizu, Atsushi; Koya, Ikuko (2015-10-03). "Long-Term Safety Issues of iPSC-Based Cell Therapy in a Spinal Cord Injury Model: Oncogenic Transformation with Epithelial-Mesenchymal Transition". *Stem Cell Reports* 4 (3): 360–373. doi:10.1016/j.stemcr.2015.01.006. ISSN 2213-6711. PMC 4375796. PMID 25684226.

[51] Aoi, T.; Yae, K.; Nakagawa, M.; Ichisaka, T.; Okita, K.; Takahashi, K.; Chiba, T.; Yamanaka, S. (2008). "Generation of Pluripotent Stem Cells from Adult Mouse Liver and Stomach Cells". *Science* 321 (5889): 699–702. doi:10.1126/science.1154884. PMID 18276851.

[52] Ivan Gutierrez-Aranda.; et al. (2010). "Human Induced Pluripotent Stem Cells Develop Teratoma More Efficiently and Faster than Human Embryonic Stem Cells Regardless of the Site of Injection". *Stem Cells* 28 (9): 1568–1570. doi:10.1002/stem.471. PMC 2996086. PMID 20641038.

[53] Zhao, Tongbiao; Zhen-Ning, Zhang,Zhili Rong, Yang Xu (13 May 2011). "Immunogenicity of induced pluripotent stem cells". *Nature* 474 (7350): 212–215. doi:10.1038/nature10135. PMID 21572395. Cite uses deprecated parameter |coauthors= (help);

[54] Araki, Ryoko; Mashiro Uda; Yuko Hoki; Misato Sunayama; Miki Nakamura; Shunsuke Ando; Mayumi Sugiura; Hisashi Ideno; Akemi Shimada; Akira Nifuji; Masumi Abe (7 February 2013). "Negligible immunogenicity of terminally differentiated cells derived from induced pluripotent or embryonic stem cells". *Nature* 494 (494): 100–106. doi:10.1038/nature11807. PMID 23302801.

[55] Okano Hideyuki, Nakamura Masaya, Yoshida Kenji; et al. (Feb 2013). "& Kyoko Miura (2013) Steps Toward Safe Cell Therapy Using Induced Pluripotent Stem Cells". *Circulation Research* 112: 523–533. doi:10.1161/CIRCRESAHA.111.256149. PMID 23371901.

[56] Boland, MY; Hazen, Jennifer L.; Nazor, Kristopher L.; Rodriguez, Alberto R.; Gifford, Wesley; Martin, Greg; Kupriyanov, Sergey; Baldwin, Kristin K. (2009). "Adult mice generated from induced pluripotent stem cells". *Nature* 461 (7260): 91–4. doi:10.1038/nature08310. PMID 19672243.

[57] Shalom-Feuerstein R et al. Impaired epithelial differentiation of induced pluripotent stem cells from EEC patients is rescued by APR-246/PRIMA-1MET. P.N.A.S. 2012 http://minus.com/lbmC3TVGDx350s

[58] Park, IH; Arora, N; Huo, H; Maherali, N; Ahfeldt, T; Shimamura, A; Lensch, MW; Cowan, C; Hochedlinger, K; Daley, GQ (Sep 5, 2008). "Disease-specific induced pluripotent stem cells.". *Cell* **134** (5): 877–86. doi:10.1016/j.cell.2008.07.041. PMC 2633781. PMID 18691744.

[59] Freedman, BS; Lam, AQ; Sundsbak, JL; Iatrino, R; Su, X; Koon, SJ; Wu, M; Daheron, L; Harris, PC; Zhou, J; Bonventre, JV (October 2013). "Reduced ciliary polycystin-2 in induced pluripotent stem cells from polycystic kidney disease patients with PKD1 mutations.". *Journal of the American Society of Nephrology : JASN* **24** (10): 1571–86. doi:10.1681/ASN.2012111089. PMID 24009235.

[60] Grskovic, M; Javaherian, A; Strulovici, B; Daley, GQ (Nov 11, 2011). "Induced pluripotent stem cells--opportunities for disease modelling and drug discovery.". *Nature reviews. Drug discovery* **10** (12): 915–29. doi:10.1038/nrd3577. PMID 22076509.

[61] Gerlin, Andrea (5 December 2012). "Roche, Pfizer, Sanofi Plan $72.7 Million Stem-Cell Bank". Bloomberg.com. Retrieved 23 December 2012.

[62] Shinnawi, Rami; Huber, I; Maizels, L; Shaheen, N; Gepstein, A; Arbel, G; Tijsen, A; Gepstein, L (2015). "Monitoring human-induced pluripotent stem cell-derived cardiomyocytes with genetically encoded calcium and voltage fluorescent reporters.". *Stem Cell Reports* **5** (4): 582–596. doi:10.1016/j.stemcr.2015.08.009. PMID 26372632.

[63] Baker, Monya (3 July 2013). "Miniature human liver grown in mice". Nature.com. Retrieved 19 July 2013.

[64] Takebe, Takanori; Sekine, Keisuke; Enomura, Masahiro; Koike, Hiroyuki; Kimura, Masaki; Ogaeri, Takurnori; Zhang, Ran-Ran; Ueno, Yasuharu; Zheng, Yun-Wen (3 July 2013). "Vascularized and functional human liver from an iPSC-derived organ bud transplant". *Nature* **499**: 481–484. doi:10.1038/nature12271.

[65] Mini-Livers May Reduce Animal Testing

[66] Mullin, Emily (28 January 2014). "Researchers repair retinas in mice with virus-free stem cells". fiercebiotech.com. Retrieved 17 February 2014.

[67] Zambidis, Elias; Lutty, Gerard; Park, Tea Soon; Bhutto, Imran; et al. (2014). "Vascular Progenitors From Cord Blood-Derived Induced Pluripotent Stem Cells Possess Augmented Capacity for Regenerating Ischemic Retinal Vasculature". *Circulation* (American Heart Association) **129** (3): 359–372. doi:10.1161/CIRCULATIONAHA.113.003000.

[68] Hailiang T, et al. (2013). "Tracking Induced Pluripotent Stem Cells-Derived Neural Stem Cells in the Central Nervous System of Rats and Monkeys". *Cellular Reprogramming* **15**: 435–442.

[69] "First transfusions of "manufactured" blood planned for 2016". Gizmag.com. Retrieved 2014-04-23.

[70] Riken Center for Developmental Biology. "Information on proposed pilot study of the safety and feasibility of transplantation of autologous hiPSC-derived retinal pigment epithelium (RPE) cell sheets in patients with neovascular age-related macular degeneration". *Research*. Archived from the original on 26 June 2013. Retrieved 23 July 2013.

[71] Gallagher, James (19 July 2013). "Pioneering adult stem cell trial approved by Japan". *BBC News*. Retrieved 23 July 2013.

14.9 External links

- Center for iPS Cell Research and Application, Kyoto University

- Takahashi, Kazutoshi; Tanabe, Koji; Ohnuki, Mari; Narita, Megumi; Ichisaka, Tomoko; Tomoda, Kiichiro; Yamanaka, Shinya (2007). "Induction of Pluripotent Stem Cells from Adult Human Fibroblasts by Defined Factors". *Cell* **131** (5): 861–872. doi:10.1016/j.cell.2007.11.019. PMID 18035408.

- Takahashi, Kazutoshi; Yamanaka, Shinya (2006). "Induction of Pluripotent Stem Cells from Mouse Embryonic and Adult Fibroblast Cultures by Defined Factors". *Cell* **126** (4): 663–676. doi:10.1016/j.cell.2006.07.024. PMID 16904174.

- With few factors, adult cells take on character of embryonic stem cells

- Generating iPS Cells from MEFS through Forced Expression of Sox-2, Oct-4, c-Myc, and Klf4

- 2 Minute Video from BSCRF about Induced Pluripotent Stem Cells

- 20Minute Video / The Discovery and Future of Induced Pluripotent Stem (iPS) Cells by Dr. Yamanaka January 8, 2008

- A blog focusing specifically on iPS cells and research, biotech, and patient-oriented issues

- Fact sheet on reprogramming

- Detailed protocols for reprogramming and for analysis of iPSCs

- University of Oxford practical workshop on pluripotent stem cell technology

Chapter 15

Induced stem cells

Induced stem cells (iSC) are stem cells artificially derived from somatic, reproductive, pluripotent or other cell types by deliberate epigenetic reprogramming. They are classified as either totipotent (iTC), pluripotent (iPSC) or progenitor (multipotent—iMSC, also called an induced multipotent progenitor cell—iMPC) or unipotent -- (iUSC) according to their developmental potential and degree of dedifferentiation. Progenitors are obtained by so-called direct reprogramming or directed differentiation and are also called induced somatic stem cells.

Three techniques are widely recognized:[1]

- Transplantation of nuclei taken from somatic cells into an oocyt (egg cell) lacking its own nucleus (removed in lab) [2][3][4][5]

- Fusion of somatic cells with pluripotent stem cells[6] and

- Modification of somatic cells, inducing its transformation into a stem cell, using: the genetic material encoding reprogramming protein factors,[7][8][9] recombinant proteins;[10] microRNA,[11][12][13][14][15] a synthetic, self-replicating polycistronic RNA,[16] and low-molecular weight biologically active substances.[17][18][19]

15.1 Natural processes of induction

In 1895 Thomas Morgan removed one of the two frog blastomeres and found that amphibians are able to form whole embryo from the remaining part. This meant that the cells can change their differentiation pathway. Later, in 1924, Spemann and Mangold demonstrated the key importance of cell–cell inductions during animal development.[20] The reversible transformation of cells of one differentiated cell type to another is called metaplasia.[21] This transition can be a part of the normal maturation process, or caused by an inducing stimulus. For example: transformation of

iris cells to lens cells in the process of maturation and transformation of retinal pigment epithelium cells into the neural retina during regeneration in adult newt eyes. This process allows the body to replace cells not suitable to new conditions with more suitable new cells. In Drosophila imaginal discs, cells have to choose from a limited number of standard discrete differentiation states. The fact that transdetermination (change of the path of differentiation) often occurs for a group of cells rather than single cells shows that it is induced rather than part of maturation.[22]

The researchers were able to identify the minimal conditions and factors that would be sufficient for starting the cascade of molecular and cellular processes to instruct pluripotent cells to organize the embryo. They showed that opposing gradients of bone morphogenetic protein (BMP) and Nodal, two transforming growth factor family members that act as morphogens, are sufficient to induce molecular and cellular mechanisms required to organize, *in vivo* or *in vitro*, uncommitted cells of the zebrafish blastula animal pole into a well-developed embryo.[23]

Some types of mature, specialized adult cells can naturally revert to stem cells. For example, "chief" cells express the stem cell marker Troy. While they normally produce digestive fluids for the stomach, they can revert into stem cells to make temporary repairs to stomach injuries, such as a cut or damage from infection. Moreover, they can make this transition even in the absence of noticeable injuries and are capable of replenishing entire gastric units, in essence serving as quiescent "reserve" stem cells.[24] Differentiated airway epithelial cells can revert into stable and functional stem cells in vivo.[25]

After injury, mature terminally differentiated kidney cells dedifferentiate into more primordial versions of themselves, and then differentiate into the cell types needing replacement in the damaged tissue[26] Macrophages can self-renew by local proliferation of mature differentiated cells.[27][28] In newts, muscle tissue is regenerated from specialized muscle cells that dedifferentiate and forget the type of cell they had been. This capacity to regenerate does not decline with age and may be linked to their ability to make

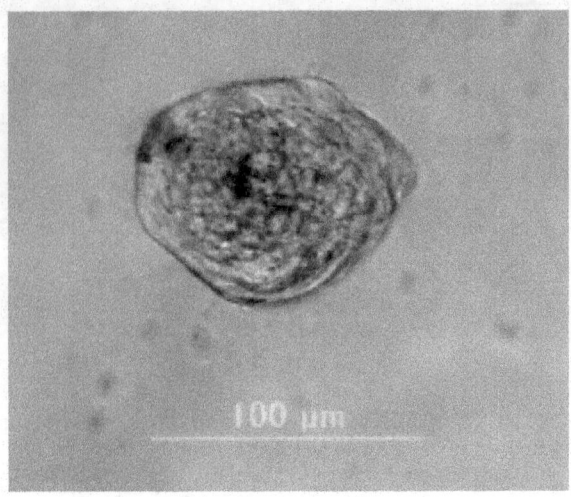

Cluster forming of pluripotent Muse/Stem cell

new stem cells from muscle cells on demand.[29]

A variety of nontumorigenic stem cells display the ability to generate multiple cell types. For instance, multilineage-differentiating stress-enduring (Muse) cells are stress-tolerant adult human stem cells that can self-renew. They form characteristic cell clusters in suspension culture that express a set of genes associated with pluripotency and can differentiate into endodermal, ectodermal and mesodermal cells both in vitro and in vivo.[30][31][32][33][34]

Other well-documented examples of transdifferentiation and their significance in development and regeneration were described in detail.[35][36]

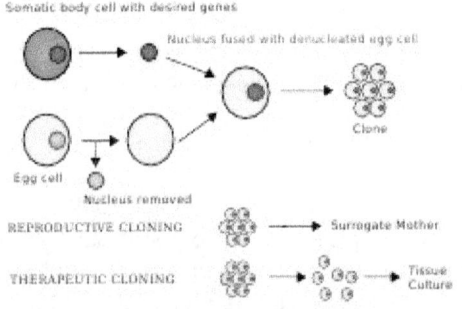

Induced totipotent cells usually can be obtained by reprogramming somatic cells by somatic-cell nuclear transfer (SCNT).

15.2 Induced totipotent cells

15.2.1 SCNT-mediated

Induced totipotent cells can be obtained by reprogramming somatic cells with somatic-cell nuclear transfer (SCNT). The process involves sucking out the nucleus of a somatic (body) cell and injecting it into an oocyte that has had its nucleus removed.[3][5][37][38]

Using an approach based on the protocol outlined by Tachibana et al.,[3] hESCs can be generated by SCNT using dermal fibroblasts nuclei from both a middle-aged 35-year-old male and an elderly, 75-year-old male, suggesting that age-associated changes are not necessarily an impediment to SCNT-based nuclear reprogramming of human cells.[39] Such reprogramming of somatic cells to a pluripotent state holds huge potentials for regenerative medicine. Unfortunately, the cells generated by this technology, potentially are not completely protected from the immune system of the patient (donor of nuclei), because they have the same mitochondrial DNA, as a donor of oocytes, instead of the patients mitochondrial DNA. This reduces their value as a source for autologous stem cell transplantation therapy, as for the present, it is not clear whether it can induce an immune response of the patient upon treatment.

Induced androgenetic haploid embryonic stem cells can be used instead of sperm for cloning. These cells, synchronized in M phase and injected into the oocyte can produce viable offspring.[40]

These developments, together with data on the possibility of unlimited oocytes from mitotically active reproductive stem cells,[41] offer the possibility of industrial production of transgenic farm animals. Repeated recloning of viable mice through a SCNT method that includes a histone deacetylase inhibitor, trichostatin, added to the cell culture medium,[42] show that it may be possible to reclone animals indefinitely with no visible accumulation of reprogramming or genomic errors[43] However, research into technologies to develop sperm and egg cells from stem cells raises bioethical issues.[44]

Such technologies may also have far-reaching clinical applications for overcoming cytoplasmic defects in human oocytes.[3][45] For example, the technology could prevent inherited mitochondrial disease from passing to future generations. Mitochondrial genetic material is passed from mother to child. Mutations can cause diabetes, deafness, eye disorders, gastrointestinal disorders, heart disease, dementia and other neurological diseases. The nucleus from one human egg has been transferred to another, including its mitochondria, creating a cell that could be regarded as having two mothers. The eggs were then fertilised, and the resulting embryonic stem cells carried the swapped mitochondrial DNA.[46] As evidence that the technique is safe author of this method points to the existence of the healthy monkeys that are now more than four years old — and are the product of mitochondrial transplants across different genetic backgrounds.[47]

In late-generation telomerase-deficient (Terc−/−) mice, SCNT-mediated reprogramming mitigates telomere dysfunction and mitochondrial defects to a greater extent than iPSC-based reprogramming.[48]

Other cloning and totipotent transformation achievements have been described.[49]

15.2.2   Obtained without SCNT

Recently some researchers succeeded to get the totipotent cells without the aid of SCNT. Totipotent cells were obtained using the epigenetic factors such as oocyte germinal isoform of histone.[50] Reprogramming in vivo, by transitory induction of the four factors Oct4, Sox2, Klf4 and c-Myc in mice, confers totipotency features. Intraperitoneal injection of such in vivo iPS cells generates embryo-like structures that express embryonic and extraembryonic (trophectodermal) markers.[51]

15.3   Rejuvenation to iPSc

malignant teratocarcinoma cells, confirming the cells' pluripotency.[54][55][56] It turned out that teratocarcinoma cells are able to maintain a culture of pluripotent embryonic stem cell in an undifferentiated state, by supplying the culture medium with various factors.[57] In the 1980s, it became clear that transplanting pluripotent/embryonic stem cells into the body of adult mammals, usually leads to the formation of teratomas, which can then turn into a malignant tumor teratocarcinoma.[58] However, putting teratocarcinoma cells into the embryo at the blastocyst stage, caused them to become incorporated in the inner cell mass and often produced a normal chimeric (i.e. composed of cells from different organisms) animal.[59][60][61] This indicated that the cause of the teratoma is a dissonance - mutual miscommunication between young donor cells and surrounding adult cells (the recipient's so-called "niche").

In August 2006, Japanese researchers circumvented the need for an oocyte, as in SCNT. By reprograming mouse embryonic fibroblasts into pluripotent stem cells via the ectopic expression of four transcription factors, namely Oct4, Sox2, Klf4 and c-Myc, they proved that the overexpression of a small number of factors can push the cell to transition to a new stable state that is associated with changes in the activity of thousands of genes.[7]

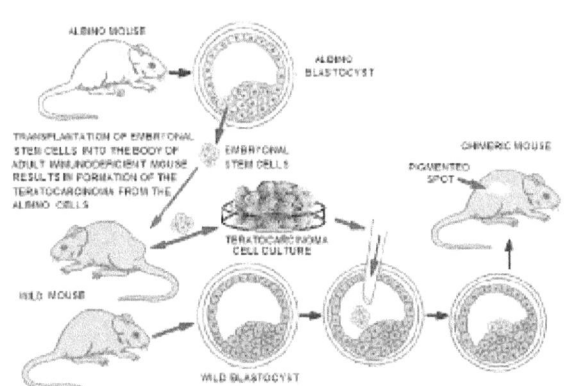

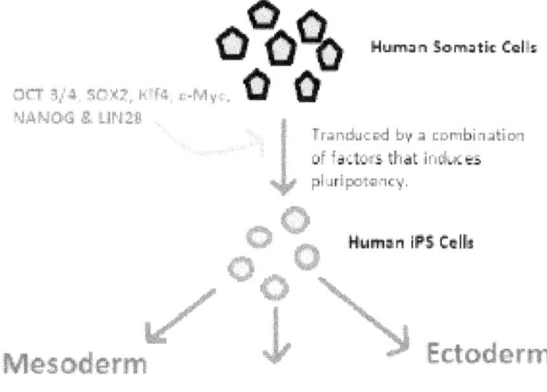

Transplantation of pluripotent/embryonic stem cells into the body of adult mammals, usually leads to the formation of teratomas, which can then turn into a malignant tumor teratocarcinoma. However, putting teratocarcinoma cells into the embryo at the blastocyst stage, caused them to become incorporated in the cell mass and often produced a normal healthy chimeric (i.e. composed of cells from different organisms) animal

Main article: Induced pluripotent stem cells

iPSc were first obtained in the form of transplantable teratocarcinoma induced by grafts taken from mouse embryos.[52] Teratocarcinoma formed from somatic cells.[53] Genetically mosaic mice were obtained from

Human somatic cells are made pluripotent by transducing them with factors that induces pluripotency (OCT 3/4, SOX2, Klf4, c-Myc, NANOG and LIN28). This results in the production of IPS cells, which can differentiate into any cells of the three embryonic germ layers (Mesoderm, Endoderm, Ectoderm).

Reprogramming mechanisms are thus linked, rather than independent and are centered on a small number of genes.[62] IPSC properties are very similar to ESCs.[63] iPSCs have been shown to support the development of all-iPSC mice using a tetraploid (4n) embryo,[64] the most stringent assay for developmental potential. However, some genetically normal iPSCs failed to produce all-iPSC mice because of aberrant epigenetic silencing of the imprinted

Dlk1-Dio3 gene cluster.[18]

An important advantage of iPSC over ESC is that they can be derived from adult cells, rather than from embryos. Therefore, it became possible to obtain iPSC from adult and even elderly patients.[9][65][66]

Reprogramming somatic cells to iPSC leads to rejuvenation. It was found that reprogramming leads to telomere lengthening and subsequent shortening after their differentiation back into fibroblast-like derivatives.[67] Thus, reprogramming leads to the restoration of embryonic telomere length,[68] and hence increases the potential number of cell divisions otherwise limited by the Hayflick limit.[69]

However, because of the dissonance between rejuvenated cells and the surrounding niche of the recipient's older cells, the injection of his own iPSC usually leads to an immune response,[70] which can be used for medical purposes,[71] or the formation of tumors such as teratoma.[72] The reason has been hypothesized to be that some cells differentiated from ESC and iPSC in vivo continue to synthesize embryonic protein isoforms.[73] So, the immune system might detect and attack cells that are not cooperating properly.

A small molecule called MitoBloCK-6 can force the pluripotent stem cells to die by triggering apoptosis (via cytochrome c release across the mitochondrial outer membrane) in human pluripotent stem cells, but not in differentiated cells. Shortly after differentiation, daughter cells became resistant to death. When MitoBloCK-6 was introduced to differentiated cell lines, the cells remained healthy. The key to their survival, was hypothesized to be due to the changes undergone by pluripotent stem cell mitochondria in the process of cell differentiation. This ability of MitoBloCK-6 to separate the pluripotent and differentiated cell lines has the potential to reduce the risk of teratomas and other problems in regenerative medicine.[74]

In 2012 other small molecules (selective cytotoxic inhibitors of human pluripotent stem cells—hPSCs) were identified that prevented human pluripotent stem cells from forming teratomas in mice. The most potent and selective compound of them (PluriSIn #1) inhibits stearoyl-coA desaturase (the key enzyme in oleic acid biosynthesis), which finally results in apoptosis. With the help of this molecule the undifferentiated cells can be selectively removed from culture.[75][76] An efficient strategy to selectively eliminate pluripotent cells with teratoma potential is targeting pluripotent stem cell-specific antiapoptotic factor(s) (i.e., survivin or Bcl10). A single treatment with chemical survivin inhibitors (e.g., quercetin or YM155) can induce selective and complete cell death of undifferentiated hPSCs and is claimed to be sufficient to prevent teratoma formation after transplantation.[77] However, it is unlikely that any kind of preliminary clearance,[78] is able to secure the replanting iPSC or ESC. After the selective removal of pluripotent cells, they re-emerge quickly by reverting differentiated cells into stem cells, which leads to tumors.[79] This may be due to the disorder of let-7 regulation of its target Nr6a1 (also known as Germ cell nuclear factor - GCNF), an embryonic transcriptional repressor of pluripotency genes that regulates gene expression in adult fibroblasts following micro-RNA miRNA loss.[80]

Teratoma formation by pluripotent stem cells may be caused by low activity of PTEN enzyme, reported to promote the survival of a small population (0,1-5% of total population) of highly tumorigenic, aggressive, teratoma-initiating embryonic-like carcinoma cells during differentiation. The survival of these teratoma-initiating cells is associated with failed repression of Nanog as well as a propensity for increased glucose and cholesterol metabolism.[81] These teratoma-initiating cells also expressed a lower ratio of p53/p21 when compared to non-tumorigenic cells.[82] In connection with the above safety problems, the use iPSC for cell therapy is still limited.[83] However, they can be used for a variety of other purposes - including the modeling of disease,[84] screening (selective selection) of drugs, toxicity testing of various drugs.[85]

Small molecule modulators of stem-cell fate.

It is interesting to note that the tissue grown from iPSCs, placed in the "chimeric" embryos in the early stages of mouse development, practically do not cause an immune response (after the embryos have grown into adult mice) and are suitable for autologous transplantation[86] At the same time, full reprogramming of adult cells in vivo within tissues by transitory induction of the four factors Oct4, Sox2, Klf4 and c-Myc in mice results in teratomas emerging from multiple organs.[51] Furthermore, partial reprogramming of cells toward pluripotency in vivo in mice demonstrates that incomplete reprogramming entails epigenetic changes (failed repression of Polycomb targets and altered DNA methylation) in cells that drive cancer development.[87]

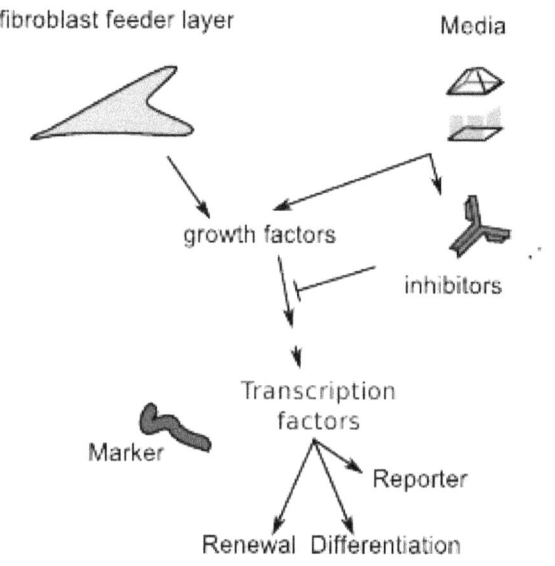

Cell culture example of a small molecule as a tool instead of a protein, in cell culture to obtain a pancreatic lineage from mesodermal stem cells the retinoic acid signalling pathway must be activated while the sonic hedgehog pathway inhibited, which can be done by adding to the media anti-shh antibodies, Hedgehog interacting protein or cyclopamine, the first two are protein and the last a small molecule.[88]

15.3.1   Mogrify algorithm

Determining the unique set of cellular factors that is needed to be manipulated for each cell conversion is a long and costly process that involved much trial and error. As a result, this first step of identifying the key set of cellular factors for cell conversion is the major obstacle researchers face in the field of cell reprogramming. An international team of researchers have developed an algorithm, called Mogrify(1), that can predict the optimal set of cellular factors required to convert one human cell type to another. When tested, Mogrify was able to accurately predict the set of cellular factors required for previously published cell conversions correctly. To further validate Mogrify's predictive ability, the team conducted two novel cell conver-

sions in the laboratory using human cells, and these were successful in both attempts solely using the predictions of Mogrify.[89][90][91] Mogrify has been made available online for other researchers and scientists.

15.3.2   Chemical inducement

Main article: Chemical biology § Chemical approaches to stem-cell biology

By using solely small molecules, Deng Hongkui and colleagues demonstrated that endogenous "master genes" are enough for cell fate reprogramming. They induced a pluripotent state in adult cells from mice using seven small-molecule compounds.[17] The effectiveness of the method is quite high: it was able to convert 0.02% of the adult tissue cells into iPSCs, which is comparable to the gene insertion conversion rate. The authors note that the mice generated from CiPSCs were "100% viable and apparently healthy for up to 6 months". So, this chemical reprogramming strategy has potential use in generating functional desirable cell types for clinical applications.[92][93]

In 2015th year a robust chemical reprogramming system was established with a yield up to 1,000-fold greater than that of the previously reported protocol. So, chemical reprogramming became a promising approach to manipulate cell fates.[94]

15.3.3   Differentiation from induced teratoma

The fact that human iPSCs capable of forming teratomas not only in humans but also in some animal body, in particular in mice or pigs, allowed to develop a method for differentiation of iPSCs in vivo. For this purpose, iPSCs with an agent for inducing differentiation into target cells are injected to genetically modified pig or mouse that has suppressed immune system activation on human cells. The formed teratoma is cut out and used for the isolation of the necessary differentiated human cells[95] by means of monoclonal antibody to tissue-specific markers on the surface of these cells. This method has been successfully used for the production of functional myeloid, erythroid, and lymphoid human cells suitable for transplantation (yet only to mice).[96] Mice engrafted with human iPSC teratoma-derived hematopoietic cells produced human B and T cells capable of functional immune responses. These results offer hope that in vivo generation of patient customized cells is feasible, providing materials that could be useful for transplantation, human antibody generation, and drug screening applications. Using MitoBloCK-6 [74] and / or

PluriSIn # 1 the differentiated progenitor cells can be further purified from teratoma forming pluripotent cells. The fact, that the differentiation takes place even in the teratoma niche, offers hope that the resulting cells are sufficiently stable to stimuli able to cause their transition back to the dedifferentiated (pluripotent) state, and therefore safe. A similar in vivo differentiation system, yielding engraftable hematopoietic stem cells from mouse and human iPSCs in teratoma-bearing animals in combination with a maneuver to facilitate hematopoiesis, was described by Suzuki et al.[97] They noted that neither leukemia nor tumors were observed in recipients after intravenous injection of iPSC-derived hematopoietic stem cells into irradiated recipients. Moreover, this injection resulted in multilineage and long-term reconstitution of the hematolymphopoietic system in serial transfers. Such system provides a useful tool for practical application of iPSCs in the treatment of hematologic and immunologic diseases.[98]

For further development of this method animal in which is grown the human cell graft, for example mouse, must have so modified genome that all its cells express and have on its surface human SIRPα.[99] To prevent rejection after transplantation to the patient of the allogenic organ or tissue, grown from the pluripotent stem cells in vivo in the animal, these cells should express two molecules: CTLA4-Ig, which disrupts T cell costimulatory pathways, and PD-L1, which activates T cell inhibitory pathway.[100]

See also: US 20130058900 patent.

15.3.4 Differentiated cell types

Retinal cells

In the near-future, clinical trials designed to demonstrate the safety of the use of iPSCs for cell therapy of the people with age-related macular degeneration, a disease causing blindness through retina damaging, will begin. There are several articles describing methods for producing retinal cells from iPSCs[101] [102] and how to use them for cell therapy.[103][104] Reports of iPSC-derived retinal pigmented epithelium transplantation showed enhanced visual-guided behaviors of experimental animals for 6 weeks after transplantation.[105] However, clinical trials have been successful: ten patients suffering from retinitis pigmentosa have had their eyesight restored—including a woman who had only 17 percent of her vision left. [106]

Lung and airway epithelial cells

Chronic lung diseases such as idiopathic pulmonary fibrosis and cystic fibrosis or chronic obstructive pulmonary disease and asthma are leading causes of morbidity and mortality worldwide with a considerable human, societal, and financial burden. So there is an urgent need for effective cell therapy and lung tissue engineering.[107][108] Several protocols have been developed for generation of the most cell types of the respiratory system, which may be useful for deriving patient-specific therapeutic cells.[109][110][111][112][113]

Reproductive cells

Some lines of iPSCs have the potentiality to differentiate into male germ cells and oocyte-like cells in an appropriate niche (by culturing in retinoic acid and porcine follicular fluid differentiation medium or seminiferous tubule transplantation). Moreover, iPSC transplantation make a contribution to repairing the testis of infertile mice, demonstrating the potentiality of gamete derivation from iPSCs in vivo and in vitro.[114]

15.4 Induced progenitor stem cells

15.4.1 Direct transdifferentiation

The risk of cancer and tumors creates the need to develop methods for safer cell lines suitable for clinical use. An alternative approach is so-called "direct reprogramming" - transdifferentiation of cells without passing through the pluripotent state.[115][116][117][118][119][120] The basis for this approach was that 5-azacytidine - a DNA demethylation reagent - can cause the formation of myogenic, chondrogenic and adipogeni clones in the immortal cell line of mouse embryonic fibroblasts[121] and that the activation of a single gene, later named MyoD1, is sufficient for such reprogramming.[122] Compared with iPSC whose reprogramming requires at least two weeks, the formation of induced progenitor cells sometimes occurs within a few days and the efficiency of reprogramming is usually many times higher. This reprogramming does not always require cell division.[123] The cells resulting from such reprogramming are more suitable for cell therapy because they do not form teratomas.[120]

Single transcription factor transdifferentiation

Originally only early embryonic cells could be coaxed into changing their identity. Mature cells are resistant to changing their identity once they've committed to a specific kind. However, brief expression of a single transcription factor, the ELT-7 GATA factor, can convert the identity of fully differentiated, specialized non-endodermal cells of the pharynx into fully differentiated intestinal cells in intact larvae and adult roundworm Caenorhabditis elegans with no requirement for a dedifferentiated intermediate.[124]

Transdifferentiation with CRISPR-mediated activator

The cell fate can be effectively manipulated by directly activating of specific endogenous gene expression with CRISPR-mediated activator. When dCas9 (that has been modified so that it no longer cuts DNA, but still can be guided to specific sequences and to bind to them) is combined with transcription activators, it can precisely manipulate endogenous gene expression. Using this method, Wei et al., enhanced the expression of endogenous Cdx2 and Gata6 genes by CRISPR-mediated activators, thus directly converted mouse embryonic stem cells into two extraembryonic lineages, i.e., typical trophoblast stem cells and extraembryonic endoderm cells.[125]

15.4.2   Phased process modeling regeneration

Another way of reprogramming is the simulation of the processes that occur during amphibian limb regeneration. In urodele amphibians, an early step in limb regeneration is skeletal muscle fiber dedifferentiation into a cellulate that proliferates into limb tissue. However, sequential small molecule treatment of the muscle fiber with myoseverin, reversine (the aurora B kinase inhibitor) and some other chemicals: BIO (glycogen synthase-3 kinase inhibitor), lysophosphatidic acid (pleiotropic activator of G-protein-coupled receptors), SB203580 (p38 MAP kinase inhibitor), or SQ22536 (adenylyl cyclase inhibitor) causes the formation of new muscle cell types as well as other cell types such as precursors to fat, bone and nervous system cells.[126]

15.4.3   Antibody-based transdifferentiation

The researchers discovered that GCSF-mimicking antibody can activate a growth-stimulating receptor on marrow cells in a way that induces marrow stem cells that normally develop into white blood cells to become neural progenitor cells. The technique[127] enables researchers to search large libraries of antibodies and quickly select the ones with a desired biological effect.[128]

15.4.4   Conditionally reprogrammed cells

Schlegel and Liu[129] demonstrated that the combination of feeder cells[130][131][132] and a Rho kinase inhibitor (Y-27632)[133][134] induces normal and tumor epithelial cells from many tissues to proliferate indefinitely in vitro. This

process occurs without the need for transduction of exogenous viral or cellular genes. These cells have been termed "Conditionally Reprogrammed Cells (CRC)". The induction of CRCs is rapid and results from reprogramming of the entire cell population. CRCs do not express high levels of proteins characteristic of iPSCs or embryonic stem cells (ESCs) (e.g., Sox2, Oct4, Nanog, or Klf4). This induction of CRCs is reversible and removal of Y-27632 and feeders allows the cells to differentiate normally.[129][135][136] CRC technology can generate 2×10^6 cells in 5 to 6 days from needle biopsies and can generate cultures from cryopreserved tissue and from fewer than four viable cells. CRCs retain a normal karyotype and remain nontumorigenic. This technique also efficiently establishes cell cultures from human and rodent tumors.[129][137][138]

The ability to rapidly generate many tumor cells from small biopsy specimens and frozen tissue provides significant opportunities for cell-based diagnostics and therapeutics (including chemosensitivity testing) and greatly expands the value of biobanking.[129][137][138] Using CRC technology, researchers were able to identify an effective therapy for a patient with a rare type of lung tumor.[139] Engleman's group [140] describes a pharmacogenomic platform that facilitates rapid discovery of drug combinations that can overcome resistance using CRC system. In addition, the CRC method allows for the genetic manipulation of epithelial cells ex vivo and their subsequent evaluation in vivo in the same host. While initial studies revealed that co-culturing epithelial cells with Swiss 3T3 cells J2 was essential for CRC induction, with transwell culture plates, physical contact between feeders and epithelial cells is not required for inducing CRCs, and more importantly that irradiation of the feeder cells is required for this induction. Consistent with the transwell experiments, conditioned medium induces and maintains CRCs, which is accompanied by a concomitant increase of cellular telomerase activity. The activity of the conditioned medium correlates directly with radiation-induced feeder cell apoptosis. Thus, conditional reprogramming of epithelial cells is mediated by a combination of Y-27632 and a soluble factor(s) released by apoptotic feeder cells.[141]

Riegel et al.[142] demonstrate that mouse ME cells isolated from normal mammary glands or from mouse mammary tumor virus (MMTV)-Neu–induced mammary tumors, can be cultured indefinitely as conditionally reprogrammed cells (CRCs). Cell surface progenitor-associated markers are rapidly induced in normal mouse ME-CRCs relative to ME cells. However, the expression of certain mammary progenitor subpopulations, such as CD49f+ ESA+ CD44+, drops significantly in later passages. Nevertheless, mouse ME-CRCs grown in a three-dimensional extracellular matrix gave rise to mammary acinar structures. ME-CRCs isolated from MMTV-Neu transgenic mouse mammary tumors express high levels of HER2/neu, as well as tumor-initiating cell markers, such as CD44+, CD49f+, and ESA+ (EpCam). These patterns of expression are sustained in later CRC passages. Early and late passage ME-CRCs from MMTV-Neu tumors that were implanted in the mammary fat pads of syngeneic or nude mice developed vascular tumors that metastasized within 6 weeks of transplantation. Importantly, the histopathology of these tumors was indistinguishable from that of the parental tumors that develop in the MMTV-Neu mice. Application of the CRC system to mouse mammary epithelial cells provides an attractive model system to study the genetics and phenotype of normal and transformed mouse epithelium in a defined culture environment and in vivo transplant studies.

A different approach to CRC is to inhibit CD47 - a membrane protein that is the thrombospondin-1 receptor. Loss of CD47 permits sustained proliferation of primary murine endothelial cells, increases asymmetric division, and enables these cells to spontaneously reprogram to form multipotent embryoid body-like clusters. CD47 knockdown acutely increases mRNA levels of c-Myc and other stem cell transcription factors in cells in vitro and in vivo. Thrombospondin-1 is a key environmental signal that inhibits stem cell self-renewal via CD47. Thus, CD47 antagonists enable cell self-renewal and reprogramming by overcoming negative regulation of c-Myc and other stem cell transcription factors.[143] In vivo blockade of CD47 using an antisense morpholino increases survival of mice exposed to lethal total body irradiation due to increased proliferative capacity of bone marrow-derived cells and radioprotection of radiosensitive gastrointestinal tissues.[144]

Lineage-specific enhancers

Differentiated macrophages can self-renew in tissues and expand long-term in culture.[27] Under certain conditions macrophages can divide without losing features they have acquired while specializing into immune cells - which is usually not possible with differentiated cells. The macrophages achieve this by activating a gene network similar to one found in embryonic stem cells. Single-cell analysis revealed that, *in vivo*, proliferating macrophages can derepress a macrophage-specific enhancer repertoire associated with a gene network controlling self-renewal. This happened when concentrations of two transcription factors named MafB and c-Maf were naturally low or were inhibited for a short time. Genetic manipulations that turned off MafB and c-Maf in the macrophages caused the cells to start a self-renewal program. The similar network also controls embryonic stem cell self-renewal but is associated with distinct embryonic stem cell-specific enhancers.[28]

Hence macrophages isolated from MafB- and c-Maf-double deficient mice divide indefinitely; the self-renewal depends on c-Myc and Klf4.[145]

15.4.5 Indirect lineage conversion

Indirect lineage conversion is a reprogramming methodology in which somatic cells transition through a plastic intermediate state of partially reprogrammed cells (pre-iPSC), induced by brief exposure to reprogramming factors, followed by differentiation in a specially developed chemical environment (artificial niche).[146]

This method could be both more efficient and safer, since it does not seem to produce tumors or other undesirable genetic changes, and results in much greater yield than other methods. However, the safety of these cells remains questionable. Since lineage conversion from pre-iPSC relies on the use of iPSC reprogramming conditions, a fraction of the cells could acquire pluripotent properties if they do not stop the de-differentiation process in vitro or due to further de-differentiation in vivo.[147]

15.4.6 Outer membrane glycoprotein

A common feature of pluripotent stem cells is the specific nature of protein glycosylation of their outer membrane. That distinguishes them from most nonpluripotent cells, although not white blood cells.[148] The glycans on the stem cell surface respond rapidly to alterations in cellular state and signaling and are therefore ideal for identifying even minor changes in cell populations. Many stem cell markers are based on cell surface glycan epitopes including the widely used markers SSEA-3, SSEA-4, Tra 1-60, and Tra 1-81.[149] Suila Heli et al.[150] speculate that in human stem cells extracellular O-GlcNAc and extracellular O-LacNAc, play a crucial role in the fine tuning of Notch signaling pathway - a highly conserved cell signaling system, that regulates cell fate specification, differentiation, left–right asymmetry, apoptosis, somitogenesis, angiogenesis, and plays a key role in stem cell proliferation (reviewed by Perdigoto and Bardin[151] and Jafar-Nejad et al.[152])

Changes in outer membrane protein glycosylation are markers of cell states connected in some way with pluripotency and differentiation.[153] The glycosylation change is apparently not just the result of the initialization of gene expression, but perform as an important gene regulator involved in the acquisition and maintenance of the undifferentiated state.[154]

For example, activation of glycoprotein ACA,[155] linking glycosylphosphatidylinositol on the surface of the progenitor cells in human peripheral blood, induces increased expression of genes Wnt, Notch-1, BMI1 and HOXB4 through a signaling cascade PI3K/Akt/mTor/PTEN, and promotes the formation of a self-renewing population of hematopoietic stem cells.[156]

Furthermore, dedifferentiation of progenitor cells induced by ACA-dependent signaling pathway leads to ACA-induced pluripotent stem cells, capable of differentiating in vitro into cells of all three germ layers.[157] The study of lectins' ability to maintain a culture of pluripotent human stem cells has led to the discovery of lectin Erythrina crista-galli (ECA), which can serve as a simple and highly effective matrix for the cultivation of human pluripotent stem cells.[158]

15.4.7   Reprogramming through a physical approach

Cell adhesion protein E-cadherin is indispensable for a robust pluripotent phenotype.[159] During reprogramming for iPS cell generation, N-cadherin can replace function of E-cadherin.[160] These functions of cadherins are not directly related to adhesion because sphere morphology helps maintaining the "stemness" of stem cells.[161] Moreover, sphere formation, due to forced growth of cells on a low attachment surface, sometimes induces reprogramming. For example, neural progenitor cells can be generated from fibroblasts directly through a physical approach without introducing exogenous reprogramming factors.

Physical cues, in the form of parallel microgrooves on the surface of cell-adhesive substrates, can replace the effects of small-molecule epigenetic modifiers and significantly improve reprogramming efficiency. The mechanism relies on the mechanomodulation of the cells' epigenetic state. Specifically, "decreased histone deacetylase activity and upregulation of the expression of WD repeat domain 5 (WDR5)—a subunit of H3 methyltranferase—by microgrooved surfaces lead to increased histone H3 acetylation and methylation". Nanofibrous scaffolds with aligned fibre orientation produce effects similar to those produced by microgrooves, suggesting that changes in cell morphology may be responsible for modulation of the epigenetic state.[162]

Substrate rigidity is an important biophysical cue influencing neural induction and subtype specification. For example, soft substrates promote neuroepithelial conversion while inhibiting neural crest differentiation of hESCs in a BMP4-dependent manner. Mechanistic studies revealed a multi-targeted mechanotransductive process involving mechanosensitive Smad phosphorylation and nucleocytoplasmic shuttling, regulated by rigidity-dependent Hippo/YAP activities and actomyosin cytoskeleton integrity and contractility.[163]

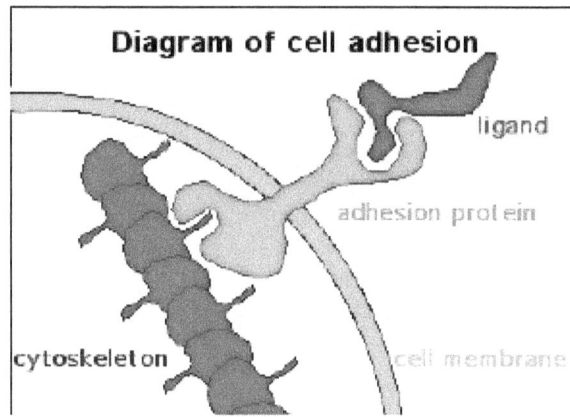

Role of cell adhesions in neural development. Image courtesy of Wikipedia user JWSchmidt under the GNU Free Documentation License

Mouse embryonic stem cells (mESCs) undergo self-renewal in the presence of the cytokine leukemia inhibitory factor (LIF). Following LIF withdrawal, mESCs differentiate, accompanied by an increase in cell–substratum adhesion and cell spreading. Restricted cell spreading in the absence of LIF by either culturing mESCs on chemically defined, weakly adhesive biosubstrates, or by manipulating the cytoskeleton allowed the cells to remain in an undifferentiated and pluripotent state. The effect of restricted cell spreading on mESC self-renewal is not mediated by increased intercellular adhesion, as inhibition of mESC adhesion using a function blocking anti E-cadherin antibody or siRNA does not promote differentiation.[164] Possible mechanisms of stem cell fate predetermination by physical interactions with the extracellular matrix have been described.[165][166]

A new method has been developed that turns cells into stem cells faster and more efficiently by 'squeezing' them using 3D microenvironment stiffness and density of the surrounding gel. The technique can be applied to a large number of cells to produce stem cells for medical purposes on an industrial scale.[167][168]

Cells involved in the reprogramming process change morphologically as the process proceeds. This results in physical difference in adhesive forces among cells. Substantial differences in 'adhesive signature' between pluripotent stem cells, partially reprogrammed cells, differentiated progeny and somatic cells allowed to develop separation process for isolation of pluripotent stem cells in microfluidic devices,[169] which is:

1. fast (separation takes less than 10 minutes);

2. efficient (separation results in a greater than 95 percent pure iPS cell culture);

3. innocuous (cell survival rate is greater than 80 percent and the resulting cells retain normal transcriptional profiles, differentiation potential and karyotype).

Stem cells possess mechanical memory (they remember past physical signals)—with the Hippo signaling pathway factors:[170] Yes-associated protein (YAP) and transcriptional coactivator with PDZ-binding domain (TAZ) acting as an intracellular mechanical rheostat—that stores information from past physical environments and influences the cells' fate.[171][172]

15.4.8 Neural stem cells

Stroke and many neurodegenerative disorders such as Parkinson's disease, Alzheimer's disease, amyotrophic lateral sclerosis need cell replacement therapy. The successful use of converted neural cells (cNs) in transplantations open a new avenue to treat such diseases.[173] Nevertheless, induced neurons (iNs), directly converted from fibroblasts are terminally committed and exhibit very limited proliferative ability that may not provide enough autologous donor cells for transplantation.[174] Self-renewing induced neural stem cells (iNSCs) provide additional advantages over iNs for both basic research and clinical applications.[118][119][120][175][176]

For example, under specific growth conditions, mouse fibroblasts can be reprogrammed with a single factor, Sox2, to form iNSCs that self-renew in culture and after transplantation can survive and integrate without forming tumors in mouse brains.[177] INSCs can be derived from adult human fibroblasts by non-viral techniques, thus offering a safe method for autologous transplantation or for the development of cell-based disease models.[176]

Neural chemically induced progenitor cells (ciNPCs) can be generated from mouse tail-tip fibroblasts and human urinary somatic cells without introducing exogenous factors, but - by a chemical cocktail, namely VCR (V, VPA, an inhibitor of HDACs; C, CHIR99021, an inhibitor of GSK-3 kinases and R, RepSox, an inhibitor of TGF beta signaling pathways), under a physiological hypoxic condition.[178] Alternative cocktails with inhibitors of histone deacetylation, glycogen synthase kinase, and TGF-β pathways (where: sodium butyrate (NaB) or Trichostatin A (TSA) could replace VPA, Lithium chloride (LiCl) or lithium carbonate (Li2CO3) could substitute CHIR99021, or Repsox may be replaced with SB-431542 or Tranilast) show similar efficacies for ciNPC induction.[178]

Multiple methods of direct transformation of somatic cells into induced neural stem cells have been described.[179]

Proof of principle experiments demonstrate that it is possible to convert transplanted human fibroblasts and human astrocytes directly in the brain that are engineered to express inducible forms of neural reprogramming genes, into neurons, when reprogramming genes (Ascl1, Brn2a and Myt1l) are activated after transplantation using a drug.[180]

Astrocytes—the most common neuroglial brain cells, which contribute to scar formation in response to injury—can be directly reprogrammed in vivo to become functional neurons that formed networks in mice without the need of cell transplantation.[181] The researchers followed the mice for nearly a year to look for signs of tumor formation and reported finding none. The same researchers have turned scar-forming astrocytes into progenitor cells called neuroblasts that regenerated into neurons in the injured adult spinal cord.[182]

15.4.9 Oligodendrocyte precursor cells

Without myelin to insulate neurons, nerve signals quickly lose power. Diseases that attack myelin, such as multiple sclerosis, result in nerve signals that cannot propagate to nerve endings, and as a consequence lead to cognitive, motor and sensory problems. Transplantation of oligodendrocyte precursor cells (OPCs), which can successfully create myelin sheaths around nerve cells, is a promising potential therapeutic response. Direct lineage conversion of mouse and rat fibroblasts into oligodendroglial cells provides a potential source of OPCs. Conversion by forced expression of both eight[183] or of the three[184] transcription factors Sox10, Olig2 and Zfp536, may provide such cells.

15.4.10 Cardiomyocytes

Cell-based in vivo therapies may provide a transformative approach to augment vascular and muscle growth and to prevent non-contractile scar formation by delivering transcription factors[115] or microRNAs[114] to the heart.[185] Cardiac fibroblasts, which represent 50% of the cells in the mammalian heart, can be reprogrammed into cardiomyocyte-like cells in vivo by local delivery of cardiac core transcription factors (GATA4, MEF2C, TBX5 and for improved reprogramming plus ESRRG, MESP1, Myocardin and ZFPM2) after coronary ligation.[115][186] These results implicated therapies that can directly remuscularize the heart without cell transplantation. However, the efficiency of such reprogramming turned out to be very low and the phenotype of received cardiomyocyte-like cells does not resemble those of a mature normal cardiomyocyte. Furthermore, transplantation of cardiac transcription factors into injured murine hearts resulted in poor cell survival and minimal expression of cardiac genes.[187]

Meanwhile, advances in the methods of obtaining cardiac

myocytes in vitro occurred.[188][189] Efficient cardiac differentiation of human iPS cells gave rise to progenitors that were retained within infarcted rat hearts, and reduced remodeling of the heart after ischemic damage.[190]

Furthermore, ischemic cardiomyopathy in the murine infarction model was targeted by iPS cell transplantation. It synchronized failing ventricles, offering a regenerative strategy to achieve resynchronization and protection from decompensation by dint of improved left ventricular conduction and contractility, reduced scarring and reversal of structural remodelling.[191] One protocol generated populations of up to 98% cardiomyocytes from hPSCs simply by modulating the canonical Wnt signaling pathway at defined time points in during differentiation, using readily accessible small molecule compounds.[192]

Discovery of the mechanisms controlling the formation of cardiomyocytes led to the development of the drug ITD-1, which effectively clears the cell surface from TGF-β receptor type II and selectively inhibits intracellular TGF-β signaling. It thus selectively enhances the differentiation of uncommitted mesoderm to cardiomyocytes, but not to vascular smooth muscle and endothelial cells.[193]

One project seeded decellularized mouse hearts with human iPSC-derived multipotential cardiovascular progenitor cells. The introduced cells migrated, proliferated and differentiated in situ into cardiomyocytes, smooth muscle cells and endothelial cells to reconstruct the hearts. In addition, the heart's extracellular matrix (the substrate of heart scaffold) signalled the human cells into becoming the specialised cells needed for proper heart function. After 20 days of perfusion with growth factors, the engineered heart tissues started to beat again and were responsive to drugs.[194]

See also: review[195]

15.4.11 Rejuvenation of the muscle stem cell

The elderly often suffer from progressive muscle weakness and regenerative failure owing in part to elevated activity of the p38α and p38β mitogen-activated kinase pathway in senescent skeletal muscle stem cells. Subjecting such stem cells to transient inhibition of p38α and p38β in conjunction with culture on soft hydrogel substrates rapidly expands and rejuvenates them that result in the return of their strength.[196]

In geriatric mice, resting satellite cells lose reversible quiescence by switching to an irreversible pre-senescence state, caused by derepression of p16INK4a (also called Cdkn2a). On injury, these cells fail to activate and expand, even in a youthful environment. p16INK4a silencing in geriatric satellite cells restores quiescence and muscle regenerative

functions.[197]

Myogenic progenitors for potential use in disease modeling or cell-based therapies targeting skeletal muscle could also be generated directly from induced pluripotent stem cells using free-floating spherical culture (EZ spheres) in a culture medium supplemented with high concentrations (100 ng/ml) of fibroblast growth factor-2 (FGF-2) and epidermal growth factor.[198]

15.4.12 Hepatocytes

Unlike current protocols for deriving hepatocytes from human fibroblasts, Saiyong Zhu et al., (2014)[199] did not generate iPSCs but, using small molecules, cut short reprogramming to pluripotency to generate an induced multipotent progenitor cell (iMPC) state from which endoderm progenitor cells and subsequently hepatocytes (iMPC-Heps) were efficiently differentiated. After transplantation into an immune-deficient mouse model of human liver failure, iMPC-Heps proliferated extensively and acquired levels of hepatocyte function similar to those of human primary adult hepatocytes. iMPC-Heps did not form tumours, most probably because they never entered a pluripotent state.

These results establish the feasibility of significant liver repopulation of mice with human hepatocytes generated in vitro, which removes a long-standing roadblock on the path to autologous liver cell therapy.

15.4.13 Insulin-producing cells

Complications of Diabetes mellitus such as cardiovascular diseases, retinopathy, neuropathy, nephropathy, and peripheral circulatory diseases depend on sugar dysregulation due to lack of insulin from pancreatic beta cells and can be lethal if they are not treated. One of the promising approaches to understand and cure diabetes is to use pluripotent stem cells (PSCs), including embryonic stem cells (ESCs) and induced PCSs (iPSCs).[200] Unfortunately, human PSC-derived insulin-expressing cells resemble human fetal β cells rather than adult β cells. In contrast to adult β cells, fetal β cells seem functionally immature, as indicated by increased basal glucose secretion and lack of glucose stimulation and confirmed by RNA-seq of whose transcripts.[201]

An alternative strategy is the conversion of fibroblasts towards distinct endodermal progenitor cell populations and, using cocktails of signalling factors, successful differentiation of these endodermal progenitor cells into functional beta-like cells both in vitro and in vivo.[202]

Overexpression of the three transcription factors, **PDX1**

cell types into a beta cell-like state.[203] An accessible and abundant source of functional insulin-producing cells is intestine. PMN expression in human intestinal "organoids" stimulates the conversion of intestinal epithelial cells into β-like cells possibly acceptable for transplantation.[204]

15.4.14 Nephron Progenitors

Adult proximal tubule cells were directly transcriptionally reprogrammed to nephron progenitors of the embryonic kidney, using a pool of six genes of instructive transcription factors (SIX1, SIX2, OSR1, Eyes absent homolog 1(EYA1), Homeobox A11 (HOXA11) and Snail homolog 2 (SNAI2)) that activated genes consistent with a cap mesenchyme/nephron progenitor phenotype in the adult proximal tubule cell line.[205] The generation of such cells may lead to cellular therapies for adult renal disease. Embryonic kidney organoids placed into adult rat kidneys can undergo onward development and vascular development.[206]

15.4.15 Blood vessel cells

As blood vessels age, they often become abnormal in structure and function, thereby contributing to numerous age-associated diseases including myocardial infarction, ischemic stroke and atherosclerosis of arteries supplying the heart, brain and lower extremities. So, an important goal is to stimulate vascular growth for the collateral circulation to prevent the exacerbation of these diseases. Induced Vascular Progenitor Cells (iVPCs) are useful for cell-based therapy designed to stimulate coronary collateral growth. They were generated by partially reprogramming endothelial cells.[146] The vascular commitment of iVPCs is related to the epigenetic memory of endothelial cells, which engenders them as cellular components of growing blood vessels. That is why, when iVPCs were implanted into myocardium, they engrafted in blood vessels and increased coronary collateral flow better than iPSCs, mesenchymal stem cells, or native endothelial cells.[207]

Ex vivo genetic modification can be an effective strategy to enhance stem cell function. For example, cellular therapy employing genetic modification with Pim-1 kinase (a downstream effector of Akt, which positively regulates neovasculogenesis) of bone marrow–derived cells[208] or human cardiac progenitor cells, isolated from failing myocardium[209] results in durability of repair, together with the improvement of functional parameters of myocardial hemodynamic performance.

Stem cells extracted from fat tissue after liposuction may be coaxed into becoming progenitor smooth muscle cells (iPVSMCs) found in arteries and veins.[210]

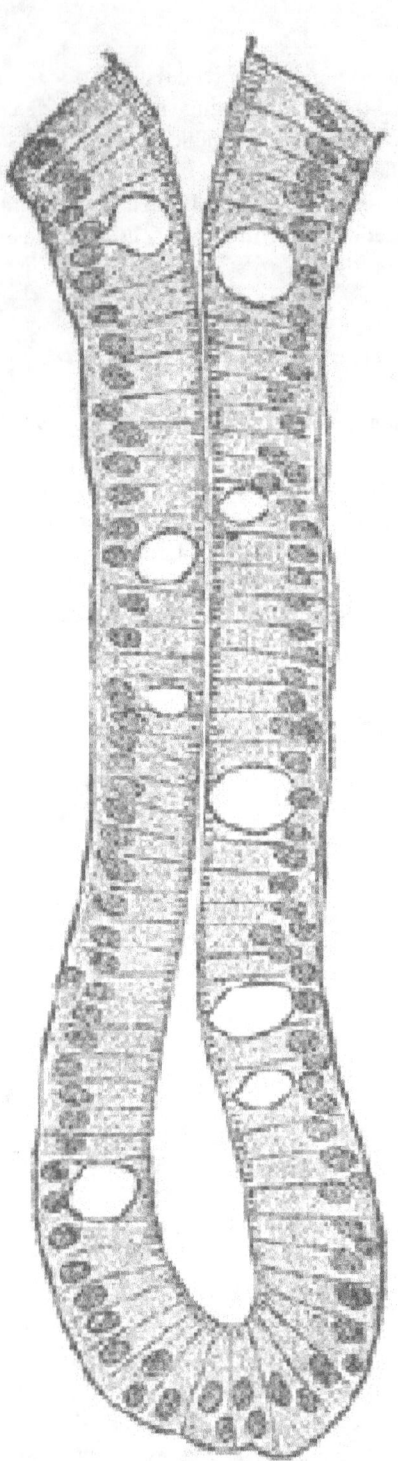

An intestinal crypt - an accessible and abundant source of intestinal epithelial cells for conversion into β-like cells.

(required for pancreatic bud outgrowth and beta-cell maturation), **NGN3** (required for endocrine precursor cell formation) and **MAFA** (for beta-cell maturation) combination (called PNM) can lead to the transformation of some

The 2D culture system of human iPS cells[211] in conjunction with triple marker selection (CD34 (a surface glycophosphoprotein expressed on developmentally early embryonic fibroblasts), NP1 (receptor - neuropilin 1) and KDR (kinase insert domain-containing receptor)) for the isolation of vasculogenic precursor cells from human iPSC, generated endothelial cells that, after transplantation, formed stable, functional mouse blood vessels in vivo, lasting for 280 days.[212]

To treat infarction, it is important to prevent the formation of fibrotic scar tissue. This can be achieved in vivo by transient application of paracrine factors that redirect native heart progenitor stem cell contributions from scar tissue to cardiovascular tissue. For example, in a mouse myocardial infarction model, a single intramyocardial injection of human vascular endothelial growth factor A mRNA (VEGF-A modRNA), modified to escape the body's normal defense system, results in long-term improvement of heart function due to mobilization and redirection of epicardial progenitor cells toward cardiovascular cell types.[213]

15.4.16   Blood stem cells

Red blood cells

RBC transfusion is necessary for many patients. However, to date the supply of RBCs remains labile. In addition, transfusion risks infectious disease transmission. A large supply of safe RBCs generated in vitro would help to address this issue. Ex vivo erythroid cell generation may provide alternative transfusion products to meet present and future clinical requirements.[214][215] Red blood cells (RBC)s generated in vitro from mobilized CD34 positive cells have normal survival when transfused into an autologous recipient.[216] RBC produced in vitro contained exclusively fetal hemoglobin (HbF), which rescues the functionality of these RBCs. In vivo the switch of fetal to adult hemoglobin was observed after infusion of nucleated erythroid precursors derived from iPSCs.[217] Although RBCs do not have nuclei and therefore can not form a tumor, their immediate erythroblasts precursors have nuclei. The terminal maturation of erythroblasts into functional RBCs requires a complex remodeling process that ends with extrusion of the nucleus and the formation of an enucleated RBC.[218] Cell reprogramming often disrupts enucleation. Transfusion of in vitro-generated RBCs or erythroblasts does not sufficiently protect against tumor formation.

The aryl hydrocarbon receptor (AhR) pathway (which has been shown to be involved in the promotion of cancer cell development) plays an important role in normal blood cell development. AhR activation in human hematopoietic progenitor cells (HPs) drives an unprecedented expansion of

HPs, megakaryocyte- and erythroid-lineage cells.[219] See also Concise Review:[220][221]

Platelets

Platelets help prevent hemorrhage in thrombocytopenic patients and patients with thrombocythemia. A significant problem for multitransfused patients is refractoriness to platelet transfusions. Thus, the ability to generate platelet products ex vivo and platelet products lacking HLA antigens in serum-free media would have clinical value. An RNA interference-based mechanism used a lentiviral vector to express short-hairpin RNAi targeting β2-microglobulin transcripts in CD34-positive cells. Generated platelets demonstrated an 85% reduction in class I HLA antigens. These platelets appeared to have normal function in vitro[222]

One clinically-applicable strategy for the derivation of functional platelets from human iPSC involves the establishment of stable immortalized megakaryocyte progenitor cell lines (imMKCLs) through doxycycline-dependent overexpression of BMI1 and BCL-XL. The resulting imMKCLs can be expanded in culture over extended periods (4–5 months), even after cryopreservation. Halting the overexpression (by the removal of doxycycline from the medium) of c-MYC, BMI1 and BCL-XL in growing imMKCLs led to the production of CD42b+ platelets with functionality comparable to that of native platelets on the basis of a range of assays in vitro and in vivo.[223]

Immune cells

A specialised type of white blood cell, known as cytotoxic T lymphocytes (CTLs), are produced by the immune system and are able to recognise specific markers on the surface of various infectious or tumour cells, causing them to launch an attack to kill the harmful cells. Thence, immunotherapy with functional antigen-specific T cells has potential as a therapeutic strategy for combating many cancers and viral infections.[224] However, cell sources are limited, because they are produced in small numbers naturally and have a short lifespan.

A potentially efficient approach for generating antigen-specific CTLs is to revert mature immune T cells into iPSCs, which possess indefinite proliferative capacity in vitro, and after their multiplication to coax them to redifferentiate back into T cells.[225][226][227]

Another method combines iPSC and chimeric antigen receptor (CAR) [228] technologies to generate human T cells targeted to CD19, an antigen expressed by malignant B cells, in tissue culture.[229] This approach of generating therapeutic human T cells may be useful for cancer im-munotherapy and other medical applications.

Invariant natural killer T (iNKT) cells have great clinical potential as adjuvants for cancer immunotherapy. iNKT cells act as innate T lymphocytes and serve as a bridge between the innate and acquired immune systems. They augment anti-tumor responses by producing interferon-gamma (IFN-γ).[230] The approach of collection, repro-gramming/dedifferentiation, re-differentiation and injec-tion has been proposed for related tumor treatment.[231]

Dendritic cells (DC) are specialized to control T-cell re-sponses. DC with appropriate genetic modifications may survive long enough to stimulate antigen-specific CTL and after that be completely eliminated. DC-like antigen-presenting cells obtained from human induced pluripotent stem cells can serve as a source for vaccination therapy.[232]

CCAAT/enhancer binding protein-α (C/EBPα) induces transdifferentiation of B cells into macrophages at high efficiencies[233] and enhances reprogramming into iPS cells when co-expressed with transcription factors Oct4, Sox2, Klf4 and Myc.[234] with a 100-fold increase in iPS cell reprogramming efficiency, involving 95% of the population.[235] Furthermore, C/EBPa can convert selected human B cell lymphoma and leukemia cell lines into macrophage-like cells at high efficiencies, impairing the cells' tumor-forming capacity.[236]

15.4.17 Thymic epithelial cells rejuvenation

The thymus is the first organ to deteriorate as people age. This shrinking is one of the main reasons the immune system becomes less effective with age. Diminished ex-pression of the thymic epithelial cell transcription factor FOXN1 has been implicated as a component of the mech-anism regulating age-related involution.[237][238]

Clare Blackburn and colleagues show that established age-related thymic involution can be reversed by forced upreg-ulation of just one transcription factor - FOXN1 in the thymic epithelial cells in order to promote rejuvenation, proliferation and differentiation of these cells into fully functional thymic epithelium.[239] This rejuvenation and increased proliferation was accompanied by upregulation of genes that promote cell cycle progression (cyclin D1, ΔNp63, FgfR2IIIb) and that are required in the thymic ep-ithelial cells to promote specific aspects of T cell develop-ment (Dll4, Kitl, Ccl25, Cxcl12, Cd40, Cd80, Ctsl, Pax1).

15.4.18 Mesenchymal stem cells

Induction

mesenchymal stem/stromal cells (MSCs) are under investi-gation for applications in cardiac, renal, neural, joint and bone repair, as well as in inflammatory conditions and hemopoietic cotransplantation.[240] This is because of their immunosuppressive properties and ability to differentiate into a wide range of mesenchymal-lineage tissues. MSCs are typically harvested from adult bone marrow or fat, but these require painful invasive procedures and are low-frequency sources, making up only 0.001%–0.01% of bone marrow cells and 0.05% in liposuction aspirates.[241] Of concern for autologous use, in particular in the elderly most in need of tissue repair, MSCs decline in quantity and qual-ity with age.[240][242][243]

IPSCs could be obtained by the cells rejuvenation of even centenarians.[9][39] Because iPSCs can be harvested free of ethical constraints and culture can be expanded indefinitely, they are an advantageous source of MSCs.[244] IPSC treat-ment with SB-431542 leads to rapid and uniform MSC gen-eration from human iPSCs. (SB-431542 is an inhibitor of activin/TGF- pathways by blocking phosphorylation of ALK4, ALK5, and ALK7 receptors.) These iPS-MSCs may lack teratoma-forming ability, display a normal sta-ble karyotype in culture and exhibit growth and differenti-ation characteristics that closely resemble those of primary MSCs. It has potential for in vitro scale-up, enabling MSC-based therapies.[245] MSC derived from iPSC have the ca-pacity to aid periodontal regeneration and are a promising source of readily accessible stem cells for use in the clinical treatment of periodontitis.[246][247]

Besides cell therapy in vivo, the culture of human mes-enchymal stem cells can be used in vitro for mass-production of exosomes, which are ideal vehicles for drug delivery.[248]

Dedifferentiated adipocytes

Adipose tissue, because of its abundance and relatively less invasive harvest methods, represents a source of mesenchy-mal stem cells (MSCs). Unfortunately, liposuction aspirates are only 0.05% MSCs.[241] However, a large amount of ma-ture adipocytes, which in general have lost their prolifera-tive abilities and therefore are typically discarded, can be easily isolated from the adipose cell suspension and dedif-ferentiated into lipid-free fibroblast-like cells, named dedif-ferentiated fat (DFAT) cells. DFAT cells re-establish active proliferation ability and express multipotent capacities.[249] Compared with adult stem cells, DFAT cells show unique advantages in abundance, isolation and homogeneity. Un-der proper induction culture in vitro or proper environ-ment in vivo, DFAT cells could demonstrate adipogenic, osteogenic, chondrogenic, and myogenic potentials. They could also exhibit perivascular characteristics and elicit neovascularization.[250][251][252]

15.4.19 Chondrogenic cells

Cartilage is the connective tissue responsible for frictionless joint movement. Its degeneration ultimately results in complete loss of joint function in the late stages of osteoarthritis. As an avascular and hypocellular tissue, cartilage has a limited capacity for self-repair. Chondrocytes are the only cell type in cartilage, in which they are surrounded by the extracellular matrix that they secrete and assemble.

One method of producing cartilage is to induce it from iPS cells.[253] Alternatively, it is possible to convert fibroblasts directly into induced chondrogenic cells (iChon) without an intermediate iPS cell stage, by inserting three reprogramming factors (c-MYC, KLF4, and SOX9).[254] Human iChon cells expressed marker genes for chondrocytes (type II collagen) but not fibroblasts.

Implanted into defects created in the articular cartilage of rats, human iChon cells survived to form cartilaginous tissue for at least four weeks, with no tumors. The method makes use of c-MYC, which is thought to have a major role in tumorigenesis and employs a retrovirus to introduce the reprogramming factors, excluding it from unmodified use in human therapy.[225][227][255]

15.5 Sources of cells for reprogramming

The most frequently used sources for reprogramming are blood cells[256][257][258][259] and fibroblasts, obtained by biopsy of the skin,[260] but taking cells from urine is less invasive.[261] The latter method does not require a biopsy or blood sampling. As of 2013, urine-derived stem cells had been differentiated into endothelial, osteogenic, chondrogenic, adipogenic, skeletal myogenic and neurogenic lineages, without forming teratomas.[262] Therefore, their epigenetic memory is suited to reprogramming into iPS cells. However, few cells appear in urine, only low conversion efficiencies had been achieved and the risk of bacterial contamination is relatively high.

Another promising source of cells for reprogramming are mesenchymal stem cells derived from human hair follicles.[263]

The origin of somatic cells used for reprogramming may influence the efficiency of reprogramming,[264][265] the functional properties of the resulting induced stem cells[266] and the ability to form tumors.[267]

IPSCs retain an epigenetic memory of their tissue of origin, which impacts their differentiation potential.[255][266][268][269][270][271] This epigenetic memory does not necessarily manifest itself at the pluripotency stage – iPSCs derived from different tissues exhibit proper morphology, express pluripotency markers and are able to differentiate into the three embryonic layers in vitro and in vivo. However, this epigenetic memory may manifest during re-differentiation into specific cell types that require the specific loci bearing residual epigenetic marks.

15.6 See also

- Transdifferentiation
- Examples of in vitro transdifferentiation by lineage-instructive approach
- Examples of in vitro transdifferentiation by initial epigenetic activation phase approach
- Examples of in vivo transdifferentiation by lineage-instructive approach
- Injury induced stem cell niches
- Transcription factors
- Growth factors
- Pioneer factors
- Cellular differentiation
- CAF-1

15.7 Notes

15.7.1 References for further reading

- Tabar, V.; Studer, L. (2014). "Pluripotent stem cells in regenerative medicine: challenges and recent progress". *Nature Reviews Genetics* **15** (2): 82–92. doi:10.1038/nrg3563. PMID 24434846.
- Tan, Y.; Ooi, S.; Wang, L. (2014). "Immunogenicity and Tumorigenicity of Pluripotent Stem Cells and their Derivatives: Genetic and Epigenetic Perspectives". *Current stem cell research & therapy* **9** (1): 63–72. doi:10.2174/1574888x113086660068.
- Yamanaka, Shinya (2012). "Induced Pluripotent Stem Cells: Past, Present, and Future". *Cell Stem Cell* **10** (6): 678–684. doi:10.1016/j.stem.2012.05.005. PMID 22704507.

- Takahashi, K.; Yamanaka, S. (2013). "Induced pluripotent stem cells in medicine and biology". *Development* **140** (12): 2457–61. doi:10.1242/dev.092551. PMID 23715538.

- Asuelime, Grace E.; Shi, Yanhong (2012). "A case of cellular alchemy: lineage reprogramming and its potential in regenerative medicine". *J Mol Cell Biol* **4**: 190–196. doi:10.1093/jmcb/mjs005. PMC 3408064. PMID 22371436.

- Lensch, M. W.; Mummery, C. L. (2013). "From Stealing Fire to Cellular Reprogramming: A Scientific History Leading to the 2012 Nobel Prize". *Stem Cell Reports* **1** (1): 5–17. doi:10.1016/j.stemcr.2013.05.001. PMID 24052937.

- Issue, Special (2013). "Induced Pluripotent Stem Cells". *Genomics, Proteomics & Bioinformatics* **11** (5): 257–334.

- Lin, Ji; Mei-rong Li, Dong-dong Ti; Wei-dong, Han (2013). "Microenvironment-evoked cell lineage conversion: Shifting the focus from internal reprogramming to external forcing". *Ageing Research Reviews* **12**: 29–38. doi:10.1016/j.arr.2012.04.002.

- Takahashi, K (2014). "Cellular Reprogramming". *Cold Spring Harb Perspect Biol* **6**: a018606. doi:10.1101/cshperspect.a018606.

- Nobel Prize in Physiology or Medicine 2012 Awarded for Discovery That Mature Cells Can Be Reprogrammed to Become Pluripotent

- Hussein, Samer MI; Nagy, Andras A (2012). "Progress made in the reprogramming field: new factors, new strategies and a new outlook". *Current Opinion in Genetics & Development* **22** (5): 435–443. doi:10.1016/j.gde.2012.08.007.

- Zhang, Yemin; Yao, Lin; Yu, Xiya; Ou, Jun; Hui, Ning; Liu, Shanrong (2012). "A poor imitation of a natural process: A call to reconsider the iPSC engineering technique". *Cell Cycle* **11** (24): 4536–4544. doi:10.4161/cc.22575.

- Sancho-Martinez, Ignacio; Hee Baek, Sung; Juan Carlos, Izpisua Belmonte (2012). "Lineage conversion methodologies meet the reprogramming toolbox" (PDF). *Nature Cell Biology* **14** (9): 892–899. doi:10.1038/ncb2567. PMID 22945254.

- Mochiduki, Y.; Okita, K. (2012). "Methods for iPS cell generation for basic research and clinical applications". *Biotechnology Journal* **7** (6): 789–797. doi:10.1002/biot.201100356. PMID 22378737.

- Madonna, Rosalinda (2012). "Human-Induced Pluripotent Stem Cells: In Quest of Clinical Applications". *Molecular Biotechnology* **52** (2): 193–203. doi:10.1007/s12033-012-9504-0. PMID 22302314.

- Lorenzo, M.; Fleischer, A.; Bachiller, D. (2012). "Generation of Mouse and Human Induced Pluripotent Stem Cells (iPSC) from Primary Somatic Cells". *Stem Cell Reviews and Reports* **9**: 435–450. doi:10.1007/s12015-012-9412-5. PMID 23104133.

- Detailed protocols for reprogramming and for analysis of iPSCs

- Buganim, Y.; Faddah, D. A.; Jaenisch, R. (2013). "Mechanisms and models of somatic cell reprogramming". *Nature Reviews Genetics* **14** (6): 427–439. doi:10.1038/nrg3473. PMC 4060150. PMID 23681063.

15.8　References

[1] Yamanaka, S.; Blau, H. M. (2010). "Nuclear reprogramming to a pluripotent state by three approaches". *Nature* **465** (7299): 704–12. doi:10.1038/nature09229. PMC 2901154. PMID 20535199.

[2] Gurdon J. B. and Ian Wilmut (2011) Nuclear Transfer to Eggs and Oocytes Cold Spring Harb Perspect Biol; 3: a002659

[3] Tachibana, M.; Amato, P.; Sparman, M.; Gutierrez, N. M.; Tippner-Hedges, R.; Ma, H.; Kang, E.; Fulati, A.; Lee, H. S.; Sritanaudomchai, H.; Masterson, K.; Larson, J.; Eaton, D.; Sadler-Fredd, K.; Battaglia, D.; Lee, D.; Wu, D.; Jensen, J.; Patton, P.; Gokhale, S.; Stouffer, R. L.; Wolf, D.; Mitalipov, S. (2013). "Human Embryonic Stem Cells Derived by Somatic Cell Nuclear Transfer". *Cell* **153** (6): 1228–38. doi:10.1016/j.cell.2013.05.006. PMID 23683578.

[4] Noggle, S.; Fung, H. L.; Gore, A.; Martinez, H.; Satriani, K. C.; Prosser, R.; Oum, K.; Paull, D.; Druckenmiller, S.; Freeby, M.; Greenberg, E.; Zhang, K.; Goland, R.; Sauer, M. V.; Leibel, R. L.; Egli, D. (2011). "Human oocytes reprogram somatic cells to a pluripotent state". *Nature* **478** (7367): 70–5. doi:10.1038/nature10397. PMID 21979046.

[5] Pan, G.; Wang, T.; Yao, H.; Pei, D. (2012). "Somatic cell reprogramming for regenerative medicine: SCNT vs. IPS cells". *BioEssays* **34** (6): 472–6. doi:10.1002/bies.201100174. PMID 22419173.

[6] Do, J. T.; Han, D. W.; Gentile, L; Sobek-Klocke, I; Stehling, M; Lee, H. T.; Schöler, H. R. (2007). "Erasure of cellular memory by fusion with pluripotent cells". *Stem Cells* **25** (4): 1013–20. doi:10.1634/stemcells.2006-0691. PMID 17218392.

[7] Takahashi, K.; Tanabe, K.; Ohnuki, M.; Narita, M.; Ichisaka, T.; Tomoda, K.; Yamanaka, S. (2007). "Induction of Pluripotent Stem Cells from Adult Human Fibroblasts by Defined Factors". *Cell* **131** (5): 861–872. doi:10.1016/j.cell.2007.11.019. PMID 18035408.

[8] Wang, W.; Yang, J.; Liu, H.; Lu, D.; Chen, X.; Zenonos, Z.; Campos, L. S.; Rad, R.; Guo, G.; Zhang, S.; Bradley, A.; Liu, P. (2011). "Rapid and efficient reprogramming of somatic cells to induced pluripotent stem cells by retinoic acid receptor gamma and liver receptor homolog 1". *Proceedings of the National Academy of Sciences* **108** (45): 18283–18288. doi:10.1073/pnas.1100893108.

[9] Lapasset, L.; Milhavet, O.; Prieur, A.; Besnard, E.; Babled, A.; Aït-Hamou, N.; Leschik, J.; Pellestor, F.; Ramirez, J.-M.; De Vos, J.; Lehmann, S.; Lemaitre, J.-M. (2011). "Rejuvenating senescent and centenarian human cells by reprogramming through the pluripotent state". *Genes & Development* **25** (21): 2248–2253. doi:10.1101/gad.173922.111.

[10] Zhou, H.; Wu, S.; Joo, J. Y.; Zhu, S.; Han, D. W.; Lin, T.; Trauger, S.; Bien, G.; Yao, S.; Zhu, Y.; Siuzdak, G.; Schöler, H. R.; Duan, L.; Ding, S. (2009). "Generation of Induced Pluripotent Stem Cells Using Recombinant Proteins". *Cell Stem Cell* **4** (5): 381–4. doi:10.1016/j.stem.2009.04.005. PMID 19398399.

[11] Li, Z.; Rana, T. M. (2012). "Using MicroRNAs to Enhance the Generation of Induced Pluripotent Stem Cells". *Current Protocols in Stem Cell Biology*. doi:10.1002/9780470151808.sc04a04s20. ISBN 0470151803.

[12] Anokye-Danso, F.; Trivedi, C. M.; Juhr, D.; Gupta, M.; Cui, Z.; Tian, Y.; Zhang, Y.; Yang, W.; Gruber, P. J.; Epstein, J. A.; Morrisey, E. E. (2011). "Highly Efficient miRNA-Mediated Reprogramming of Mouse and Human Somatic Cells to Pluripotency". *Cell Stem Cell* **8** (4): 376–88. doi:10.1016/j.stem.2011.03.001. PMID 21474102.

[13] Miyoshi, N.; Ishii, H.; Nagano, H.; Haraguchi, N.; Dewi, D. L.; Kano, Y.; Nishikawa, S.; Tanemura, M.; Mimori, K.; Tanaka, F.; Saito, T.; Nishimura, J.; Takemasa, I.; Mizushima, T.; Ikeda, M.; Yamamoto, H.; Sekimoto, M.; Doki, Y.; Mori, M. (2011). "Reprogramming of Mouse and Human Cells to Pluripotency Using Mature MicroRNAs". *Cell Stem Cell* **8** (6): 633–8. doi:10.1016/j.stem.2011.05.001. PMID 21620789.

[14] Jayawardena, T. M.; Egemnazarov, B.; Finch, E. A.; Zhang, L.; Payne, J. A.; Pandya, K.; Zhang, Z.; Rosenberg, P.; Mirotsou, M.; Dzau, V. J. (2012). "MicroRNA-Mediated in Vitro and in Vivo Direct Reprogramming of Cardiac Fibroblasts to Cardiomyocytes". *Circulation Research* **110** (11): 1465–73. doi:10.1161/CIRCRESAHA.112.269035. PMID 22539765.

[15] Bao, X.; Zhu, X.; Liao, B.; Benda, C.; Zhuang, Q.; Pei, D.; Qin, B.; Esteban, M. A. (2013). "MicroRNAs in somatic cell reprogramming". *Current Opinion in Cell Biology* **25** (2): 208–14. doi:10.1016/j.ceb.2012.12.004. PMID 23332905.

[16] Yoshioka, N.; Gros, E.; Li, H. R.; Kumar, S.; Deacon, D. C.; Maron, C.; Muotri, A. R.; Chi, N. C.; Fu, X. D.; Yu, B. D.; Dowdy, S. F. (2013). "Efficient Generation of Human iPSCs by a Synthetic Self-Replicative RNA". *Cell Stem Cell* **13** (2): 246–54. doi:10.1016/j.stem.2013.06.001. PMID 23910086.

[17] Hou, P.; Li, Y.; Zhang, X.; Liu, C.; Guan, J.; Li, H.; Zhao, T.; Ye, J.; Yang, W.; Liu, K.; Ge, J.; Xu, J.; Zhang, Q.; Zhao, Y.; Deng, H. (2013). "Pluripotent Stem Cells Induced from Mouse Somatic Cells by Small-Molecule Compounds". *Science* **341** (6146): 651–4. doi:10.1126/science.1239278. PMID 23868920.
Efe, J. A.; Ding, S. (2011). "The evolving biology of small molecules: Controlling cell fate and identity". *Philosophical Transactions of the Royal Society B: Biological Sciences* **366** (1575): 2208–2221. doi:10.1098/rstb.2011.0006.

[18] Stadtfeld, M.; Apostolou, E.; Ferrari, F.; Choi, J.; Walsh, R. M.; Chen, T.; Ooi, S. S. K.; Kim, S. Y.; Bestor, T. H.; Shioda, T.; Park, P. J.; Hochedlinger, K. (2012). "Ascorbic acid prevents loss of Dlk1-Dio3 imprinting and facilitates generation of all–iPS cell mice from terminally differentiated B cells". *Nature Genetics* **44** (4): 398–405, S1–2. doi:10.1038/ng.1110. PMID 22387999.

[19] Pandian, G. N.; Sugiyama, H. (2012). "Programmable genetic switches to control transcriptional machinery of pluripotency". *Biotechnology Journal* **7** (6): 798–809. doi:10.1002/biot.201100361. PMID 22588775.
Pandian, G. N.; Nakano, Y.; Sato, S.; Morinaga, H.; Bando, T.; Nagase, H.; Sugiyama, H. (2012). "A synthetic small molecule for rapid induction of multiple pluripotency genes in mouse embryonic fibroblasts". *Scientific Reports* **2**: 544. doi:10.1038/srep00544. PMC 3408130. PMID 22848790.

[20] De Robertis, Edward M. (2006). "Spemanns *organizer and self-regulation in amphibian embryos*". *Nature Reviews Molecular Cell Biology* **7**: 296–302. doi:10.1038/nrm1855. PMC 2464568. PMID 16482093.

[21] Slack, J. M. W. (2009). "Metaplasia and somatic cell reprogramming". *The Journal of Pathology* **217** (2): 161–8. doi:10.1002/path.2442. PMID 18855879.

[22] Wei, G.; Schubiger, G.; Harder, F.; Müller, A. M. (2000). "Stem Cell Plasticity in Mammals and Transdetermination in *Drosophila*: Common Themes?". *Stem Cells* **18** (6): 409–14. doi:10.1634/stemcells.18-6-409. PMID 11072028.
Worley, M. I.; Setiawan, L.; Hariharan, I. K. (2012). "Regeneration and Transdetermination in *Drosophila Imaginal* Discs". *Annual Review of Genetics* **46**: 289–310. doi:10.1146/annurev-genet-110711-155637. PMID 22934642.

[23] Xu, Peng-Fei; Houssin, Nathalie; Ferri-Lagneau, Karine F.; Thisse, Bernard; Thisse, Christine (2014). "Construction of a Vertebrate Embryo from Two Opposing Morphogen Gradients". *Science* **344** (6179): 87–89. doi:10.1126/science.1248252.

[24] Stange, D. E.; Koo, B. K.; Huch, M.; Sibbel, G.; Basak, O.; Lyubimova, A.; Kujala, P.; Bartfeld, S.; Koster, J.; Geahlen, J. H.; Peters, P. J.; Van Es, J. H.; Van De Wetering, M.; Mills, J. C.; Clevers, H. (2013). "Differentiated Troy+ Chief Cells Act as Reserve Stem Cells to Generate All Lineages of the Stomach Epithelium". *Cell* **155** (2): 357–68. doi:10.1016/j.cell.2013.09.008. PMC 4094146. PMID 24120136.

[25] Tata, P. R.; Mou, H.; Pardo-Saganta, A.; Zhao, R.; Prabhu, M.; Law, B. M.; Vinarsky, V.; Cho, J. L.; Breton, S.; Sahay, A.; Medoff, B. D.; Rajagopal, J. (2013). "Dedifferentiation of committed epithelial cells into stem cells in vivo". *Nature*. doi:10.1038/nature12777.

[26] Kusaba, T.; Lalli, M.; Kramann, R.; Kobayashi, A.; Humphreys, B. D. (2013). "Differentiated kidney epithelial cells repair injured proximal tubule". *Proceedings of the National Academy of Sciences* **111** (4): 1527–1532. doi:10.1073/pnas.1310653110.

[27] Sieweke, Michael H.; Allen, Judith E. (2013). "Beyond Stem Cells: Self-Renewal of Differentiated Macrophages". *Science* **342** (6161): 1242974. doi:10.1126/science.1242974. PMID 24264994.

[28] Soucie, E. L.; Weng, Z.; Geirsdóttir, L.; et al. (2016). "Lineage-specific enhancers activate self-renewal genes in macrophages and embryonic stem cells". *Science* **351**: aad5510. doi:10.1126/science.aad5510.

[29] Sandoval-Guzmán, T.; Wang, H.; Khattak, S.; Schuez, M.; Roensch, K.; Nacu, E.; Tazaki, A.; Joven, A.; Tanaka, E. M.; Simon, A. S. (2014). "Fundamental Differences in Dedifferentiation and Stem Cell Recruitment during Skeletal Muscle Regeneration in Two Salamander Species". *Cell Stem Cell* **14** (2): 174–87. doi:10.1016/j.stem.2013.11.007. PMID 24268695.

[30] Kuroda, Y; Wakao, S; Kitada, M; Murakami, T; Nojima, M; Dezawa, M (2013). "Isolation, culture and evaluation of multilineage-differentiating stress-enduring (Muse) cells". *Nat Protoc* **8** (7): 1391–415. doi:10.1038/nprot.2013.076. PMID 23787896.

[31] Kuroda, Y.; Kitada, M.; Wakao, S.; et al. (2010). "Unique multipotent cells in adult human mesenchymal cell populations". *PNAS* **107** (19): 8639–8643. doi:10.1073/pnas.0911647107. PMC 2889306. PMID 20421459.

[32] Ogura, F; Wakao, S; Kuroda, Y; Tsuchiyama, K; Bagheri, M; Heneidi, S; Chazenbalk, G; Aiba, S; Dezawa, M (2014). "Human Adipose Tissue Possesses a Unique Population of Pluripotent Stem Cells with Nontumorigenic and Low Telomerase Activities: Potential Implications in Regenerative Medicine". *Stem Cells Dev.* **23**: 717–28. doi:10.1089/scd.2013.0473. PMID 24256547.

[33] Heneidi, S; Simerman, AA; Keller, E; Singh, P; Li, X; et al. (2013). "Awakened by Cellular Stress: Isolation and Characterization of a Novel Population of Pluripotent Stem Cells Derived from Human Adipose Tissue". *PLoS ONE* **8** (6): e64752. doi:10.1371/journal.pone.0064752. PMC 3673968. PMID 23755141.

[34] Shigemoto, T; Kuroda, Y; Wakao, S; Dezawa, M (2013). "A Novel Approach to Collecting Satellite Cells From Adult Skeletal Muscles on the Basis of Their Stress Tolerance". *Stem Cells Trans Med* **2** (7): 488–498. doi:10.5966/sctm.2012-0130. PMC 3697816. PMID 23748608.

[35] Sisakhtnezhad, S.; Matin, M. M. (2012). "Transdifferentiation: A cell and molecular reprogramming process". *Cell and Tissue Research* **348** (3): 379–96. doi:10.1007/s00441-012-1403-y. PMID 22526624.

[36] Allyson J. Merrell & Ben Z. Stanger (2016). Adult cell plasticity in vivo: de-differentiation and transdifferentiation are back in style. Nature Reviews Molecular Cell Biology. doi:10.1038/nrm.2016.24

[37] Dinnyes, Andras; Tian, Xiuchun Cindy; Oback, Bj"orn (17 April 2013). Robert A. Meyers, ed. *Nuclear Transfer for Cloning Animals.* John Wiley & Sons. pp. 299–344. ISBN 978-3-527-66854-0.
Jullien, J.; Pasque, V.; Halley-Stott, R. P.; Miyamoto, K.; Gurdon, J. B. (2011). "Mechanisms of nuclear reprogramming by eggs and oocytes: A deterministic process?". *Nature Reviews Molecular Cell Biology* **12** (7): 453–9. doi:10.1038/nrm3140. PMID 21697902.
Campbell, K. H. S. (2002). "A background to nuclear transfer and its applications in agriculture and human therapeutic medicine". *Journal of Anatomy* **200** (3): 267–275. doi:10.1046/j.1469-7580.2002.00035.x.

[38] US 8,647,872 patent

[39] Chung, Y. G.; Eum, J. H.; Lee, J. E.; Shim, S. H.; Sepilian, V.; Hong, S. W.; Lee, Y.; Treff, N. R.; Choi, Y. H.; Kimbrel, E. A.; Dittman, R. E.; Lanza, R.; Lee, D. R. (2014). "Human Somatic Cell Nuclear Transfer Using Adult Cells". *Cell Stem Cell* **14** (6): 777–80. doi:10.1016/j.stem.2014.03.015. PMID 24746675.

[40] Yang, H.; Shi, L.; Wang, B. A.; Liang, D.; Zhong, C.; Liu, W.; Nie, Y.; Liu, J.; Zhao, J.; Gao, X.; Li, D.; Xu, G. L.; Li, J. (2012). "Generation of Genetically Modified Mice by Oocyte Injection of Androgenetic Haploid Embryonic Stem Cells". *Cell* **149** (3): 605–17. doi:10.1016/j.cell.2012.04.002. PMID 22541431.

[41] Hayashi, K.; Ogushi, S.; Kurimoto, K.; Shimamoto, S.; Ohta, H.; Saitou, M. (2012). "Offspring from Oocytes Derived from in Vitro Primordial Germ Cell-like Cells in Mice". *Science* **338** (6109): 971–975. doi:10.1126/science.1226889. PMID 23042295.

[42] Kishigami, S.; Mizutani, E.; Ohta, H.; Hikichi, T.; Thuan, N. V.; Wakayama, S.; Bui, H. T.; Wakayama, T. (2006). "Significant improvement of mouse cloning technique by treatment with trichostatin a after somatic nuclear transfer". *Biochemical and Biophysical Research Communications*

340 (1): 183–9. doi:10.1016/j.bbrc.2005.11.164. PMID 16356478.

[43] Wakayama, S.; Kohda, T.; Obokata, H.; Tokoro, M.; Li, C.; Terashita, Y.; Mizutani, E.; Nguyen, V. T.; Kishigami, S.; Ishino, F.; Wakayama, T. (2013). "Successful Serial Recloning in the Mouse over Multiple Generations". *Cell Stem Cell* **12** (3): 293–7. doi:10.1016/j.stem.2013.01.005. PMID 23472871.

[44] Official website of the Presidential Commission for the Study of Bioethical Issues

[45] Paull, D.; Emmanuele, V.; Weiss, K. A.; Treff, N.; Stewart, L.; Hua, H.; Zimmer, M.; Kahler, D. J.; Goland, R. S.; Noggle, S. A.; Prosser, R.; Hirano, M.; Sauer, M. V.; Egli, D. (2012). "Nuclear genome transfer in human oocytes eliminates mitochondrial DNA variants". *Nature* **493** (7434): 632–7. doi:10.1038/nature11800. PMID 23254936.

[46] Tachibana, M.; Amato, P.; Sparman, M.; Woodward, J.; Sanchis, D. M.; Ma, H.; Gutierrez, N. M.; Tippner-Hedges, R.; Kang, E.; Lee, H. S.; Ramsey, C.; Masterson, K.; Battaglia, D.; Lee, D.; Wu, D.; Jensen, J.; Patton, P.; Gokhale, S.; Stouffer, R.; Mitalipov, S. (2012). "Towards germline gene therapy of inherited mitochondrial diseases". *Nature* **493** (7434): 627–31. doi:10.1038/nature11647. PMID 23103867.

[47] Check Hayden, E. (2013). "Regulators weigh benefits of 'three-parent' fertilization". *Nature* **502** (7471): 284–5. doi:10.1038/502284a. PMID 24132269.

[48] Le, R.; Kou, Z.; Jiang, Y.; Li, M.; Huang, B.; Liu, W.; Li, H.; Kou, X.; He, W.; Rudolph, K. L.; Ju, Z.; Gao, S. (2014). "Enhanced Telomere Rejuvenation in Pluripotent Cells Reprogrammed via Nuclear Transfer Relative to Induced Pluripotent Stem Cells". *Cell Stem Cell* **14** (1): 27–39. doi:10.1016/j.stem.2013.11.005. PMID 24268696.

[49] Cibelli, Jose; Lanza, Robert; Campbell, Keith H.S.; West, Michael D. (14 September 2002). *Principles of Cloning*. Academic Press. ISBN 978-0-08-049215-5.

[50] Shinagawa, T.; Takagi, T.; Tsukamoto, D.; Tomaru, C.; Huynh, L. M.; Sivaraman, P.; Kumarevel, T.; Inoue, K.; Nakato, R.; Katou, Y.; Sado, T.; Takahashi, S.; Ogura, A.; Shirahige, K.; Ishii, S. (2014). "Histone Variants Enriched in Oocytes Enhance Reprogramming to Induced Pluripotent Stem Cells". *Cell Stem Cell* **14** (2): 217–27. doi:10.1016/j.stem.2013.12.015. PMID 24506885.

[51] Abad, M. A.; Mosteiro, L.; Pantoja, C.; Cañamero, M.; Rayon, T.; Ors, I.; Graña, O.; Megias, D.; Domínguez, O.; Martinez, D.; Manzanares, M.; Ortega, S.; Serrano, M. (2013). "Reprogramming in vivo produces teratomas and iPS cells with totipotency features". *Nature* **502** (7471): 340–5. doi:10.1038/nature12586. PMID 24025773. Naik, Gautam (2013-09-11). "New Promise for Stem Cells - WSJ.com". Online.wsj.com. Retrieved 2014-01-30.

[52] Stevens, L. C. (1970). "The development of transplantable teratocarcinomas from intratesticular grafts of pre- and postimplantation mouse embryos". *Developmental Biology* **21** (3): 364–82. doi:10.1016/0012-1606(70)90130-2. PMID 5436899.

[53] Mintz, B; Cronmiller, C; Custer, R. P. (1978). "Somatic cell origin of teratocarcinomas". *Proceedings of the National Academy of Sciences of the United States of America* **75** (6): 2834–8. doi:10.1073/pnas.75.6.2834. PMC 392659. PMID 275854.

[54] Mintz, B; Illmensee, K (1975). "Normal genetically mosaic mice produced from malignant teratocarcinoma cells". *Proc Natl Acad Sci U S A* **72** (9): 3585–3589. doi:10.1073/pnas.72.9.3585.

[55] Martin, G. R.; Evans, M. J. (1975). "Differentiation of clonal lines of teratocarcinoma cells: formation of embryoid bodies in vitro". *Proc. Natl. Acad. Sci. USA* **72** (4): 1441–1445. doi:10.1073/pnas.72.4.1441.

[56] Illmensee, K; Mintz, B (1976). "Totipotency and normal differentiation of single teratocarcinoma cells cloned by injection into blastocysts". *Proceedings of the National Academy of Sciences of the United States of America* **73** (2): 549–53. doi:10.1073/pnas.73.2.549. PMC 335947. PMID 1061157.

[57] Martin, GR (1981). "Isolation of a pluripotent cell line from early mouse embryos cultured in medium conditioned by teratocarcinoma stem cells". *Proc. Natl. Acad. Sci. USA* **78** (12): 7634–7638. doi:10.1073/pnas.78.12.7634. PMC 349323. PMID 6950406.

[58] Martin, G. R. (1980). "Teratocarcinomas and mammalian embryogenesis". *Science* **209** (4458): 768–76. doi:10.1126/science.6250214. PMID 6250214.

[59] Papaioannou, V. E.; Gardner, R. L.; McBurney, M. W.; Babinet, C; Evans, M. J. (1978). "Participation of cultured teratocarcinoma cells in mouse embryogenesis". *Journal of embryology and experimental morphology* **44**: 93–104. PMID 650144.

[60] GRAHAM, C. F. (January 1977). *Teratocarcinoma cells and normal mouse embryogenesiseditor=Michael I. Sherman*. MIT Press. ISBN 978-0-262-19158-6.

[61] ILLMENSEE, K. (14 June 2012). L. B. Russell, ed. *Reversion of malignancy and normalized differentiation of teratocarcinoma cells in chimeric mice*. Springer London, Limited. pp. 3–24. ISBN 978-1-4684-3392-0.

[62] Stuart, H. T.; Van Oosten, A. L.; Radzisheuskaya, A.; Martello, G.; Miller, A.; Dietmann, S.; Nichols, J.; Silva, J. C. R. (2014). "NANOG Amplifies STAT3 Activation and They Synergistically Induce the Naive Pluripotent Program". *Current Biology* **24** (3): 340–6. doi:10.1016/j.cub.2013.12.040. PMID 24462001.

[63] Choi, Jiho; Lee, Soohyun; Mallard, William; et al. (2015). "A comparison of genetically matched cell lines reveals the equivalence of human iPSCs and ESCs". *Nature Biotechnology* **33**: 1173–1181. doi:10.1038/nbt.3388.

[64] Boland, M. J.; Hazen, J. L.; Nazor, K. L.; Rodriguez, A. R.; Gifford, W; Martin, G; Kupriyanov, S; Baldwin, K. K. (2009). "Adult mice generated from induced pluripotent stem cells". *Nature* **461** (7260): 91–4. doi:10.1038/nature08310. PMID 19672243.
Kang, L.; Wang, J.; Zhang, Y.; Kou, Z.; Gao, S. (2009). "IPS Cells Can Support Full-Term Development of Tetraploid Blastocyst-Complemented Embryos". *Cell Stem Cell* **5** (2): 135–8. doi:10.1016/j.stem.2009.07.001. PMID 19631602.

[65] Yagi, T.; Kosakai, A.; Ito, D.; Okada, Y.; Akamatsu, W.; Nihei, Y.; Nabetani, A.; Ishikawa, F.; Arai, Y.; Hirose, N.; Okano, H.; Suzuki, N. (2012). "Establishment of Induced Pluripotent Stem Cells from Centenarians for Neurodegenerative Disease Research". *PLoS ONE* **7** (7): e41572. doi:10.1371/journal.pone.0041572. PMID 22848530.

[66] Rohani, L.; Johnson, A. A.; Arnold, A.; Stolzing, A. (2014). "The aging signature: A hallmark of induced pluripotent stem cells?". *Aging Cell* **13** (1): 2–7. doi:10.1111/acel.12182. PMC 4326871. PMID 24256351.

[67] Yehezkel, S; Rebibo-Sabbah, A; Segev, Y; Tzukerman, M; Shaked, R; Huber, I; Gepstein, L; Skorecki, K; Selig, S (2011). "Reprogramming of telomeric regions during the generation of human induced pluripotent stem cells and subsequent differentiation into fibroblast-like derivatives". *Epigenetics* **6** (1): 63–75. doi:10.4161/epi.6.1.13390. PMC 3052915. PMID 20861676.

[68] West, M. D.; Vaziri, H (2010). "Back to immortality: The restoration of embryonic telomere length during induced pluripotency". *Regenerative Medicine* **5** (4): 485–8. doi:10.2217/rme.10.51. PMID 20632849.

[69] Marión, R. M.; Blasco, M. A. (2010). "Telomere rejuvenation during nuclear reprogramming". *Current Opinion in Genetics & Development* **20** (2): 190–6. doi:10.1016/j.gde.2010.01.005. PMID 20176474.
Gourronc, F. A.; Klingelhutz, A. J. (2012). "Therapeutic opportunities: Telomere maintenance in inducible pluripotent stem cells". *Mutation Research/Fundamental and Molecular Mechanisms of Mutagenesis* **730** (1–2): 98–105. doi:10.1016/j.mrfmmm.2011.05.008. PMC 3179558. PMID 21605571.

[70] Zhao, T.; Zhang, Z. N.; Rong, Z.; Xu, Y. (2011). "Immunogenicity of induced pluripotent stem cells". *Nature* **474** (7350): 212–5. doi:10.1038/nature10135. PMID 21572395.

[71] Dhodapkar, M. V.; Dhodapkar, K. M. (2011). "Spontaneous and therapy-induced immunity to pluripotency genes in humans: Clinical implications, opportunities and challenges". *Cancer Immunology, Immunotherapy*

60 (3): 413–8. doi:10.1007/s00262-010-0944-8. PMC 3574640. PMID 21104412.

[72] Gutierrez-Aranda, I.; Ramos-Mejia, V.; Bueno, C.; Munoz-Lopez, M.; Real, P. J.; Mácia, A.; Sanchez, L.; Ligero, G.; Garcia-Parez, J. L.; Menendez, P. (2010). "Human Induced Pluripotent Stem Cells Develop Teratoma More Efficiently and Faster Than Human Embryonic Stem Cells Regardless the Site of Injection". *Stem Cells* **28** (9): 1568–1570. doi:10.1002/stem.471. PMC 2996086. PMID 20641038.

[73] Chang, C. J.; Mitra, K.; Koya, M.; Velho, M.; Desprat, R.; Lenz, J.; Bouhassira, E. E. (2011). "Production of Embryonic and Fetal-Like Red Blood Cells from Human Induced Pluripotent Stem Cells". *PLoS ONE* **6** (10): e25761. doi:10.1371/journal.pone.0025761. PMID 22022444.

[74] Dabir, D. V.; Hasson, S. A.; Setoguchi, K.; Johnson, M. E.; Wongkongkathep, P.; Douglas, C. J.; Zimmerman, J.; Damoiseaux, R.; Teitell, M. A.; Koehler, C. M. (2013). "A Small Molecule Inhibitor of Redox-Regulated Protein Translocation into Mitochondria". *Developmental Cell* **25** (1): 81–92. doi:10.1016/j.devcel.2013.03.006. PMC 3726224. PMID 23597483.

[75] Ben-David, Uri; Gan, Qing-Fen; Golan-Lev, Tamar; et al. (2013). "Selective Elimination of Human Pluripotent Stem Cells by an Oleate Synthesis Inhibitor Discovered in a High-Throughput Screen". *Cell Stem Cell* **12** (2): 167–179. doi:10.1016/j.stem.2012.11.015. PMID 23318055.

[76] Lou, K. J. (2013). "Small molecules vs. Teratomas". *Science-Business eXchange* **6** (7). doi:10.1038/scibx.2013.158.

[77] Lee, M. -O.; Moon, S. H.; Jeong, H. -C.; Yi, J. -Y.; Lee, T. -H.; Shim, S. H.; Rhee, Y. -H.; Lee, S. -H.; Oh, S. -J.; Lee, M. -Y.; Han, M. -J.; Cho, Y. S.; Chung, H. -M.; Kim, K. -S.; Cha, H. -J. (2013). "Inhibition of pluripotent stem cell-derived teratoma formation by small molecules". *Proceedings of the National Academy of Sciences* **110** (35): E3281. doi:10.1073/pnas.1303669110.

[78] Tang, C; Weissman, I. L.; Drukker, M (2012). "The safety of embryonic stem cell therapy relies on teratoma removal". *Oncotarget* **3** (1): 7–8. doi:10.18632/oncotarget.434. PMC 3292887. PMID 22294556.

[79] Chaffer, C. L.; Brueckmann, I.; Scheel, C.; Kaestli, A. J.; Wiggins, P. A.; Rodrigues, L. O.; Brooks, M.; Reinhardt, F.; Su, Y.; Polyak, K.; Arendt, L. M.; Kuperwasser, C.; Bierie, B.; Weinberg, R. A. (2011). "Normal and neoplastic non-stem cells can spontaneously convert to a stem-like state". *Proceedings of the National Academy of Sciences* **108** (19): 7950–7955. doi:10.1073/pnas.1102454108.
"Residual Undifferentiated Cells During Differentiation of Induced Pluripotent Stem Cells In Vitro and In Vivo". *Stem Cells and Development* **21**: 521–529. doi:10.1089/scd.2011.0131.

[80] Gurtan, A. M.; Ravi, A.; Rahl, P. B.; Bosson, A. D.; Jnbaptiste, C. K.; Bhutkar, A.; Whittaker, C. A.; Young,

R. A.; Sharp, P. A. (2013). "Let-7 represses Nr6a1 and a mid-gestation developmental program in adult fibroblasts". *Genes & Development* **27** (8): 941–954. doi:10.1101/gad.215376.113.
Wang, H.; Wang, X.; Xu, X.; Zwaka, T. P.; Cooney, A. J. (2013). "Epigenetic Reprogramming of the Germ Cell Nuclear Factor Gene is Required for Proper Differentiation of Induced Pluripotent Cells". *Stem Cells* **31** (12): 2659–66. doi:10.1002/stem.1367. PMID 23495137.

[81] Lindgren, A. G.; Natsuhara, K.; Tian, E.; Vincent, J. J.; Li, X.; Jiao, J.; Wu, H.; Banerjee, U.; Clark, A. T. (2011). "Loss of Pten Causes Tumor Initiation Following Differentiation of Murine Pluripotent Stem Cells Due to Failed Repression of Nanog". *PLoS ONE* **6** (1): e16478. doi:10.1371/journal.pone.0016478. PMC 3029365. PMID 21304588.

[82] Grad, I; Hibaoui, Y; Jaconi, M; Chicha, L; Bergström-Tengzelius, R; Sailani, M. R.; Pelte, M. F.; Dahoun, S; Mitsiadis, T. A.; Töhönen, V; Bouillaguet, S; Antonarakis, S. E.; Kere, J; Zucchelli, M; Hovatta, O; Feki, A (2011). "NANOG priming before full reprogramming may generate germ cell tumours". *European cells & materials* **22**: 258–74; discussio 274. PMID 22071697.

[83] Okano, H.; Nakamura, M.; Yoshida, K.; Okada, Y.; Tsuji, O.; Nori, S.; Ikeda, E.; Yamanaka, S.; Miura, K. (2013). "Steps Toward Safe Cell Therapy Using Induced Pluripotent Stem Cells". *Circulation Research* **112** (3): 523–33. doi:10.1161/CIRCRESAHA.111.256149. PMID 23371901.
Cunningham, J. J.; Ulbright, T. M.; Pera, M. F.; Looijenga, L. H. J. (2012). "Lessons from human teratomas to guide development of safe stem cell therapies". *Nature Biotechnology* **30** (9): 849–57. doi:10.1038/nbt.2329. PMID 22965062.

[84] Bellin, M.; Marchetto, M. C.; Gage, F. H.; Mummery, C. L. (2012). "Induced pluripotent stem cells: The new patient?". *Nature Reviews Molecular Cell Biology* **13** (11): 713–26. doi:10.1038/nrm3448. PMID 23034453.
Sandoe, J.; Eggan, K. (2013). "Opportunities and challenges of pluripotent stem cell neurodegenerative disease models". *Nature Neuroscience* **16** (7): 780–9. doi:10.1038/nn.3425. PMID 23799470.

[85] Takahashi, K.; Yamanaka, S. (2013). "Induced pluripotent stem cells in medicine and biology". *Development* **140** (12): 2457–61. doi:10.1242/dev.092551. PMID 23715538.
Fu, X; Xu, Y (2012). "Challenges to the clinical application of pluripotent stem cells: Towards genomic and functional stability". *Genome Medicine* **4** (6): 55. doi:10.1186/gm354. PMC 3698533. PMID 22741526.

[86] Araki, R.; Uda, M.; Hoki, Y.; Sunayama, M.; Nakamura, M.; Ando, S.; Sugiura, M.; Ideno, H.; Shimada, A.; Nifuji, A.; Abe, M. (2013). "Negligible immunogenicity of terminally differentiated cells derived from induced pluripotent or embryonic stem cells". *Nature* **494** (7435): 100–4. doi:10.1038/nature11807. PMID 23302801.
Wahlestedt, M.; Norddahl, G. L.; Sten, G.; Ugale, A.;

Frisk, M. -A. M.; Mattsson, R.; Deierborg, T.; Sigvardsson, M.; Bryder, D. (2013). "An epigenetic component of hematopoietic stem cell aging amenable to reprogramming into a young state". *Blood* **121** (21): 4257–64. doi:10.1182/blood-2012-11-469080. PMID 23476050.

[87] Ohnishi, K.; Semi, K.; Yamamoto, T.; Shimizu, M.; Tanaka, A.; Mitsunaga, K.; Yamada, Y. (2014). "Premature Termination of Reprogramming In Vivo Leads to Cancer Development through Altered Epigenetic Regulation". *Cell* **156** (4): 663–677. doi:10.1016/j.cell.2014.01.005. PMID 24529372.

[88] Mfopou JK, De Groote V, Xu X, Heimberg H, Bouwens L. (May 2007). "Sonic hedgehog and other soluble factors from differentiating embryoid bodies inhibit pancreas development". *Stem Cells* **25** (5): 1156–65. doi:10.1634/stemcells.2006-0720. PMID 17272496.

[89] Mapping out cell conversion

[90] Owen, Rackham; Gough, Julian (2016). "A predictive computational framework for direct reprogramming between human cell types". *Nature Genetics*. doi:10.1038/ng.3487.

[91] New Algorithm May Someday Enable Scientists to Regrow Limbs and Replace Damaged Organs

[92] De Los, Angeles; Daley, G. Q. (2013). "A chemical logic for reprogramming to pluripotency". *Cell Research* **23**: 1337–1338. doi:10.1038/cr.2013.119.

[93] Federation, A. J.; Bradner, J. E.; Meissner, A. (2013). "The use of small molecules in somatic-cell reprogramming". *Trends in Cell Biology* **24**: 179–187. doi:10.1016/j.tcb.2013.09.011. PMID 24183602.

[94] Zhao, Y.; Zhao, T.; Guan, J. (2015). "A XEN-like State Bridges Somatic Cells to Pluripotency during Chemical Reprogramming". *Cell* **163**: 1678–1691. doi:10.1016/j.cell.2015.11.017.

[95] NAKAUCHI Hiromitsu, KAMIYA Akihide, SUZUKI Nao, ITO Keiichi, YAMAZAKI Satoshi (2011) METHOD FOR PRODUCING CELLS INDUCED TO DIFFERENTIATE FROM PLURIPOTENT STEM CELLS PATENT COOPERATION TREATY APPLICATION, patno: WO2011071085 (A1) — 2011-06-16 (C12N5/07)

[96] Amabile, G.; Welner, R. S.; Nombela-Arrieta, C.; d'Alise, A. M.; Di Ruscio, A.; Ebralidze, A. K.; Kraytsberg, Y.; Ye, M.; Kocher, O.; Neuberg, D. S.; Khrapko, K.; Silberstein, L. E.; Tenen, D. G. (2012). "In vivo generation of transplantable human hematopoietic cells from induced pluripotent stem cells". *Blood* **121** (8): 1255–64. doi:10.1182/blood-2012-06-434407. PMID 23212524.

[97] Suzuki, N.; Yamazaki, S.; Yamaguchi, T.; Okabe, M.; Masaki, H.; Takaki, S.; Otsu, M.; Nakauchi, H. (2013). "Generation of Engraftable Hematopoietic Stem Cells from Induced Pluripotent Stem Cells by Way of Teratoma Formation". *Molecular Therapy* **21** (7): 1424–31. doi:10.1038/mt.2013.71. PMID 23670574.

[98] Chou, B. K.; Ye, Z.; Cheng, L. (2013). "Generation and Homing of iPSC-Derived Hematopoietic Cells in Vivo". *Molecular Therapy* **21** (7): 1292–3. doi:10.1038/mt.2013.129. PMID 23812546.

[99] Yamauchi, T; Takenaka, K; Urata, S; Shima, T; Kikushige, Y; Tokuyama, T; Iwamoto, C; Nishihara, M; Iwasaki, H; Miyamoto, T; Honma, N; Nakao, M; Matozaki, T; Akashi, K (2013). "Polymorphic Sirpa is the genetic determinant for NOD-based mouse lines to achieve efficient human cell engraftment". *Blood* **121** (8): 1316–25. doi:10.1182/blood-2012-06-440354. PMID 23293079.

[100] Rong, Z.; Wang, M.; Hu, Z.; Stradner, M.; Zhu, S.; Kong, H.; Yi, H.; Goldrath, A.; Yang, Y. G.; Xu, Y.; Fu, X. (2014). "An Effective Approach to Prevent Immune Rejection of Human ESC-Derived Allografts". *Cell Stem Cell* **14** (1): 121–30. doi:10.1016/j.stem.2013.11.014. PMC 4023958. PMID 24388175.

[101] Hirami, Y.; Osakada, F.; Takahashi, K.; Okita, K.; Yamanaka, S.; Ikeda, H.; Yoshimura, N.; Takahashi, M. (2009). "Generation of retinal cells from mouse and human induced pluripotent stem cells". *Neuroscience Letters* **458** (3): 126–31. doi:10.1016/j.neulet.2009.04.035. PMID 19379795.

[102] Buchholz, D. E.; Hikita, S. T.; Rowland, T. J.; Friedrich, A. M.; Hinman, C. R.; Johnson, L. V.; Clegg, D. O. (2009). "Derivation of Functional Retinal Pigmented Epithelium from Induced Pluripotent Stem Cells". *Stem Cells* **27** (10): 2427–34. doi:10.1002/stem.189. PMID 19658190.

[103] Yang, Jin; Nong, Eva; Tsang, Stephen H (2013). "Induced pluripotent stem cells and retinal degeneration treatment". *Expert Rev. Ophthalmol* **8** (1): 5–8. doi:10.1586/EOP.12.75.

[104] Fields., Mark A.; Hwang., John; Gong., Jie; Cai., Hui; Del Priore, Lucian (9 December 2012). Stephen Tsang, ed. *The Eye as a Target Organ for Stem Cell Therapy*. Springer. pp. 1–30. ISBN 978-1-4614-5493-9.

[105] Li, Y; Tsai, YT; Hsu, CW; et al. (2012). "Long-term safety and efficacy of human induced pluripotent stem cell (iPS) grafts in a preclinical model of retinitis pigmentosa". *Mol. Med.* **18** (1): 1312–1319. doi:10.2119/molmed.2012.00242. PMC 3521789. PMID 22895806.

[106] Stem cell therapy for RP is now offered at St. Luke's Medical Center.

[107] Tzouvelekis, A.; Ntolios, P.; Bouros, D. (2013). "Stem Cell Treatment for Chronic Lung Diseases". *Respiration* **85** (3): 179–92. doi:10.1159/000346525. PMID 23364286.

[108] Wagner, Darcy E.; Bonvillain, Ryan W.; Jensen, Todd; Girard, Eric D.; Bunnell, Bruce A.; Finck, Christine M.; Hoffman5, Andrew M.; Weiss2, Daniel J. (Jul 2013). "Can stem cells be used to generate new lungs? Ex vivo lung bioengineering with decellularized whole lung scaffolds". *Respirology* **18** (6): 895–911. doi:10.1111/resp.12102. PMID 23614471.

[109] Wong, A. P.; Rossant, J. (2013). "Generation of Lung Epithelium from Pluripotent Stem Cells". *Current pathobiology reports* **1** (2): 137–145. doi:10.1007/s40139-013-0016-9. PMC 3646155. PMID 23662247.

[110] Mou, H.; Zhao, R.; Sherwood, R.; Ahfeldt, T.; Lapey, A.; Wain, J.; Sicilian, L.; Izvolsky, K.; Lau, F. H.; Musunuru, K.; Cowan, C.; Rajagopal, J. (2012). "Generation of Multipotent Lung and Airway Progenitors from Mouse ESCs and Patient-Specific Cystic Fibrosis iPSCs". *Cell Stem Cell* **10** (4): 385–97. doi:10.1016/j.stem.2012.01.018. PMID 22482504.

[111] Ghaedi, M.; Calle, E. A.; Mendez, J. J.; Gard, A. L.; Balestrini, J.; Booth, A.; Bove, P. F.; Gui, L.; White, E. S.; Niklason, L. E. (2013). "Human iPS cell-derived alveolar epithelium repopulates lung extracellular matrix". *Journal of Clinical Investigation* **123** (11): 4950–62. doi:10.1172/JCI68793. PMID 24135142.

[112] Ghaedi, M.; Mendez, J. J.; Bove, P. F.; Sivarapatna, A.; Raredon, M. S. B.; Niklason, L. E. (2014). "Alveolar epithelial differentiation of human induced pluripotent stem cells in a rotating bioreactor". *Biomaterials* **35** (2): 699–710. doi:10.1016/j.biomaterials.2013.10.018. PMID 24144903.

[113] Huang, S. X. L.; Islam, M. N.; O'Neill, J.; Hu, Z.; Yang, Y. G.; Chen, Y. W.; Mumau, M.; Green, M. D.; Vunjak-Novakovic, G.; Bhattacharya, J.; Snoeck, H. W. (2013). "Efficient generation of lung and airway epithelial cells from human pluripotent stem cells". *Nature Biotechnology* **32** (1): 84–91. doi:10.1038/nbt.2754. PMC 4101921. PMID 24291815.

[114] Niu, Z.; Hu, Y.; Chu, Z.; Yu, M.; Bai, Y.; Wang, L.; Hua, J. (2013). "Germ-like cell differentiation from induced pluripotent stem cells (iPSCs)". *Cell Biochemistry and Function* **31** (1): 12–9. doi:10.1002/cbf.2924. PMID 23086862. Yang, S.; Bo, J.; Hu, H.; Guo, X.; Tian, R.; Sun, C.; Zhu, Y.; Li, P.; Liu, P.; Zou, S.; Huang, Y.; Li, Z. (2012). "Derivation of male germ cells from induced pluripotent stem cells in vitro and in reconstituted seminiferous tubules". *Cell Proliferation* **45** (2): 91–100. doi:10.1111/j.1365-2184.2012.00811.x. PMID 22324506. Panula, S.; Medrano, J. V.; Kee, K.; Bergstrom, R.; Nguyen, H. N.; Byers, B.; Wilson, K. D.; Wu, J. C.; Simon, C.; Hovatta, O.; Reijo Pera, R. A. (2010). "Human germ cell differentiation from fetal- and adult-derived induced pluripotent stem cells". *Human Molecular Genetics* **20** (4): 752–62. doi:10.1093/hmg/ddq520. PMID 21131292.

[115] Qian, L.; Huang, Y.; Spencer, C. I.; Foley, A.; Vedantham, V.; Liu, L.; Conway, S. J.; Fu, J. D.; Srivastava, D. (2012). "In vivo reprogramming of murine cardiac fibroblasts into induced cardiomyocytes". *Nature* **485** (7400): 593–8. doi:10.1038/nature11044. PMID 22522929.

[116] Szabo, E.; Rampalli, S.; Risueño, R. M.; Schnerch, A.; Mitchell, R.; Fiebig-Comyn, A.; Levadoux-Martin, M.; Bhatia, M. (2010). "Direct conversion of human fibroblasts to multilineage blood progenitors". *Nature* **468** (7323): 521–526. doi:10.1038/nature09591. PMID 21057492.

[117] Efe, J. A.; Hilcove, S.; Kim, J.; Zhou, H.; Ouyang, K.; Wang, G.; Chen, J.; Ding, S. (2011). "Conversion of mouse fibroblasts into cardiomyocytes using a direct reprogramming strategy". *Nature Cell Biology* **13** (3): 215–222. doi:10.1038/ncb2164. PMID 21278734.

[118] Lujan, E.; Chanda, S.; Ahlenius, H.; Sudhof, T. C.; Wernig, M. (2012). "Direct conversion of mouse fibroblasts to self-renewing, tripotent neural precursor cells". *Proceedings of the National Academy of Sciences* **109** (7): 2527–2532. doi:10.1073/pnas.1121003109.

[119] Thier, M.; Wörsdörfer, P.; Lakes, Y. B.; Gorris, R.; Herms, S.; Opitz, T.; Seiferling, D.; Quandel, T.; Hoffmann, P.; Nöthen, M. M.; Brüstle, O.; Edenhofer, F. (2012). "Direct Conversion of Fibroblasts into Stably Expandable Neural Stem Cells". *Cell Stem Cell* **10** (4): 473–9. doi:10.1016/j.stem.2012.03.003. PMID 22445518.

[120] Han, D. W.; Tapia, N.; Hermann, A.; Hemmer, K.; Höing, S.; Araúzo-Bravo, M. J.; Zaehres, H.; Wu, G.; Frank, S.; Moritz, S. R.; Greber, B.; Yang, J. H.; Lee, H. T.; Schwamborn, J. C.; Storch, A.; Schöler, H. R. (2012). "Direct Reprogramming of Fibroblasts into Neural Stem Cells by Defined Factors". *Cell Stem Cell* **10** (4): 465–72. doi:10.1016/j.stem.2012.02.021. PMID 22445517.

[121] Taylor, S. M.; Jones, P. A. (1979). "Multiple new phenotypes induced in 10T1/2 and 3T3 cells treated with 5-azacytidine". *Cell* **17** (4): 771–9. doi:10.1016/0092-8674(79)90317-9. PMID 90553.

[122] Lassar, A. B.; Paterson, B. M.; Weintraub, H (1986). "Transfection of a DNA locus that mediates the conversion of 10T1/2 fibroblasts to myoblasts". *Cell* **47** (5): 649–56. doi:10.1016/0092-8674(86)90507-6. PMID 2430720.
Davis, R. L.; Weintraub, H.; Lassar, A. B. (1987). "Expression of a single transfected cDNA converts fibroblasts to myoblasts". *Cell* **51** (6): 987–1000. doi:10.1016/0092-8674(87)90585-x. PMID 3690668.
Weintraub, H; Tapscott, S. J.; Davis, R. L.; Thayer, M. J.; Adam, M. A.; Lassar, A. B.; Miller, A. D. (1989). "Activation of muscle-specific genes in pigment, nerve, fat, liver, and fibroblast cell lines by forced expression of MyoD". *Proceedings of the National Academy of Sciences of the United States of America* **86** (14): 5434–8. doi:10.1073/pnas.86.14.5434. PMC 297637. PMID 2748593.

[123] Vierbuchen, T.; Wernig, M. (2011). "Direct lineage conversions: Unnatural but useful?". *Nature Biotechnology* **29** (10): 892–907. doi:10.1038/nbt.1946. PMID 21997635.

[124] Riddle, M. R.; Weintraub, A.; Nguyen, K. C. Q.; Hall, D. H.; Rothman, J. H. (2013). "Transdifferentiation and remodeling of post-embryonic C. Elegans cells by a single transcription factor". *Development* **140** (24): 4844–9. doi:10.1242/dev.103010. PMID 24257624.

[125] Wei, S; Zou, Q; Lai, S; et al. (2016). "Conversion of embryonic stem cells into extraembryonic lineages by CRISPR-mediated activators". *Sci Rep.* **6**: 19648. doi:10.1038/srep19648.

[126] Jung, D. W.; Williams, D. R. (2011). "Novel Chemically Defined Approach to Produce Multipotent Cells from Terminally Differentiated Tissue Syncytia". *ACS Chemical Biology* **6** (6): 553–62. doi:10.1021/cb2000154. PMID 21322636.
"In This Issue". *ACS Chemical Biology* **7** (4): 619. 2012. doi:10.1021/cb300127f.

[127] Zhang, Hongkai; Wilson, Ian A.; Lerner, Richard A. (2012). "Selection of antibodies that regulate phenotype from intracellular combinatorial antibody libraries". *PNAS* **109** (39): 15728–15733. doi:10.1073/pnas.1214275109. PMID 23019357.

[128] Antibody that Transforms Bone Marrow Stem Cells Directly into Brain Cells
Xie, Jia; Zhang, Hongkai; Yea, Kyungmoo; Lerner, Richard A. (2013). "Autocrine signaling based selection of combinatorial antibodies that transdifferentiate human stem cells". *PNAS* **110**: 8099–8104. doi:10.1073/pnas.1306263110.

[129] Liu, X.; Ory, V.; Chapman, S.; Yuan, H.; Albanese, C.; Kallakury, B.; Timofeeva, O. A.; Nealon, C.; Dakic, A.; Simic, V.; Haddad, B. R.; Rhim, J. S.; Dritschilo, A.; Riegel, A.; McBride, A.; Schlegel, R. (2012). "ROCK Inhibitor and Feeder Cells Induce the Conditional Reprogramming of Epithelial Cells". *The American Journal of Pathology* **180** (2): 599–607. doi:10.1016/j.ajpath.2011.10.036. PMID 22189618.

[130] Rheinwald, J. G.; Green, H (1975). "Serial cultivation of strains of human epidermal keratinocytes: The formation of keratinizing colonies from single cells". *Cell* **6** (3): 331–43. doi:10.1016/S0092-8674(75)80001-8. PMID 1052771.

[131] Hiew, Y.-L. (2011) Examining the biological consequences of DNA damage caused by irradiated J2-3T3 fibroblast feeder cells and HPV16: characterisation of the biological functions of Mll. Doctoral thesis, UCL (University College London)

[132] Szumiel, I. (2012). "Radiation hormesis: Autophagy and other cellular mechanisms". *International Journal of Radiation Biology* **88** (9): 619–28. doi:10.3109/09553002.2012.699698. PMID 22702489.

[133] Kurosawa, H. (2012). "Application of Rho-associated protein kinase (ROCK) inhibitor to human pluripotent stem cells". *Journal of Bioscience and Bioengineering* **114** (6): 577–81. doi:10.1016/j.jbiosc.2012.07.013. PMID 22898436.

[134] Terunuma, A.; Limgala, R. P.; Park, C. J.; Choudhary, I.; Vogel, J. C. (2010). "Efficient Procurement of Epithelial Stem Cells from Human Tissue Specimens Using a Rho-Associated Protein Kinase Inhibitor Y-27632". *Tissue Engineering Part A* **16** (4): 1363–8. doi:10.1089/ten.tea.2009.0339. PMID 19912046.

[135] Sandra Chapman, Xuefeng Liu, Craig Meyers, Richard Schlegel, and Alison A. McBride. (2010) Human keratinocytes are efficiently immortalized by a Rho kinase inhibitor

[136] Suprynowicz, F. A.; Upadhyay, G.; Krawczyk, E.; Kramer, S. C.; Hebert, J. D.; Liu, X.; Yuan, H.; Cheluvaraju, C.; Clapp, P. W.; Boucher, R. C.; Kamonjoh, C. M.; Randell, S. H.; Schlegel, R. (2012). "Conditionally reprogrammed cells represent a stem-like state of adult epithelial cells". *Proceedings of the National Academy of Sciences* **109** (49): 20035–20040. doi:10.1073/pnas.1213241109.

[137] Agarwal, S.; Rimm, D. L. (2012). "Making Every Cell Like He *La*". *The American Journal of Pathology* **180** (2): 443–5. doi:10.1016/j.ajpath.2011.12.001. PMID 22192626.

[138] Lisanti, MP; Tanowitz, HB (Apr 2012). "Translational discoveries, personalized medicine, and living biobanks of the future". *The American Journal of Pathology* **180** (4): 1334–6. doi:10.1016/j.ajpath.2012.02.003. PMID 22453029.

[139] Yuan, H.; Myers, S.; Wang, J.; Zhou, D.; Woo, J. A.; Kallakury, B.; Ju, A.; Bazylewicz, M.; Carter, Y. M.; Albanese, C.; Grant, N.; Shad, A.; Dritschilo, A.; Liu, X.; Schlegel, R. (2012). "Use of Reprogrammed Cells to Identify Therapy for Respiratory Papillomatosis". *New England Journal of Medicine* **367** (13): 1220–7. doi:10.1056/NEJMoa1203055. PMID 23013073.

[140] Crystal, AS; Shaw, AT; Sequist, LV; Friboulet, L; Niederst, MJ; Lockerman, EL; Frias, RL; Gainor, JF; Amzallag, A; Greninger, P; Lee, D; Kalsy, A; Gomez-Caraballo, M; Elamine, L; Howe, E; Hur, W; Lifshits, E; Robinson, HE; Katayama, R; Faber, AC; Awad, MM; Ramaswamy, S; Mino-Kenudson, M; Iafrate, AJ; Benes, CH; Engelman, JA (2014). "Patient-derived models of acquired resistance can identify effective drug combinations for cancer". *Science* **346** (6216): 1480–6. doi:10.1126/science.1254721. PMID 25394791.

[141] Palechor-Ceron, N.; Suprynowicz, F. A.; Upadhyay, G.; Dakic, A.; Minas, T.; Simic, V.; Johnson, M.; Albanese, C.; Schlegel, R.; Liu, X. (2013). "Radiation Induces Diffusible Feeder Cell Factor(s) That Cooperate with ROCK Inhibitor to Conditionally Reprogram and Immortalize Epithelial Cells". *The American Journal of Pathology* **183** (6): 1862–70. doi:10.1016/j.ajpath.2013.08.009. PMID 24096078.

[142] Saenz, F. R.; Ory, V.; AlOtaiby, M.; Rosenfield, S.; Furlong, M.; Cavalli, L. R.; Johnson, M. D.; Liu, X.; Schlegel, R.; Wellstein, A.; Riegel, A. T. (2014). "Conditionally Reprogrammed Normal and Transformed Mouse Mammary Epithelial Cells Display a Progenitor-Cell–Like Phenotype". *PLoS ONE* **9** (5): e97666. doi:10.1371/journal.pone.0097666. PMID 24831228.

[143] Kaur, Sukhbir; Soto-Pantoja, David R.; Stein, Erica V.; et al. (2013). "Thrombospondin-1 Signaling through CD47 Inhibits Self-renewal by Regulating c-Myc and Other Stem Cell Transcription Factors". *Scientific Reports* **3**: 1673. doi:10.1038/srep01673. PMC 3628113. PMID 23591719.

[144] Soto-Pantoja, D. R.; Ridnour, L. A.; Wink, D. A.; Roberts, D. D. (2013). "Blockade of CD47 increases survival of mice exposed to lethal total body irradiation". *Scientific Reports* **3**. doi:10.1038/srep01038.

[145] Gentek, R.; Molawi, K.; Sieweke, M. H. (2014). "Tissue macrophage identity and self-renewal". *Immunological Reviews* **262**: 56–73. doi:10.1111/imr.12224.

[146] Kurian, Leo; Sancho-Martinez, Ignacio; Nivet, Emmanuel; Juan; Izpisua Belmonte, Carlos (2012). "Conversion of human fibroblasts to angioblast-like progenitor cells". *Nature Methods* **10**: 77–83. doi:10.1038/nmeth.2255.

[147] Morris, S. A.; Daley, G. Q. (2013). "A blueprint for engineering cell fate: current technologies to reprogram cell identity". *Cell research* **23** (1): 33–48. doi:10.1038/cr.2013.1. PMID 23277278.

[148] Wang, Y. C.; Nakagawa, M.; Garitaonandia, I.; Slavin, I.; Altun, G.; Lacharite, R. M.; Nazor, K. L.; Tran, H. T.; Lynch, C. L.; Leonardo, T. R.; Liu, Y.; Peterson, S. E.; Laurent, L. C.; Yamanaka, S.; Loring, J. F. (2011). "Specific lectin biomarkers for isolation of human pluripotent stem cells identified through array-based glycomic analysis". *Cell Research* **21** (11): 1551–63. doi:10.1038/cr.2011.148. PMID 21894191.

[149] Thomson, J. A.; Itskovitz-Eldor, J; Shapiro, S. S.; Waknitz, M. A.; Swiergiel, J. J.; Marshall, V. S.; Jones, J. M. (1998). "Embryonic Stem Cell Lines Derived from Human Blastocysts". *Science* **282** (5391): 1145–7. doi:10.1126/science.282.5391.1145. PMID 9804556.

[150] Suila, H.; Hirvonen, T.; Ritamo, I.; Natunen, S.; Tuimala, J.; Laitinen, S.; Anderson, H.; Nystedt, J.; Räbinä, J.; Valmu, L. (2014). "Extracellular O-Linked N-Acetylglucosamine is Enriched in Stem Cells Derived from Human Umbilical Cord Blood". *BioResearch Open Access* **3** (2): 39–44. doi:10.1089/biores.2013.0050. PMID 24804163.

[151] Perdigoto, C. N.; Bardin, A. J. (2013). "Sending the right signal: Notch and stem cells". *Biochimica et Biophysica Acta (BBA) - General Subjects* **1830** (2): 2307–2322. doi:10.1016/j.bbagen.2012.08.009.

[152] Jafar-Nejad, H.; Leonardi, J.; Fernandez-Valdivia, R. (2010). "Role of glycans and glycosyltransferases in the regulation of Notch signaling". *Glycobiology* **20** (8): 931–949. doi:10.1093/glycob/cwq053. PMID 20368670.

[153] Alisson-Silva, Frederico; Deivid; Rodrigues, Carvalho; Vairo, Leandro; et al. (2014). "and Adriane R Todeschini (2014). Evidences for the involvement of cell surface glycans in stem cell pluripotency and differentiation". *Glycobiology* **24** (5): 458–468. doi:10.1093/glycob/cwu012. PMID 24578376.

[154] Hasehira, K.; Tateno, H.; Onuma, Y.; Ito, Y.; Asashima, M.; Hirabayashi, J. (2012). "Structural and Quantitative Evidence for Dynamic Glycome Shift on Production of Induced Pluripotent Stem Cells". *Molecular & Cellular Proteomics* **11** (12): 1913–1923. doi:10.1074/mcp.M112.020586.

[155] Becker Kojic´, Z. A. (2002). "A Novel Human Erythrocyte Glycosylphosphatidylinositol (GPI)-anchored Glycoprotein ACA. ISOLATION, PURIFICATION, PRIMARY STRUCTURE DETERMINATION, AND MOLECULAR PARAMETERS OF ITS LIPID STRUCTURE". *Journal of Biological Chemistry* **277** (43): 40472–8. doi:10.1074/jbc.M202416200. PMID 12167612.

[156] Becker-Kojić, Z. A.; Ureña-Peralta, J. R.; Saffrich, R; Rodriguez-Jiménez, F. J.; Rubio, M. P.; Rios, P; Romero, A; Ho, A. D.; Stojković, M (2013). "A novel human glycoprotein ACA is an upstream regulator of human hematopoiesis". *Bulletin of experimental biology and medicine* **155** (4): 536–51. doi:10.1007/s10517-013-2195-0. PMID 24143385.

[157] Becker-Kojić, ZA; Ureña-Peralta, JR; Zipanćić, I.; Stojković, M. (2013). "Activation of surface glycoprotein ACA induced pluripotent hematopoietic progenitor cells" (PDF). *Cell Technologies in Biology and Medicine* **9** (2): 85–101.

[158] Mikkola, M. (2013) Human pluripotent stem cells: glycomic approaches for culturing and characterization.ISBN 978-952-10-8444-7

[159] Redmer, T.; Diecke, S.; Grigoryan, T.; Quiroga-Negreira, A.; Birchmeier, W.; Besser, D. (2011). "E-cadherin is crucial for embryonic stem cell pluripotency and can replace OCT4 during somatic cell reprogramming". *EMBO Reports* **12** (7): 720–6. doi:10.1038/embor.2011.88. PMID 21617704.

[160] Bedzhov, I.; Alotaibi, H.; Basilicata, M. F.; Ahlborn, K.; Liszewska, E.; Brabletz, T.; Stemmler, M. P. (2013). "Adhesion, but not a specific cadherin code, is indispensable for ES cell and induced pluripotency". *Stem Cell Research* **11** (3): 1250–63. doi:10.1016/j.scr.2013.08.009. PMID 24036274.

[161] Su, G.; Zhao, Y.; Wei, J.; Xiao, Z.; Chen, B.; Han, J.; Chen, L.; Guan, J.; Wang, R.; Dong, Q.; Dai, J. (2013). "Direct conversion of fibroblasts into neural progenitor-like cells by forced growth into 3D spheres on low attachment surfaces". *Biomaterials* **34** (24): 5897–906. doi:10.1016/j.biomaterials.2013.04.040. PMID 23680365.

[162] Downing, T. L.; Soto, J.; Morez, C.; Houssin, T.; Fritz, A.; Yuan, F.; Chu, J.; Patel, S.; Schaffer, D. V.; Li, S. (2013). "Biophysical regulation of epigenetic state and

cell reprogramming". *Nature Materials* **12** (12): 1154–62. doi:10.1038/nmat3777. PMID 24141451.

[163] Sun, Yubing; Koh; Aw Yong, Meng; Villa-Diaz, Luis G.; Fu, Jianping (2014). "Hippo/YAP-mediated rigidity-dependent motor neuron differentiation of human pluripotent stem cells". *Nature Materials* **13**: 599–604. doi:10.1038/nmat3945.

[164] Murray, P.; Prewitz, M.; Hopp, I.; Wells, N.; Zhang, H.; Cooper, A.; Parry, K. L.; Short, R.; Antoine, D. J.; Edgar, D. (2013). "The self-renewal of mouse embryonic stem cells is regulated by cell–substratum adhesion and cell spreading". *The International Journal of Biochemistry & Cell Biology* **45** (11): 2698–2705. doi:10.1016/j.biocel.2013.07.001.

[165] Guilak, F.; Cohen, D. M.; Estes, B. T.; et al. (2009). "Control of stem cell fate by physical interactions with the extracellular matrix". *Cell stem cell* **5** (1): 17–26. doi:10.1016/j.stem.2009.06.016. PMC 2768283. PMID 19570510.

[166] Worley, K.; Certo, A.; Wan, L. Q. (2012). "Geometry–Force Control of Stem Cell Fate". *BioNanoScience* **3**: 43–51. doi:10.1007/s12668-012-0067-0.

[167] Caiazzo, Massimiliano; Okawa, Yuya; Ranga, Adrian; Piersigilli, Alessandra; Tabata, Yoji; Lutolf, Matthias P. (2016). "Defined three-dimensional microenvironments boost induction of pluripotency". *Nature Materials*. doi:10.1038/nmat4536.

[168] Squeezing cells into stem cells. ScienceDaily, 11 January 2016

[169] Singh, A.; Suri, S.; Lee, T.; Chilton, J. M.; Cooke, M. T.; Chen, W.; Fu, J.; Stice, S. L.; Lu, H.; McDevitt, T. C.; Garcia, A. S. J. (2013). "Adhesion strength–based, label-free isolation of human pluripotent stem cells". *Nature Methods* **10** (5): 438–44. doi:10.1038/nmeth.2437. PMID 23563795.

[170] Wang, Kainan; Degerny, Cindy; Xu, Minghong; Yang, Xiang-Jiao (2009). "YAP, TAZ, and Yorkie: A conserved family of signal-responsive transcriptional coregulators in animal development and human disease". *Biochemistry and Cell Biology* **87** (1): 77–91. doi:10.1139/O08-114. PMID 19234525.

[171] Yang, C.; Tibbitt, M.W.; Basta, L.; Anseth, K.S. (2014). "Mechanical memory and dosing influence stem cell fate". *Nature Materials* **13**: 645–652. doi:10.1038/nmat3889.

[172] Nampe, D.; Tsutsui, H. (2013). "Engineered Micromechanical Cues Affecting Human Pluripotent Stem Cell Regulations and Fate". *Journal of laboratory automation* **18** (6): 482–493. doi:10.1177/2211068213503156. PMID 24062363.

[173] Zhang, W.; Duan, S.; Li, Y.; Xu, X.; Qu, J.; Zhang, W.; Liu, G. H. (2012). "Converted neural cells: Induced to a cure?". *Protein & Cell* **3** (2): 91–97. doi:10.1007/s13238-012-2029-2.

[174] Yang, N.; Ng, Y. H.; Pang, Z. P.; Südhof, T. C.; Wernig, M. (2011). "Induced Neuronal Cells: How to Make and Define a Neuron". *Cell Stem Cell* 9 (6): 517–25. doi:10.1016/j.stem.2011.11.015. PMID 22136927.

[175] Sheng, C.; Zheng, Q.; Wu, J.; Xu, Z.; Sang, L.; Wang, L.; Guo, C.; Zhu, W.; Tong, M.; Liu, L.; Li, W.; Liu, Z. H.; Zhao, X. Y.; Wang, L.; Chen, Z.; Zhou, Q. (2012). "Generation of dopaminergic neurons directly from mouse fibroblasts and fibroblast-derived neural progenitors". *Cell Research* 22 (4): 769–72. doi:10.1038/cr.2012.32. PMID 22370632.

[176] Maucksch, C.; Firmin, E.; et al. (2012). "Non-viral generation of neural precursor-like cells from adult human fibroblasts". *J Stem Cells Regen Med* 8 (3): 1–9.

[177] Ring, K. L.; Tong, L. M.; Balestra, M. E.; Javier, R.; Andrews-Zwilling, Y.; Li, G.; Walker, D.; Zhang, W. R.; Kreitzer, A. C.; Huang, Y. (2012). "Direct Reprogramming of Mouse and Human Fibroblasts into Multipotent Neural Stem Cells with a Single Factor". *Cell Stem Cell* 11 (1): 100–9. doi:10.1016/j.stem.2012.05.018. PMC 3399516. PMID 22683203.

[178] Generation of neural progenitor cells by chemical cocktails and hypoxia Cheng, L.; Hu, W.; Qiu, B.; Zhao, J.; Yu, Y.; Guan, W.; Wang, M.; Yang, W.; Pei, G. (2014). "Generation of neural progenitor cells by chemical cocktails and hypoxia". *Cell Research* 24 (6): 665–79. doi:10.1038/cr.2014.32. PMID 24638034.

[179] Liu, G. H.; Yi, F.; Suzuki, K.; Qu, J.; Belmonte, J. C. I. (2012). "Induced neural stem cells: A new tool for studying neural development and neurological disorders". *Cell Research* 22 (7): 1087–91. doi:10.1038/cr.2012.73. PMID 22547025.

[180] Torper, O.; Pfisterer, U.; Wolf, D. A.; Pereira, M.; Lau, S.; Jakobsson, J.; Bjorklund, A.; Grealish, S.; Parmar, M. (2013). "Generation of induced neurons via direct conversion in vivo". *Proceedings of the National Academy of Sciences* 110 (17): 7038–7043. doi:10.1073/pnas.1303829110.

[181] Niu, W.; Zang, T.; Zou, Y.; Fang, S.; Smith, D. K.; Bachoo, R.; Zhang, C. L. (2013). "In vivo reprogramming of astrocytes to neuroblasts in the adult brain". *Nature Cell Biology* 15 (10): 1164–75. doi:10.1038/ncb2843. PMID 24056302.

[182] Su, Z.; Niu, W.; Liu, M. L.; Zou, Y.; Zhang, C. L. (2014). "In vivo conversion of astrocytes to neurons in the injured adult spinal cord". *Nature Communications* 5. doi:10.1038/ncomms4338.

[183] Najm, F. J.; Lager, A. M.; Zaremba, A.; Wyatt, K.; Caprariello, A. V.; Factor, D. C.; Karl, R. T.; Maeda, T.; Miller, R. H.; Tesar, P. J. (2013). "Transcription factor–mediated reprogramming of fibroblasts to expandable, myelinogenic oligodendrocyte progenitor cells". *Nature Biotechnology* 31 (5): 426–33. doi:10.1038/nbt.2561. PMID 23584611.

[184] Yang, N.; Zuchero, J. B.; Ahlenius, H.; Marro, S.; Ng, Y. H.; Vierbuchen, T.; Hawkins, J. S.; Geissler, R.; Barres, B. A.; Wernig, M. (2013). "Generation of oligodendroglial cells by direct lineage conversion". *Nature Biotechnology* 31 (5): 434–9. doi:10.1038/nbt.2564. PMID 23584610.

[185] Xu, C. (2012). "Turning cardiac fibroblasts into cardiomyocytes in vivo". *Trends in Molecular Medicine* 18 (10): 575–6. doi:10.1016/j.molmed.2012.06.009. PMID 22770847.

[186] Fu, J. D.; Stone, N. R.; Liu, L.; Spencer, C. I.; Qian, L.; Hayashi, Y.; Delgado-Olguin, P.; Ding, S.; Bruneau, B. G.; Srivastava, D. (2013). "Direct Reprogramming of Human Fibroblasts toward a Cardiomyocyte-like State". *Stem Cell Reports* 1 (3): 235–47. doi:10.1016/j.stemcr.2013.07.005. PMID 24319660.

[187] Chen, J. X.; Krane, M.; Deutsch, M. -A.; Wang, L.; Rav-Acha, M.; Gregoire, S.; Engels, M. C.; Rajarajan, K.; Karra, R.; Abel, E. D.; Wu, J. C.; Milan, D.; Wu, S. M. (2012). "Inefficient Reprogramming of Fibroblasts into Cardiomyocytes Using Gata4, Mef2c, and Tbx5". *Circulation Research* 111 (1): 50–5. doi:10.1161/CIRCRESAHA.112.270264. PMC 3390172. PMID 22581928.

[188] Burridge, P. W.; Keller, G.; Gold, J. D.; Wu, J. C. (2012). "Production of De Novo Cardiomyocytes: Human Pluripotent Stem Cell Differentiation and Direct Reprogramming". *Cell Stem Cell* 10 (1): 16–28. doi:10.1016/j.stem.2011.12.013. PMC 3255078. PMID 22226352.

[189] Wang, H.; Cao, N.; Spencer, C. I.; Nie, B.; Ma, T.; Xu, T.; Zhang, Y.; Wang, X.; Srivastava, D.; Ding, S. (2014). "Small Molecules Enable Cardiac Reprogramming of Mouse Fibroblasts with a Single Factor, Oct4". *Cell Reports* 6 (5): 951–60. doi:10.1016/j.celrep.2014.01.038. PMC 4004339. PMID 24561253.

[190] Carpenter, L.; Carr, C.; Yang, C. T.; Stuckey, D. J.; Clarke, K.; Watt, S. M. (2012). "Efficient Differentiation of Human Induced Pluripotent Stem Cells Generates Cardiac Cells That Provide Protection Following Myocardial Infarction in the Rat". *Stem Cells and Development* 21 (6): 977–86. doi:10.1089/scd.2011.0075. PMID 22182484.

[191] Yamada, S.; Nelson, T. J.; Kane, G. C.; Martinez-Fernandez, A.; Crespo-Diaz, R. J.; Ikeda, Y.; Perez-Terzic, C.; Terzic, A. (2013). "IPS Cell Intervention Rescues Wall Motion Disparity Achieving Biological Cardiac Resynchronization Post-Infarction". *The Journal of Physiology* 591 (17): 4335–4349. doi:10.1113/jphysiol.2013.252288.

[192] Lian, X.; Hsiao, C.; Wilson, G.; Zhu, K.; Hazeltine, L. B.; Azarin, S. M.; Raval, K. K.; Zhang, J.; Kamp, T. J.; Palecek, S. P. (2012). "Cozzarelli Prize Winner: Robust cardiomyocyte differentiation from human pluripotent stem cells via temporal modulation of canonical Wnt signaling". *Proceedings of the National Academy of Sciences* 109 (27): E1848. doi:10.1073/pnas.1200250109.

[193] Willems, E.; Cabral-Teixeira, J.; Schade, D.; Cai, W.; Reeves, P.; Bushway, P. J.; Lanier, M.; Walsh, C.; Kirchhausen, T.; Izpisua Belmonte, J. C.; Cashman, J.; Mercola, M. (2012). "Small Molecule-Mediated TGF-β Type II Receptor Degradation Promotes Cardiomyogenesis in Embryonic Stem Cells". *Cell Stem Cell* **11** (2): 242–52. doi:10.1016/j.stem.2012.04.025. PMID 22862949.

[194] Lu, T. Y.; Lin, B.; Kim, J.; Sullivan, M.; Tobita, K.; Salama, G.; Yang, L. (2013). "Repopulation of decellularized mouse heart with human induced pluripotent stem cell-derived cardiovascular progenitor cells". *Nature Communications* **4**. doi:10.1038/ncomms3307.

[195] Budniatzky, I.; Gepstein, L. (2014). "Concise Review: Reprogramming Strategies for Cardiovascular Regenerative Medicine: From Induced Pluripotent Stem Cells to Direct Reprogramming". *Stem cells translational medicine* **3** (4): 448–457. doi:10.5966/sctm.2013-0163. PMID 24591731.

[196] Cosgrove, B. D.; Gilbert, P. M.; Porpiglia, E.; Mourkioti, F.; Lee, S. P.; Corbel, S. Y.; Blau, H. M. (2014). "Rejuvenation of the muscle stem cell population restores strength to injured aged muscles". *Nature Medicine* **20** (3): 255–264. doi:10.1038/nm.3464. PMC 3949152. PMID 24531378.

[197] Sousa-Victor, P.; Gutarra, S.; Garcia-Prat, L.; Rodriguez-Ubreva, J.; Ortet, L.; Ruiz-Bonilla, V.; Jardi, M.; Ballestar, E.; González, S.; Serrano, A. L.; Perdiguero, E.; Muñoz-Cánoves, P. (2014). "Geriatric muscle stem cells switch reversible quiescence into senescence". *Nature* **506** (7488): 316–21. doi:10.1038/nature13013. PMID 24522534.

[198] Hosoyama; et al. "and Masatoshi Suzuki (March, 2014). Derivation of Myogenic Progenitors Directly From Human Pluripotent Stem Cells Using a Sphere-Based Culture". *Stem Cells Trans Med.* **3**: 564–574. doi:10.5966/sctm.2013-0143.

[199] Zhu, S.; Rezvani, M.; Harbell, J.; Mattis, A. N.; Wolfe, A. R.; Benet, L. Z.; Willenbring, H.; Ding, S. (2014). "Mouse liver repopulation with hepatocytes generated from human fibroblasts". *Nature* **508** (7494): 93–7. doi:10.1038/nature13020. PMID 24572354.

[200] Abdelalim, E. M.; Bonnefond, A.; Bennaceur-Griscelli, A.; Froguel, P. (2014). "Pluripotent Stem Cells as a Potential Tool for Disease Modelling and Cell Therapy in Diabetes". *Stem Cell Reviews and Reports* **10**: 327–337. doi:10.1007/s12015-014-9503-6.

[201] Hrvatin, S.; O'Donnell, C. W.; Deng, F.; et al. (2014). "Differentiated human stem cells resemble fetal, not adult, β cells". *Proceedings of the National Academy of Sciences* **111** (8): 3038–3043. doi:10.1073/pnas.1400709111.

[202] Zhu, Saiyong; Russ, Holger A.; Wang, Xiaojing; Zhang, Mingliang; Ma, Tianhua; Xu, Tao; Tang, Shibing; Hebrok, Matthias; Ding, Sheng (2016). "Human pancreatic beta-like cells converted from fibroblasts". *Nature Communications* **7**: 10080. doi:10.1038/ncomms10080.

[203] Akinci, E; Banga, A; Tungatt, K; et al. (2013). "Reprogramming of Various Cell Types to a Beta-Like State by Pdx1, Ngn3 and MafA". *PLoS ONE* **8** (11): e82424. doi:10.1371/journal.pone.0082424.

[204] Chen, Y. J.; Finkbeiner, S. R.; Weinblatt, D.; Stanger, B. Z. (2014). "De Novo Formation of Insulin-Producing "Neo-β Cell Islets" from Intestinal Crypts". *Cell Reports* **6**: 1046–1058. doi:10.1016/j.celrep.2014.02.013.

[205] Hendry, C. E.; Vanslambrouck, J. M.; Ineson, J.; Suhaimi, N.; Takasato, M.; Rae, F.; Little, M. H. (2013). "Direct Transcriptional Reprogramming of Adult Cells to Embryonic Nephron Progenitors". *Journal of the American Society of Nephrology* **24** (9): 1424–34. doi:10.1681/ASN.2012121143. PMID 23766537.

[206] Xinaris, C; Benedetti, V; Rizzo, P; et al. (2012). "In vivo maturation of functional renal organoids formed from embryonic cell suspensions" (PDF). *J Am Soc Nephrol* **23** (11): 1857–1868. doi:10.1681/ASN.2012050505. PMID 23085631.

[207] Yin, L.; Ohanyan, V.; Fen Pung, Y.; Delucia, A.; Bailey, E.; Enrick, M.; Stevanov, K.; Kolz, C. L.; Guarini, G.; Chilian, W. M. (2011). "Induction of Vascular Progenitor Cells from Endothelial Cells Stimulates Coronary Collateral Growth". *Circulation Research* **110** (2): 241–52. doi:10.1161/CIRCRESAHA.111.250126. PMID 22095729.

[208] Quijada, P.; Toko, H.; Fischer, K. M.; Bailey, B.; Reilly, P.; Hunt, K. D.; Gude, N. A.; Avitabile, D.; Sussman, M. A. (2012). "Preservation of Myocardial Structure is Enhanced by Pim-1 Engineering of Bone Marrow Cells". *Circulation Research* **111** (1): 77–86. doi:10.1161/CIRCRESAHA.112.265207. PMC 3398618. PMID 22619278.

[209] Mohsin, S.; Khan, M.; Toko, H.; Bailey, B.; Cottage, C. T.; Wallach, K.; Nag, D.; Lee, A.; Siddiqi, S.; Lan, F.; Fischer, K. M.; Gude, N.; Quijada, P.; Avitabile, D.; Truffa, S.; Collins, B.; Dembitsky, W.; Wu, J. C.; Sussman, M. A. (2012). "Human Cardiac Progenitor Cells Engineered with Pim-I Kinase Enhance Myocardial Repair". *Journal of the American College of Cardiology* **60** (14): 1278–87. doi:10.1016/j.jacc.2012.04.047. PMID 22841153.

[210] American Heart Association (2012, July 25). Adult stem cells from liposuction used to create blood vessels in the lab. ScienceDaily.

[211] Wang, Z. Z.; Au, P.; Chen, T.; Shao, Y.; Daheron, L. M.; Bai, H.; Arzigian, M.; Fukumura, D.; Jain, R. K.; Scadden, D. T. (2007). "Endothelial cells derived from human embryonic stem cells form durable blood vessels in vivo". *Nature Biotechnology* **25** (3): 317–8. doi:10.1038/nbt1287. PMID 17322871.

[212] Samuel, R.; Daheron, L.; Liao, S.; Vardam, T.; Kamoun, W. S.; Batista, A.; Buecker, C.; Schafer, R.; Han, X.; Au, P.; Scadden, D. T.; Duda, D. G.; Fukumura, D.; Jain, R. K.

(2013). "Generation of functionally competent and durable engineered blood vessels from human induced pluripotent stem cells". *Proceedings of the National Academy of Sciences* **110** (31): 12774–12779. doi:10.1073/pnas.1310675110.

[213] Zangi, L.; Lui, K. O.; von Gise, A.; Ma, Q.; Ebina, W.; Ptaszek, L. M.; Später, D.; Xu, H.; Tabebordbar, M.; Gorbatov, R.; Sena, B.; Nahrendorf, M.; Briscoe, D. M.; Li, R. A.; Wagers, A. J.; Rossi, D. J.; Pu, W. T.; Chien, K. R. (2013). "Modified mRNA directs the fate of heart progenitor cells and induces vascular regeneration after myocardial infarction". *Nature Biotechnology* **31** (10): 898–907. doi:10.1038/nbt.2682. PMID 24013197.

[214] Zeuner, A.; Martelli, F.; Vaglio, S.; Federici, G.; Whitsett, C.; Migliaccio, A. R. (2012). "Concise Review: Stem Cell-Derived Erythrocytes as Upcoming Players in Blood Transfusion". *Stem Cells* **30** (8): 1587–96. doi:10.1002/stem.1136. PMID 22644674.

[215] Hirose, S. I.; Takayama, N.; Nakamura, S.; Nagasawa, K.; Ochi, K.; Hirata, S.; Yamazaki, S.; Yamaguchi, T.; Otsu, M.; Sano, S.; Takahashi, N.; Sawaguchi, A.; Ito, M.; Kato, T.; Nakauchi, H.; Eto, K. (2013). "Immortalization of Erythroblasts by c-MYC and BCL-XL Enables Large-Scale Erythrocyte Production from Human Pluripotent Stem Cells". *Stem Cell Reports* **1** (6): 499–508. doi:10.1016/j.stemcr.2013.10.010. PMID 24371805.

[216] Giarratana, M. -C.; Rouard, H.; Dumont, A.; Kiger, L.; Safeukui, I.; Le Pennec, P. -Y.; Francois, S.; Trugnan, G.; Peyrard, T.; Marie, T.; Jolly, S.; Hebert, N.; Mazurier, C.; Mario, N.; Harmand, L.; Lapillonne, H.; Devaux, J. -Y.; Douay, L. (2011). "Proof of principle for transfusion of in vitro-generated red blood cells". *Blood* **118** (19): 5071–9. doi:10.1182/blood-2011-06-362038. PMID 21885599.

[217] Kobari, L.; Yates, F.; Oudrhiri, N.; Francina, A.; Kiger, L.; Mazurier, C.; Rouzbeh, S.; El-Nemer, W.; Hebert, N.; Giarratana, M. -C.; Francois, S.; Chapel, A.; Lapillonne, H.; Luton, D.; Bennaceur-Griscelli, A.; Douay, L. (2012). "Human induced pluripotent stem cells can reach complete terminal maturation: In vivo and in vitro evidence in the erythropoietic differentiation model". *Haematologica* **97** (12): 1795–803. doi:10.3324/haematol.2011.055566. PMID 22733021.

[218] Keerthivasan, G.; Wickrema, A.; Crispino, J. D. (2011). "Erythroblast Enucleation". *Stem Cells International* **2011**: 1–9. doi:10.4061/2011/139851.

[219] Smith, B. W.; Rozelle, S. S.; Leung, A.; Ubellacker, J.; Parks, A.; Nah, S. K.; French, D.; Gadue, P.; Monti, S.; Chui, D. H. K.; Steinberg, M. H.; Frelinger, A. L.; Michelson, A. D.; Theberge, R.; McComb, M. E.; Costello, C. E.; Kotton, D. N.; Mostoslavsky, G.; Sherr, D. H.; Murphy, G. J. (2013). "The aryl hydrocarbon receptor directs hematopoietic progenitor cell expansion and differentiation". *Blood* **122** (3): 376–85. doi:10.1182/blood-2012-11-466722. PMID 23723449.

[220] Shah, Siddharth; Huang, Xiaosong; Cheng, Linzhao (2014). "Stem Cell-Based Approaches to Red Blood Cell Production for Transfusion". *Stem Cells Trans Med* **3** (3): 346–355. doi:10.5966/sctm.2013-0054.

[221] Scientific Breakthrough as Artificial Blood is Created from Stem Cells

[222] Figueiredo, C. A.; Goudeva, L.; Horn, P. A.; Eiz-Vesper, B.; Blasczyk, R.; Seltsam, A. (2010). "Generation of HLA-deficient platelets from hematopoietic progenitor cells". *Transfusion* **50** (8): 1690–701. doi:10.1111/j.1537-2995.2010.02644.x. PMID 20412529.

[223] Nakamura, S.; Takayama, N.; Hirata, S.; Seo, H.; Endo, H.; Ochi, K.; Fujita, K. I.; Koike, T.; Harimoto, K. I.; Dohda, T.; Watanabe, A.; Okita, K.; Takahashi, N.; Sawaguchi, A.; Yamanaka, S.; Nakauchi, H.; Nishimura, S.; Eto, K. (2014). "Expandable Megakaryocyte Cell Lines Enable Clinically Applicable Generation of Platelets from Human Induced Pluripotent Stem Cells". *Cell Stem Cell* **14** (4): 535–48. doi:10.1016/j.stem.2014.01.011. PMID 24529595.

[224] Riddell, S. R.; Greenberg, P. D. (1995). "Principles for Adoptive T Cell Therapy of Human Viral Diseases". *Annual Review of Immunology* **13**: 545–86. doi:10.1146/annurev.iy.13.040195.002553. PMID 7612234.

[225] Nishimura, T.; Kaneko, S.; Kawana-Tachikawa, A.; Tajima, Y.; Goto, H.; Zhu, D.; Nakayama-Hosoya, K.; Iriguchi, S.; Uemura, Y.; Shimizu, T.; Takayama, N.; Yamada, D.; Nishimura, K.; Ohtaka, M.; Watanabe, N.; Takahashi, S.; Iwamoto, A.; Koseki, H.; Nakanishi, M.; Eto, K.; Nakauchi, H. (2013). "Generation of Rejuvenated Antigen-Specific T Cells by Reprogramming to Pluripotency and Redifferentiation". *Cell Stem Cell* **12** (1): 114–26. doi:10.1016/j.stem.2012.11.002. PMID 23290140.

[226] Vizcardo, R.; Masuda, K.; Yamada, D.; Ikawa, T.; Shimizu, K.; Fujii, S. I.; Koseki, H.; Kawamoto, H. (2013). "Regeneration of Human Tumor Antigen-Specific T Cells from iPSCs Derived from Mature CD8+ T Cells". *Cell Stem Cell* **12** (1): 31–6. doi:10.1016/j.stem.2012.12.006. PMID 23290135.

[227] Lei, F.; Haque, R.; Xiong, X.; Song, J. (2012). "Directed Differentiation of Induced Pluripotent Stem Cells towards T Lymphocytes". *Journal of Visualized Experiments* (63). doi:10.3791/3986.

[228] Sadelain, M; Brentjens, R; Rivière, I (2013). "The basic principles of chimeric antigen receptor design". *Cancer Discovery* **3** (4): 388–98. doi:10.1158/2159-8290.CD-12-0548. PMC 3667586. PMID 23550147.

[229] Themeli, M.; Kloss, C. C.; Ciriello, G.; Fedorov, V. D.; Perna, F.; Gonen, M.; Sadelain, M. (2013). "Generation of tumor-targeted human T lymphocytes from induced pluripotent stem cells for cancer therapy". *Nature Biotechnology* **31** (10): 928–33. doi:10.1038/nbt.2678. PMID 23934177.

[230] Pilones, K. A.; Aryankalayil, J.; Demaria, S. (2012). "Invariant NKT Cells as Novel Targets for Immunotherapy in Solid Tumors". *Clinical and Developmental Immunology* **2012**: 1–11. doi:10.1155/2012/720803.

[231] Watarai, H.; Yamada, D.; Fujii, S. I.; Taniguchi, M.; Koseki, H. (2012). "Induced pluripotency as a potential path towards iNKT cell-mediated cancer immunotherapy". *International Journal of Hematology* **95** (6): 624–31. doi:10.1007/s12185-012-1091-0. PMID 22592322.

[232] Haruta, M.; Tomita, Y.; Yuno, A.; Matsumura, K.; Ikeda, T.; Takamatsu, K.; Haga, E.; Koba, C.; Nishimura, Y.; Senju, S. (2012). "TAP-deficient human iPS cell-derived myeloid cell lines as unlimited cell source for dendritic cell-like antigen-presenting cells". *Gene Therapy* **20** (5): 504–13. doi:10.1038/gt.2012.59. PMID 22875043.

[233] Xie, H.; Ye, M.; Feng, R.; Graf, T. (2004). "Stepwise Reprogramming of B Cells into Macrophages". *Cell* **117** (5): 663–76. doi:10.1016/S0092-8674(04)00419-2. PMID 15163413.
Bussmann, L. H.; Schubert, A.; Vu Manh, T. P.; De Andres, L.; Desbordes, S. C.; Parra, M.; Zimmermann, T.; Rapino, F.; Rodriguez-Ubreva, J.; Ballestar, E.; Graf, T. (2009). "A Robust and Highly Efficient Immune Cell Reprogramming System". *Cell Stem Cell* **5** (5): 554–66. doi:10.1016/j.stem.2009.10.004. PMID 19896445.

[234] Hanna, J.; Markoulaki, S.; Schorderet, P.; Carey, B. W.; Beard, C.; Wernig, M.; Creyghton, M. P.; Steine, E. J.; Cassady, J. P.; Foreman, R.; Lengner, C. J.; Dausman, J. A.; Jaenisch, R. (2008). "Direct Reprogramming of Terminally Differentiated Mature B Lymphocytes to Pluripotency". *Cell* **133** (2): 250–64. doi:10.1016/j.cell.2008.03.028. PMID 18423197.

[235] Di Stefano, B.; Sardina, J. L.; Van Oevelen, C.; Collombet, S.; Kallin, E. M.; Vicent, G. P.; Lu, J.; Thieffry, D.; Beato, M.; Graf, T. (2013). "C/EBPα poises B cells for rapid reprogramming into induced pluripotent stem cells". *Nature* **506** (7487): 235–9. doi:10.1038/nature12885. PMID 24336202.

[236] Rapino, F.; Robles, E. F.; Richter-Larrea, J. A.; Kallin, E. M.; Martinez-Climent, J. A.; Graf, T. (2013). "C/EBPα Induces Highly Efficient Macrophage Transdifferentiation of B Lymphoma and Leukemia Cell Lines and Impairs Their Tumorigenicity". *Cell Reports* **3** (4): 1153–63. doi:10.1016/j.celrep.2013.03.003. PMID 23545498.

[237] Guo, J.; Feng, Y.; Barnes, P.; Huang, F. F.; Idell, S.; Su, D. M.; Shams, H. (2012). "Deletion of FoxN1 in the thymic medullary epithelium reduces peripheral T cell responses to infection and mimics changes of aging". *PLoS ONE* **7** (4): e34681. doi:10.1371/journal.pone.0034681.

[238] Sun, L.; Guo, J.; Brown, R.; Amagai, T.; Zhao, Y.; Su, D.-M. (2010). "Declining expression of a single epithelial cell-autonomous gene accelerates age-related thymic involution". *Aging Cell* **9** (3): 347–357. doi:10.1111/j.1474-9726.2010.00559.x. PMID 20156205.

[239] Bredenkamp, Nicholas; Nowell, Craig S.; Blackburn, C. Clare (2014). "Regeneration of the aged thymus by a single transcription factor". *Development* **141** (8): 1627–1637. doi:10.1242/dev.103614. PMID 24715454.

[240] Peng, Y.; Huang, S.; Cheng, B.; Nie, X.; Enhe, J.; Feng, C.; Fu, X. (2013). "Mesenchymal stem cells: A revolution in therapeutic strategies of age-related diseases". *Ageing Research Reviews* **12** (1): 103–15. doi:10.1016/j.arr.2012.04.005. PMID 22569401.

[241] Bieback, K; Kern, S; Kocaömer, A; Ferlik, K; Bugert, P (2008). "Comparing mesenchymal stromal cells from different human tissues: Bone marrow, adipose tissue and umbilical cord blood". *Bio-medical materials and engineering* **18** (1 Suppl): S71–6. PMID 18334717.

[242] Efimenko, A.; Dzhoyashvili, N.; Kalinina, N.; Kochegura, T.; Akchurin, R.; Tkachuk, V.; Parfyonova, Y. (2013). "Adipose-Derived Mesenchymal Stromal Cells from Aged Patients with Coronary Artery Disease Keep Mesenchymal Stromal Cell Properties but Exhibit Characteristics of Aging and Have Impaired Angiogenic Potential". *Stem Cells Translational Medicine* **3** (1): 32–41. doi:10.5966/sctm.2013-0014. PMC 3902283. PMID 24353175.

[243] Stolzing, A; Jones, E; McGonagle, D; Scutt, A (2008). "Age-related changes in human bone marrow-derived mesenchymal stem cells: Consequences for cell therapies". *Mechanisms of Ageing and Development* **129** (3): 163–73. doi:10.1016/j.mad.2007.12.002. PMID 18241911.

[244] Irina Eberle, Mohsen Moslem, Reinhard Henschler, Tobias Cantz (2012) Engineered MSCs from Patient-Specific iPS Cells. Advances in Biochemical Engineering Biotechnology

[245] Chen, Y. S.; Pelekanos, R. A.; Ellis, R. L.; Horne, R.; Wolvetang, E. J.; Fisk, N. M. (2012). "Small Molecule Mesengenic Induction of Human Induced Pluripotent Stem Cells to Generate Mesenchymal Stem/Stromal Cells". *Stem Cells Translational Medicine* **1** (2): 83–95. doi:10.5966/sctm.2011-0022. PMID 23197756.

[246] Hynes, K; Menicanin, D; Han, J; Marino, V; Mrozik, K; Gronthos, S; Bartold, P. M. (2013). "Mesenchymal stem cells from iPS cells facilitate periodontal regeneration". *Journal of Dental Research* **92** (9): 833–9. doi:10.1177/0022034513498258. PMID 23884555.

[247] iPSC for Dental Tissue Regeneration

[248] Lai, Ruenn Chai; Yeo, Ronne Wee Yeh; Tan, Soon Sim; Zhang, Bin; Yin, Yijun; Sze, Newman Siu Kwan; Choo, Andre; Lim, Sai Kiang (2013). "Mesenchymal Stem Cell Therapy". *Mesenchymal Stem Cell Therapy*: 39–61. doi:10.1007/978-1-62703-200-1_3. ISBN 978-1-62703-199-8. |chapter= ignored (help)
Lai, R. C.; Yeo, R. W. Y.; Tan, K. H.; Lim, S. K. (2013). "Exosomes for drug delivery — a novel application for the mesenchymal stem cell". *Biotechnology Advances* **31** (5): 543–51. doi:10.1016/j.biotechadv.2012.08.008. PMID 22959595.

Kosaka, N.; Takeshita, F.; Yoshioka, Y.; Hagiwara, K.; Katsuda, T.; Ono, M.; Ochiya, T. (2013). "Exosomal tumor-suppressive microRNAs as novel cancer therapy". *Advanced Drug Delivery Reviews* **65** (3): 376–82. doi:10.1016/j.addr.2012.07.011. PMID 22841506.

[249] Jumabay, M.; Abdmaulen, R.; Ly, A.; Cubberly, M. R.; Shahmirian, L. J.; Heydarkhan-Hagvall, S.; Dumesic, D. A.; Yao, Y.; Bostrom, K. I. (2014). "Pluripotent Stem Cells Derived from Mouse and Human White Mature Adipocytes". *Stem Cells Translational Medicine* **3** (2): 161–71. doi:10.5966/sctm.2013-0107. PMID 24396033.

[250] Poloni, A.; Maurizi, G.; Leoni, P.; Serrani, F.; Mancini, S.; Frontini, A.; Zingaretti, M. C.; Siquini, W.; Sarzani, R.; Cinti, S. (2012). "Human Dedifferentiated Adipocytes Show Similar Properties to Bone Marrow-Derived Mesenchymal Stem Cells". *Stem Cells* **30** (5): 965–74. doi:10.1002/stem.1067. PMID 22367678.

[251] Shen, J. F.; Sugawara, A; Yamashita, J; Ogura, H; Sato, S (2011). "Dedifferentiated fat cells: An alternative source of adult multipotent cells from the adipose tissues". *International Journal of Oral Science* **3** (3): 117–24. doi:10.4248/IJOS11044. PMC 3470092. PMID 21789960.

[252] Melief, S. M.; Zwaginga, J. J.; Fibbe, W. E.; Roelofs, H. (2013). "Adipose Tissue-Derived Multipotent Stromal Cells Have a Higher Immunomodulatory Capacity Than Their Bone Marrow-Derived Counterparts". *Stem Cells Translational Medicine* **2** (6): 455–63. doi:10.5966/sctm.2012-0184. PMID 23694810.

[253] Cheng, A.; Hardingham, T. E.; Kimber, S. J. (2013). "Generating Cartilage Repair from Pluripotent Stem Cells". *Tissue Engineering Part B: Reviews* **20** (4): 131030093023007. doi:10.1089/ten.teb.2012.0757.

[254] Outani, H.; Okada, M.; Yamashita, A.; Nakagawa, K.; Yoshikawa, H.; Tsumaki, N. (2013). "Direct Induction of Chondrogenic Cells from Human Dermal Fibroblast Culture by Defined Factors". *PLoS ONE* **8** (10): e77365. doi:10.1371/journal.pone.0077365. PMID 24146984.

[255] Crompton, J. G.; Rao, M.; Restifo, N. P. (2013). "Memoirs of a Reincarnated T Cell". *Cell Stem Cell* **12** (1): 6–8. doi:10.1016/j.stem.2012.12.009. PMID 23290132.

[256] Tan, H. -K.; Toh, C. -X. D.; Ma, D.; Yang, B.; Liu, T. M.; Lu, J.; Wong, C. -W.; Tan, T. -K.; Li, H.; Syn, C.; Tan, E. -L.; Lim, B.; Lim, Y. -P.; Cook, S. A.; Loh, Y. -H. (2014). "Human Finger-Prick Induced Pluripotent Stem Cells Facilitate the Development of Stem Cell Banking". *Stem Cells Translational Medicine* **3** (5): 586–98. doi:10.5966/sctm.2013-0195. PMID 24646489.

[257] Okita, K.; Yamakawa, T.; Matsumura, Y.; Sato, Y.; Amano, N.; Watanabe, A.; Goshima, N.; Yamanaka, S. (2013). "An Efficient Nonviral Method to Generate Integration-Free Human-Induced Pluripotent Stem Cells from Cord Blood and Peripheral Blood Cells". *Stem Cells* **31** (3): 458–66. doi:10.1002/stem.1293. PMID 23193063.

[258] Geti, I.; Ormiston, M. L.; Rouhani, F.; Toshner, M.; Movassagh, M.; Nichols, J.; Mansfield, W.; Southwood, M.; Bradley, A.; Rana, A. A.; Vallier, L.; Morrell, N. W. (2012). "A Practical and Efficient Cellular Substrate for the Generation of Induced Pluripotent Stem Cells from Adults: Blood-Derived Endothelial Progenitor Cells". *Stem Cells Translational Medicine* **1** (12): 855–65. doi:10.5966/sctm.2012-0093. PMID 23283547.

[259] Staerk, J.; Dawlaty, M. M.; Gao, Q.; Maetzel, D.; Hanna, J.; Sommer, C. A.; Mostoslavsky, G.; Jaenisch, R. (2010). "Reprogramming of Human Peripheral Blood Cells to Induced Pluripotent Stem Cells". *Cell Stem Cell* **7** (1): 20–4. doi:10.1016/j.stem.2010.06.002. PMC 2917234. PMID 20621045.
Park, T. S.; Huo, J. S.; Peters, A.; Talbot, C. C.; Verma, K.; Zimmerlin, L.; Kaplan, I. M.; Zambidis, E. T. (2012). "Growth Factor-Activated Stem Cell Circuits and Stromal Signals Cooperatively Accelerate Non-Integrated iPSC Reprogramming of Human Myeloid Progenitors". *PLoS ONE* **7** (8): e42838. doi:10.1371/journal.pone.0042838. PMID 22905176.

[260] Yoshikawa, K.; Naitoh, M.; Kubota, H.; Ishiko, T.; Aya, R.; Yamawaki, S.; Suzuki, S. (2013). "Multipotent stem cells are effectively collected from adult human cheek skin". *Biochemical and Biophysical Research Communications* **431** (1): 104–10. doi:10.1016/j.bbrc.2012.12.069. PMID 23268344.

[261] Zhou, T.; Benda, C.; Duzinger, S.; Huang, Y.; Li, X.; Li, Y.; Guo, X.; Cao, G.; Chen, S.; Hao, L.; Chan, Y. -C.; Ng, K. -M.; Cy Ho, J.; Wieser, M.; Wu, J.; Redl, H.; Tse, H. -F.; Grillari, J.; Grillari-Voglauer, R.; Pei, D.; Esteban, M. A. (2011). "Generation of Induced Pluripotent Stem Cells from Urine". *Journal of the American Society of Nephrology* **22** (7): 1221–8. doi:10.1681/ASN.2011010106. PMID 21636641.
Zhou, T.; Benda, C.; Dunzinger, S.; Huang, Y.; Ho, J. C.; Yang, J.; Wang, Y.; Zhang, Y.; Zhuang, Q.; Li, Y.; Bao, X.; Tse, H. F.; Grillari, J.; Grillari-Voglauer, R.; Pei, D.; Esteban, M. A. (2012). "Generation of human induced pluripotent stem cells from urine samples". *Nature Protocols* **7** (12): 2080–9. doi:10.1038/nprot.2012.115. PMID 23138349.
Wang, L.; Wang, L.; Huang, W.; Su, H.; Xue, Y.; Su, Z.; Liao, B.; Wang, H.; Bao, X.; Qin, D.; He, J.; Wu, W.; So, K. F.; Pan, G.; Pei, D. (2012). "Generation of integration-free neural progenitor cells from cells in human urine". *Nature Methods* **10** (1): 84–9. doi:10.1038/nmeth.2283. PMID 23223155.
Cai, J; Zhang, Y; Liu, P; Chen, S; Wu, X; Sun, Y; Li, A; Huang, K; et al. (2013). "Generation of tooth-like structures from integration-free human urine induced pluripotent stem cells" (PDF). *Cell Regeneration* **2**: 6. doi:10.1186/2045-9769-2-6.

[262] Bharadwaj, S.; Liu, G.; Shi, Y.; Wu, R.; Yang, B.; He, T.; Fan, Y.; Lu, X.; Zhou, X.; Liu, H.; Atala, A.; Rohozin-

ski, J.; Zhang, Y. (2013). "Multipotential differentiation of human urine-derived stem cells: Potential for therapeutic applications in urology". *Stem Cells* **31** (9): 1840–56. doi:10.1002/stem.1424. PMID 23666768.

[263] Wang, Y.; Liu, J.; Tan, X.; Li, G.; Gao, Y.; Liu, X.; Zhang, L.; Li, Y. (2012). "Induced Pluripotent Stem Cells from Human Hair Follicle Mesenchymal Stem Cells". *Stem Cell Reviews and Reports* **9** (4): 451–60. doi:10.1007/s12015-012-9420-5. PMID 23242965.

[264] Schnabel, L. V.; Abratte, C. M.; Schimenti, J. C.; Southard, T. L.; Fortier, L. A. (2012). "Genetic background affects induced pluripotent stem cell generation". *Stem Cell Research & Therapy* **3** (4): 30. doi:10.1186/scrt121.

[265] Panopoulos, A. D.; Ruiz, S.; Yi, F.; Herrerías, A. D.; Batchelder, E. M.; Belmonte, J. C. I. (2011). "Rapid and Highly Efficient Generation of Induced Pluripotent Stem Cells from Human Umbilical Vein Endothelial Cells". *PLoS ONE* **6** (5): e19743. doi:10.1371/journal.pone.0019743. PMID 21603572.

[266] Polo, J. M.; Liu, S.; Figueroa, M. E.; Kulalert, W.; Eminli, S.; Tan, K. Y.; Apostolou, E.; Stadtfeld, M.; Li, Y.; Shioda, T.; Natesan, S.; Wagers, A. J.; Melnick, A.; Evans, T.; Hochedlinger, K. (2010). "Cell type of origin influences the molecular and functional properties of mouse induced pluripotent stem cells". *Nature Biotechnology* **28** (8): 848–55. doi:10.1038/nbt.1667. PMID 20644536.

[267] Miura, K.; Okada, Y.; Aoi, T.; Okada, A.; Takahashi, K.; Okita, K.; Nakagawa, M.; Koyanagi, M.; Tanabe, K.; Ohnuki, M.; Ogawa, D.; Ikeda, E.; Okano, H.; Yamanaka, S. (2009). "Variation in the safety of induced pluripotent stem cell lines". *Nature Biotechnology* **27** (8): 743–5. doi:10.1038/nbt.1554. PMID 19590502.
Liang, Y.; Zhang, H.; Feng, Q. S.; Cai, M. B.; Deng, W.; Qin, D.; Yun, J. P.; Tsao, G. S. W.; Kang, T.; Esteban, M. A.; Pei, D.; Zeng, Y. X. (2013). "The propensity for tumorigenesis in human induced pluripotent stem cells is related with genomic instability". *Chinese Journal of Cancer* **32** (4): 205–12. doi:10.5732/cjc.012.10065. PMID 22704487.

[268] Kim, K.; Doi, A.; Wen, B.; Ng, K.; Zhao, R.; Cahan, P.; Kim, J.; Aryee, M. J.; Ji, H.; Ehrlich, L. I. R.; Yabuuchi, A.; Takeuchi, A.; Cunniff, K. C.; Hongguang, H.; McKinney-Freeman, S.; Naveiras, O.; Yoon, T. J.; Irizarry, R. A.; Jung, N.; Seita, J.; Hanna, J.; Murakami, P.; Jaenisch, R.; Weissleder, R.; Orkin, S. H.; Weissman, I. L.; Feinberg, A. P.; Daley, G. Q. (2010). "Epigenetic memory in induced pluripotent stem cells". *Nature* **467** (7313): 285–90. doi:10.1038/nature09342. PMID 20644535.

[269] Kim, K.; Zhao, R.; Doi, A.; Ng, K.; Unternaehrer, J.; Cahan, P.; Hongguang, H.; Loh, Y. H.; Aryee, M. J.; Lensch, M. W.; Li, H.; Collins, J. J.; Feinberg, A. P.; Daley, G. Q. (2011). "Donor cell type can influence the epigenome and differentiation potential of human induced pluripotent stem cells". *Nature Biotechnology* **29** (12): 1117–9. doi:10.1038/nbt.2052. PMID 22119740.

[270] Bar-Nur, O.; Russ, H. A.; Efrat, S.; Benvenisty, N. (2011). "Epigenetic Memory and Preferential Lineage-Specific Differentiation in Induced Pluripotent Stem Cells Derived from Human Pancreatic Islet Beta Cells". *Cell Stem Cell* **9** (1): 17–23. doi:10.1016/j.stem.2011.06.007. PMID 21726830.

[271] Denker, H. W. (2012). "Time to Reconsider Stem Cell Induction Strategies". *Cells* **1** (4): 1293–312. doi:10.3390/cells1041293. PMC 3901125. PMID 24710555.

15.9   Notes

Chapter 16

Cell potency

Cell potency is a cell's ability to differentiate into other cell types.[1][2] The more cell types a cell can differentiate into, the greater its potency. Potency is also described as the gene activation potential within a cell which like a continuum begins with totipotency to designate a cell with the most differentiation potential, pluripotency, multipotency, oligopotency and finally unipotency. Potency is taken from the Latin term "potens" which means "having power."

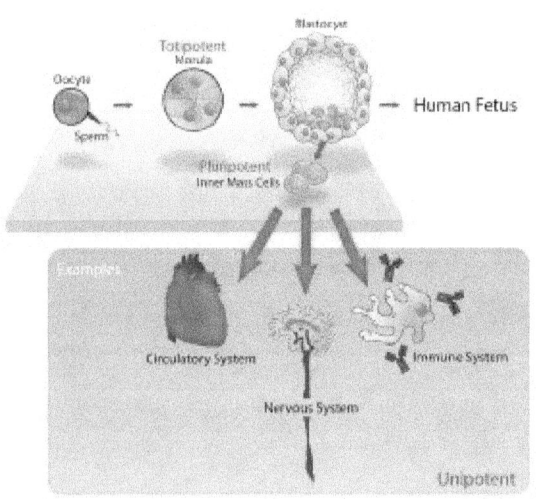

Pluripotent, embryonic stem cells originate as inner mass cells within a blastocyst. These stem cells can become any tissue in the body, excluding a placenta. Only the morula's cells are totipotent, able to become all tissues and a placenta.

16.1 Totipotency

Totipotency is the ability of a single cell to divide and produce all of the differentiated cells in an organism. Spores and Zygotes are examples of totipotent cells.[3] In the spectrum of cell potency, totipotency represents the cell with the greatest differentiation potential. *Toti* comes from the Latin *totus* which means "entirely."

It is possible for a fully differentiated cell to return to a state of totipotency.[4] This conversion to totipotency is complex, not fully understood and the subject of recent research. Research in 2011 has shown that cells may differentiate not into a fully totipotent cell, but instead into a "complex cellular variation" of totipotency.[5] Stem cells resembling totipotent blastomeres from 2-cell stage embryos can arise spontaneously in the embryonic stem cell cultures[6][7] and also can be induced to arise more frequently in vitro through down-regulation of the chromatin assembly activity of CAF-1.[8]

The human development model is one which can be used to describe how totipotent cells arise.[9] Human development begins when a sperm fertilizes an egg and the resulting fertilized egg creates a single totipotent cell, a zygote.[10] In the first hours after fertilization, this zygote divides into identical totipotent cells, which can later develop into any of the three germ layers of a human (endoderm, mesoderm, or ectoderm), into cells of the cytotrophoblast layer or syncytiotrophoblast layer of the placenta. After reaching a 16-cell stage, the totipotent cells of the morula differentiate into cells that will eventually become either the blastocyst's Inner cell mass or the outer trophoblasts. Approximately four days after fertilization and after several cycles of cell division, these totipotent cells begin to specialize. The inner cell mass, the source of embryonic stem cells, becomes pluripotent.

Research on *Caenorhabditis elegans* suggests that multiple mechanisms including RNA regulation may play a role in maintaining totipotency at different stages of development in some species.[11] Work with zebrafish and mammals suggest a further interplay between miRNA and RNA binding proteins (RBPs) in determining development differences.[12]

In September 2013, a team from the Spanish national Cancer Research Centre were able for the first time to make adult cells from mice retreat to the characteristics of embryonic stem cells thereby achieving totipotency.[13]

16.2 Pluripotency

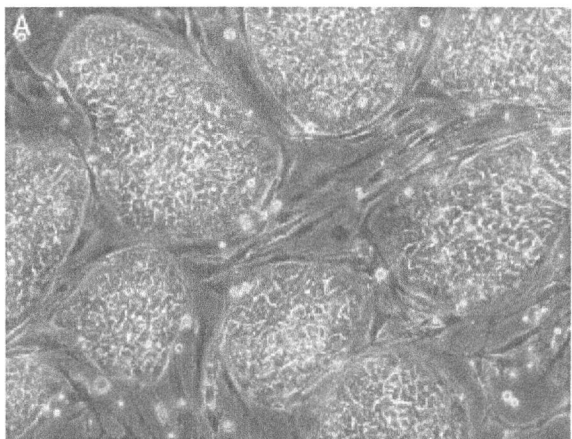

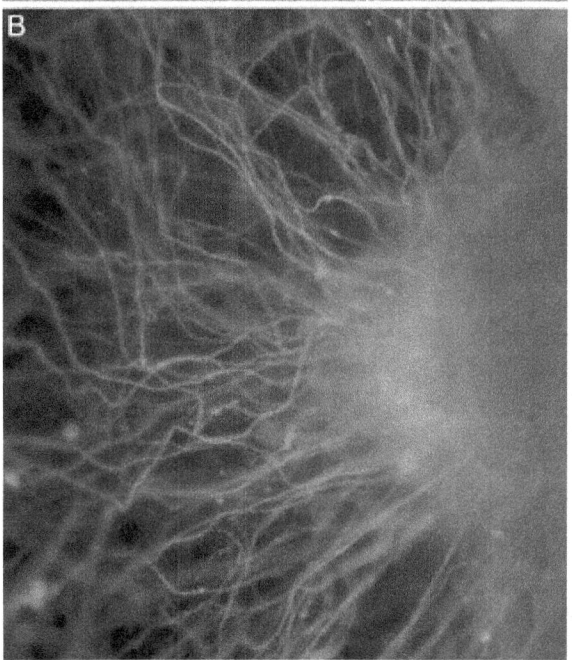

A: *Human embryonic stem cells (cell colonies that are not yet differentiated).*
B: *Nerve cells*

For substances having the capacity to produce several distinct biological responses see Pluripotency (biological compounds)
Main article: Stem cell

In cell biology, pluripotency (from the Latin plurimus, meaning *very many*, and potens, meaning *having power*)[14] refers to a stem cell that has the potential to differentiate into any of the three germ layers: endoderm (interior stomach lining, gastrointestinal tract, the lungs), mesoderm (muscle, bone, blood, urogenital), or ectoderm (epidermal tissues and nervous system).[15] However, cell pluripotency is

a continuum, ranging from the completely pluripotent cell that can form every cell of the embryo proper, e.g., embryonic stem cells and iPSCs (see below), to the incompletely or partially pluripotent cell that can form cells of all three germ layers but that may not exhibit all the characteristics of completely pluripotent cells.

16.2.1 Induced pluripotency

Main article: Induced pluripotent stem cells

Induced pluripotent stem cells, commonly abbreviated as iPS cells or iPSCs are a type of pluripotent stem cell artificially derived from a non-pluripotent cell, typically an adult somatic cell, by inducing a "forced" expression of certain genes and transcription factors.[16] These transcription factors play a key role in determining the state of these cells and also highlights the fact that these somatic cells do preserve the same genetic information as early embryonic cells.[17] The ability to induce cells into a pluripotent state was initially pioneered in 2006 using mouse fibroblasts and four transcription factors, Oct4, Sox2, Klf4 and c-Myc;[18] this technique called reprogramming earned Shinya Yamanaka and John Gurdon the Nobel Prize in Physiology or Medicine 2012.[19] This was then followed in 2007 by the successful induction of human iPSCs derived from human dermal fibroblasts using methods similar to those used for the induction of mouse cells.[20] These induced cells exhibit similar traits to those of embryonic stem cells (ESCs) but do not require the use of embryos. Some of the similarities between ESCs and iPSCs include pluripotency, morphology, self-renewal ability, a trait that implies that they can divide and replicate indefinitely, and gene expression.[21]

Epigenetic factors are also thought to be involved in the actual reprogramming of somatic cells in order to induce pluripotency. It has been theorized that certain epigenetic factors might actually work to clear the original somatic epigenetic marks in order to acquire the new epigenetic marks that are part of achieving a pluripotent state. Chromatin is also reorganized in iPSCs and becomes like that found in ESCs in that it is less condensed and therefore more accessible. Euchromatin modifications are also common which is also consistent with the state of euchromatin found in ESCs.[21]

Due to their great similarity to ESCs, iPSCs have been of great interest to the medical and research community. iPSCs could potentially have the same therapeutic implications and applications as ESCs but without the controversial use of embryos in the process, a topic of great bioethical debate. In fact, the induced pluripotency of somatic cells into undifferentiated iPS cells was originally hailed as the end of the controversial use of embryonic stem cells. However, iPSCs were found to be potentially tumorigenic, and, despite advances,[16] were never approved for clinical stage research in the United States. Setbacks such as low replication rates and early senescence have also been encountered when making iPSCs,[22] hindering their use as ESCs replacements.

Additionally, it has been determined that the somatic expression of combined transcription factors can directly induce other defined somatic cell fates (transdifferentiation); researchers identified three neural-lineage-specific transcription factors that could directly convert mouse fibroblasts (skin cells) into fully functional neurons.[23] This result challenges the terminal nature of cellular differentiation and the integrity of lineage commitment; and implies that with the proper tools, *all* cells are totipotent and may form all kinds of tissue.

Some of the possible medical and therapeutic uses for iPSCs derived from patients include their use in cell and tissue transplants without the risk of rejection that is commonly encountered. iPSCs can potentially replace animal models unsuitable as well as in-vitro models used for disease research.[24]

16.3 Multipotency

Further information: Progenitor cells

Multipotency describes progenitor cells which have the

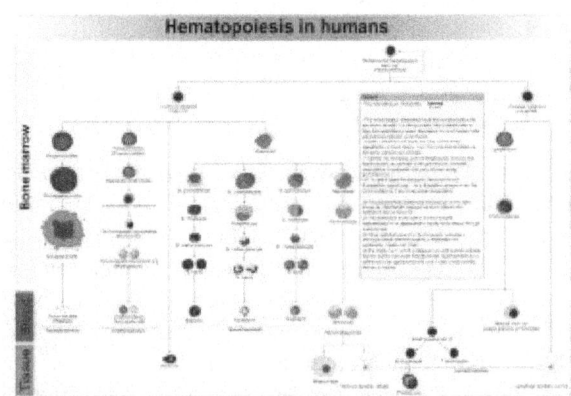

Hematopoietic stem cells are an example of multipotency. When they differentiate into myeloid or lymphoid progenitor cells, they lose potency and become oligopotent cells with the ability to give rise to all cells of its lineage.

gene activation potential to differentiate into multiple, but limited cell types. For example, a multipotent blood stem cell is a hematopoietic cell — and this cell type can differentiate itself into several types of blood cell types like lymphocytes, monocytes, neutrophils, etc., but cannot differentiate into brain cells, bone cells or other non-blood cell types.

New research related to multipotent cells suggests that multipotent cells may be capable of conversion into unrelated cell types. In one case, fibroblasts were converted into functional neurons.[23] In another case, human umbilical cord blood stem cells were converted into human neurons.[25] Research is also focusing on converting multipotent cells into pluripotent cells. [26]

Multipotent cells are found in many, but not all human cell types. Multipotent cells have been found in cord blood,[27] adipose tissue,[28] cardiac cells,[29] bone marrow, and mesenchymal stem cells (MSCs) which are found in the third molar.[30]

MSCs may prove to be a good, reliable source for stem cells because of the ease in collection of molars at 8–10 years of age and before adult dental calcification. MSCs can differentiate into osteoblasts, chondrocytes, and adipocytes.[31]

16.4 Oligopotency

In biology, oligopotency is the ability of progenitor cells to differentiate into a few cell types. It is a degree of potency. Examples of oligopotent stem cells are the lymphoid or myeloid stem cells.[1] A lymphoid cell specifically, can give rise to various blood cells such as B and T cells, however, not to a different blood cell type like a red blood cell.[32] Examples of progenitor cells are vascular stem cells that have the capacity to become both endothelial or smooth muscle cells.

16.5 Unipotency

Further information: Precursor cell

In cell biology, a unipotent cell is the concept that one stem cell has the capacity to differentiate into only one cell type. It is currently unclear if true unipotent stem cells exist. Hepatoblasts, which differentiate into hepatocytes (which constitute most of the liver) or cholangiocytes (epithelial cells of the bile duct), are bipotent.[33] A close synonym for *unipotent cell* is *precursor cell*.

16.6 See also

- Induced stem cells

16.7 References

[1] Hans R. Schöler (2007). "The Potential of Stem Cells: An Inventory". In Nikolaus Knoepffler, Dagmar Schipanski, and Stefan Lorenz Sorgner. *Human biotechnology as Social Challenge*. Ashgate Publishing, Ltd. p. 28. ISBN 978-0-7546-5755-2.

[2] "Stem Cell School: Glossary".

[3] Mitalipov S, Wolf D; Wolf (2009). "Totipotency, pluripotency and nuclear reprogramming". *Advances in Biochemical Engineering & Biotechnology* **114**: 185–99. Bibcode:2009esc..book..185M. doi:10.1007/10_2008_45. ISBN 978-3-540-88805-5. PMC 2752493. PMID 19343304.

[4] Western P (2009). "Foetal germ cells: striking the balance between pluripotency and differentiation". *Int. J. Dev. Biol.* **53** (2–3): 393–409. doi:10.1387/ijdb.082671pw. PMID 19412894.

[5] Sugimoto K, Gordon SP, Meyerowitz EM (April 2011). "Regeneration in plants and animals: dedifferentiation, transdifferentiation, or just differentiation?". *Trends Cell Biol.* **21** (4): 212–8. doi:10.1016/j.tcb.2010.12.004. PMID 21236679.

[6] Macfarlan T.S., Gifford W.D., Driscoll S., Lettieri K., Rowe H.M., Bonanomi D., Firth A., Singer O., Trono D. & Pfaff S.L. (2012) Embryonic stem cell potency fluctuates with endogenous retrovirus activity. Nature 487:57-63

[7] Morgani S.M., Canham M.A., Nichols J., Sharov A.A., Migueles R.P., Ko M.S. & Brickman J.M. (2013) Totipotent Embryonic Stem Cells Arise in Ground-State Culture Conditions. Cell Rep 3:1945-1957

[8] Ishiuchi T., Enriquez-Gasca R., Mizutani E., Boskovic A., Ziegler-Birling C., Rodriguez-Terrones D., Wakayama T., Vaquerizas J.M. & Torres-Padilla M.E.(2015) Early embryonic-like cells are induced by downregulating replication dependent chromatin assembly. Nat Struct Mol Biol. 22:662-671.

[9] Seydoux G, Braun RE (December 2006). "Pathway to totipotency: lessons from germ cells". *Cell* **127** (5): 891–904. doi:10.1016/j.cell.2006.11.016. PMID 17129777.

[10] Asch R, Simerly C, Ord T, Ord VA, Schatten G (July 1995). "The stages at which human fertilization arrests: microtubule and chromosome configurations in inseminated oocytes which failed to complete fertilization and development in humans". *Hum. Reprod.* **10** (7): 1897–906. PMID 8583008.

[11] Ciosk, R.; Depalma, Michael; Priess, James R. (10 February 2006). "Translational Regulators Maintain Totipotency in the Caenorhabditis elegans Germline". *Science* **311** (5762): 851–853. Bibcode:2006Sci...311..851C. doi:10.1126/science.1122491. PMID 16469927.

[12] Kedde M, Agami R (April 2008). "Interplay between microRNAs and RNA-binding proteins determines developmental processes". *Cell Cycle* **7** (7): 899–903. doi:10.4161/cc.7.7.5644. PMID 18414021.

[13] Serrano, Manuel (2013-09-11). "Study published in Nature is another step towards regenerative medicine" (PDF). cnio.es. Retrieved 2013-12-11.

[14] "Biology Online". Biology-Online.org. Retrieved 25 April 2013.

[15] Binder, Marc D.; Hirokawa, Nobutaka; Uwe Windhorst, eds. (2009). *Encyclopedia of neuroscience*. Berlin: Springer. ISBN 978-3540237358.

[16] Baker, Monya (2007-12-06). "Adult cells reprogrammed to pluripotency, without tumors". *Nature Reports Stem Cells*. doi:10.1038/stemcells.2007.124.

[17] Stadtfeld, M.; Hochedlinger, K. (15 October 2010). "Induced pluripotency: history, mechanisms, and applications". *Genes & Development* **24** (20): 2239–2263. doi:10.1101/gad.1963910.

[18] Takahashi, Kazutoshi; Yamanaka, Shinya (August 2006). "Induction of Pluripotent Stem Cells from Mouse Embryonic and Adult Fibroblast Cultures by Defined Factors". *Cell* **126** (4): 663–676. doi:10.1016/j.cell.2006.07.024. PMID 16904174.

[19] "The Nobel Prize in Physiology or Medicine 2012". Nobelprize.org. Nobel Media AB 2013. Web. 28 Nov 2013.

[20] Takahashi, Kazutoshi; Tanabe, Koji; Ohnuki, Mari; Narita, Megumi; Ichisaka, Tomoko; Tomoda, Kiichiro; Yamanaka, Shinya (1 November 2007). "Induction of Pluripotent Stem Cells from Adult Human Fibroblasts by Defined Factors". *Cell* **131** (5): 861–872. doi:10.1016/j.cell.2007.11.019. PMID 18035408.

[21] Liang, Gaoyang; Zhang, Yi (18 December 2012). "Embryonic stem cell and induced pluripotent stem cell: an epigenetic perspective". *Cell Research* **23** (1): 49–69. doi:10.1038/cr.2012.175. PMC 3541668. PMID 23247625.

[22] Choi, Charles. "Cell-Off: Induced Pluripotent Stem Cells Fall Short of Potential Found in Embryonic Version". Scientific American. Retrieved 25 April 2013.

[23] Vierbuchen T; Wernig M; et al. (February 2010). "Direct conversion of fibroblasts to functional neurons by defined factors". *Nature* **463** (7284): 1035–41. Bibcode:2010Natur.463.1035V. doi:10.1038/nature08797. PMC 2829121. PMID 20107439.

[24] Park, IH; Lerou, PH; Zhao, R; Huo, H; Daley, GQ (2008). "Generation of human-induced pluripotent stem cells.". *Nature protocols* **3** (7): 1180–6. doi:10.1038/nprot.2008.92. PMID 18600223.

[25] Giorgetti A; Marchetto MC; Li M; et al. (July 2012). "Cord blood-derived neuronal cells by ectopic expression of Sox2 and c-Myc". *Proc. Natl. Acad. Sci. U.S.A.* **109** (31): 12556–61. Bibcode:2012PNAS..10912556G. doi:10.1073/pnas.1209523109. PMC 3412010. PMID 22814375.

[26] Guan K; Nayernia K; Maier LS; et al. (April 2006). "Pluripotency of spermatogonial stem cells from adult mouse testis". *Nature* **440** (7088): 1199–203. Bibcode:2006Natur.440.1199G. doi:10.1038/nature04697. PMID 16565704.

[27] Yong Zhao, Theodore Mazzone (Dec 2010). "Human cord blood stem cells and the journey to a cure for type 1 diabetes". *Autoimmun Rev* **10** (2): 103–107. doi:10.1016/j.autrev.2010.08.011. PMID 20728583.

[28] Tallone T; Realini C; Böhmler A; et al. (April 2011). "Adult human adipose tissue contains several types of multipotent cells". *J Cardiovasc Transl Res* **4** (2): 200–10. doi:10.1007/s12265-011-9257-3. PMID 21327755.

[29] Beltrami AP; Barlucchi L; Torella D; et al. (September 2003). "Adult cardiac stem cells are multipotent and support myocardial regeneration". *Cell* **114** (6): 763–76. doi:10.1016/S0092-8674(03)00687-1. PMID 14505575.

[30] Ohgushi H, Arima N, Taketani T (December 2011). "[Regenerative therapy using allogeneic mesenchymal stem cells]". *Nippon Rinsho* (in Japanese) **69** (12): 2121–7. PMID 22242308.

[31] Uccelli, Antonio; Moretta, Pistoia (September 2008). "Mesenchymal stem cells in health and disease". *Nature Reviews* **8** (9): 726–36. doi:10.1038/nri2395. PMID 19172693.

[32] Ibelgaufts, Horst. "Cytokines & Cells Online Pathfinder Encyclopedia". Retrieved 25 April 2013.

[33] "hepatoblast differentiation". *GONUTS*.

Chapter 17

Zygote

For other uses, see Zygote (disambiguation).
"Fertilized egg" redirects here. For the food product, see Balut (egg).

A **zygote** (from Greek ζυγωτός zygōtos "joined" or "yoked", from ζυγοῦν zygoun "to join" or "to yoke"),[1] is a eukaryotic cell formed by a fertilization event between two gametes. The zygote's genome is a combination of the DNA in each gamete, and contains all of the genetic information necessary to form a new individual. In multicellular organisms, the zygote is the earliest developmental stage. In single-celled organisms, the zygote can divide asexually by mitosis to produce identical offspring.

Oscar Hertwig and Richard Hertwig made some of the first discoveries on animal zygote formation.

17.1 Fungi

In fungi, the sexual fusion of haploid cells is called karyogamy. The result of karyogamy is a diploid cell called a zygote or zygospore. This cell may then enter meiosis or mitosis depending on the life cycle of the species.

17.2 Plants

In plants, the zygote may be polyploid if fertilization occurs between meiotically unreduced gametes.

In land plants, the zygote is formed within a chamber called the archegonium. In seedless plants, the archegonium is usually flask-shaped, with a long hollow neck through which the sperm cell enters. As the zygote divides and grows, it does so inside the archegonium.

17.3 Humans

In human fertilization, two 1n haploid cells—an ovum (female gamete) and a sperm cell (male gamete)—combine to form a single 2n diploid cell called the zygote. DNA is then replicated in the two separate pronuclei derived from the sperm and ovum, making the zygote's chromosome number temporarily 4n diploid. After approximately 30 hours, fusion of the pronuclei and subsequent mitotic division produce two 2n diploid daughter cells called blastomeres.[2]

Between the stages of fertilization and implantation, the developing human is called the *preimplantation conceptus* or the proembryo. It is not correct to call the conceptus an *embryo*, because it will later differentiate into both intraembryonic and extraembryonic tissues,[3] and can even split to produce multiple embryos (identical twins).

After fertilization, the conceptus travels down the oviduct towards the uterus while continuing to divide[4] mitotically without actually increasing in size, in a process called cleavage.[5] After four divisions, the conceptus consists of 16 blastomeres, and it is known as the morula.[6] Through the processes of compaction, cell division, and blastulation, the conceptus takes the form of the blastocyst by the fifth day of development, just as it approaches the site of implantation.[7] When the blastocyst hatches from the zona pellucida, it can implant in the endometrial lining of the uterus and begin the embryonic stage of development.

The human zygote has been genetically edited in experiments designed to cure inherited diseases.[8]

17.4 In other species

A Chlamydomonas zygote that contains chloroplast DNA (cpDNA) from both parents, such cells generally are rare since normally cpDNA is inherited uniparental from the mt+ mating type parent. These rare biparental zygotes allowed mapping of chloroplast genes by recombination.

17.5 In Protozoa

In the Amoeba, reproduction occurs by cell division of the parent cell: first the nucleus of the parent divides into two and then the cell membrane also cleaves, becoming two "daughter" Amoebae.

17.6 See also

- Proembryo

17.7 References

[1] "English etymology of zygote". *myetymology.com*.

[2] Blastomere Encyclopædia Britannica. Encyclopædia Britannica Online. Encyclopædia Britannica Inc., 2012. Web. 06 Feb. 2012.

[3] Larsen's Human Embryology. 4th Ed. Page 4.

[4] O'Reilly, Deirdre. "Fetal development". *MedlinePlus Medical Encyclopedia* (2007-10-19). Retrieved 2009-02-15.

[5] Klossner, N. Jayne and Hatfield, Nancy. *Introductory Maternity & Pediatric Nursing*, p. 107 (Lippincott Williams & Wilkins, 2006).

[6] Neas, John F. "Human Development". *Embryology Atlas*

[7] Blackburn, Susan. *Maternal, Fetal, & Neonatal Physiology*, p. 80 (Elsevier Health Sciences 2007).

[8] Human zygote edited genetically

Chapter 18

Spore

This article is about spores in eukaryotes. For bacterial spores, see endospore. For other uses, see Spore (disambiguation).

In biology, a **spore** is a unit of asexual reproduction that

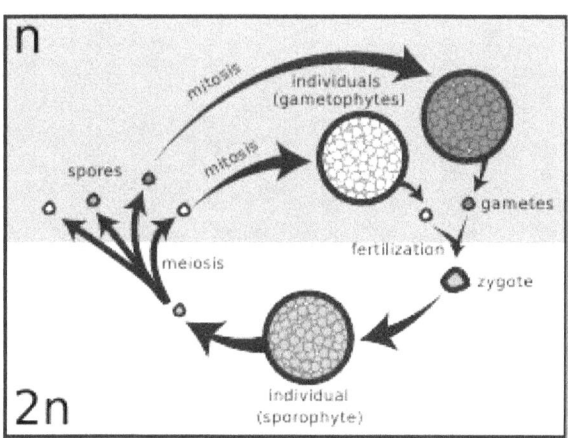

Spores produced in a sporic life cycle.

Fresh snow partially covers Rough-stalked Feather-moss (Brachythecium rutabulum), growing on a thinned hybrid black poplar (Populus x canadensis). The last stage of the moss lifecycle is shown, where the sporophytes are visible before dispersion of their spores: the calyptra (1) is still attached to the capsule (2). The tops of the gametophytes (3) can be discerned as well. Inset shows the surrounding, black poplars growing on sandy loam on the bank of a kolk, with the detail area marked.

may be adapted for dispersal and for survival, often for extended periods of time, in unfavorable conditions. By contrast, gametes are units of sexual reproduction. Spores form part of the life cycles of many plants, algae, fungi and protozoa.[1] Bacterial spores are not part of a sexual cycle but are resistant structures used for survival under unfavourable conditions. Myxozoan spores release amoebulae into their hosts for parasitic infection, but also reproduce within the hosts through the pairing of two nuclei within the plasmodium, which develops from the amoebula.[2]

Spores are usually haploid and unicellular and are produced by meiosis in the sporangium of a diploid sporophyte. Under favourable conditions the spore can develop into a new organism using mitotic division, producing a multicellular gametophyte, which eventually goes on to produce gametes. Two gametes fuse to form a zygote which develops into a new sporophyte. This cycle is known as alternation of generations.

The spores of seed plants, however, are produced internally and the megaspores, formed within the ovules and the microspores are involved in the formation of more complex structures that form the dispersal units, the seeds and pollen grains.

18.1 Definition

The term *spore* derives from the ancient Greek word σπορά *spora*, meaning "seed, sowing," related to σπόρος *sporos*, "sowing," and σπείρειν *speirein*, "to sow."

In common parlance, the difference between a "spore" and a "gamete" (both together called gonites) is that a spore will germinate and develop into a sporeling, while a gamete needs to combine with another gamete to form a zygote before developing further.

The chief difference between spores and seeds as dispersal units is that spores are unicellular, while seeds contain within them a multicellular gametophyte that produces a developing embryo, the multicellular sporophyte of the next generation. Spores germinate to give rise to haploid gametophytes, while seeds germinate to give rise to diploid sporophytes.

18.2   Classification of spore-producing organisms

Vascular plant spores are always haploid.   Vascular plants are either **homosporous (or isosporous)** or **heterosporous**. Plants that are homosporous produce spores of the same size and type. Heterosporous plants, such as seed plants, spikemosses, quillworts, and some aquatic ferns produce spores of two different sizes: the larger spore (megaspore) in effect functioning as a "female" spore and the smaller (microspore) functioning as a "male".

18.3   Classification of spores

Spores can be classified in several ways:

18.3.1   By spore-producing structure

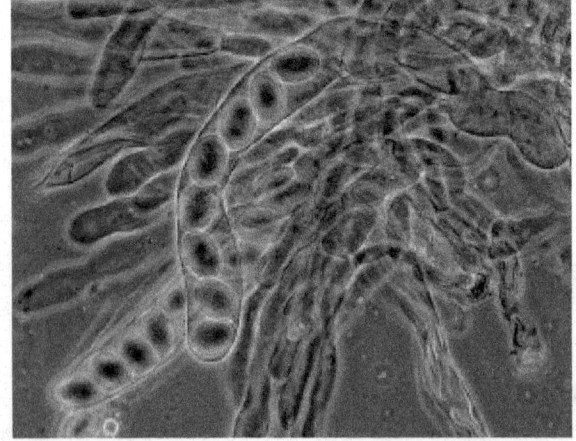

Asci of Morchella elata, *containing ascospores*

In fungi and fungus-like organisms, spores are often classified by the structure in which meiosis and spore production occurs. Since fungi are often classified according to their spore-producing structures, these spores are often characteristic of a particular taxon of the fungi.

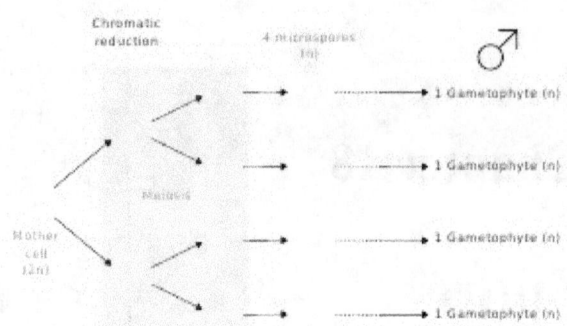

In plants, microspores, and in some cases megaspores, are formed from all four products of meiosis.

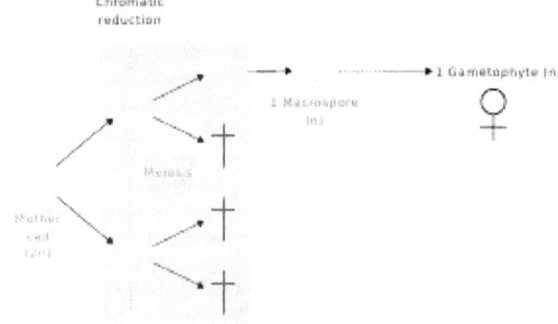

In contrast, in many seed plants and heterosporous ferns, only a single product of meiosis will become a megaspore (macrospore), with the rest degenerating.

- **Sporangiospores**: spores produced by a sporangium in many fungi such as zygomycetes.

- **Zygospores**: spores produced by a zygosporangium, characteristic of zygomycetes.

- **Ascospores**: spores produced by an ascus, characteristic of ascomycetes.

- **Basidiospores**: spores produced by a basidium, characteristic of basidiomycetes.

- **Aeciospores**: spores produced by an aecium in some fungi such as rusts or smuts.

- **Urediniospores**: spores produced by a uredinium in some fungi such as rusts or smuts.

- **Teliospores**: spores produced by a telium in some fungi such as rusts or smuts.

- **Oospores**: spores produced by an oogonium, characteristic of oomycetes.

- **Carpospores**: spores produced by a carposporophyte, characteristic of red algae.

- **Tetraspores**: spores produced by a tetrasporophyte, characteristic of red algae.

18.3.2   By function

- **Chlamydospores**: thick-walled resting spores of fungi produced to survive unfavorable conditions.

- **Parasitic fungal spores** may be classified into internal spores, which germinate within the host, and external spores, also called environmental spores, released by the host to infest other hosts.[3]

18.3.3   By origin during life cycle

- **Meiospores**: spores produced by meiosis; they are thus haploid, and give rise to a haploid daughter cell(s) or a haploid individual. Examples are the precursor cells of gametophytes of seed plants found in flowers (angiosperms) or cones (gymnosperms), and the zoospores produced from meiosis in the sporophytes of algae such as *Ulva*.

    - **Microspores**: meiospores that give rise to a male gametophyte, (pollen in seed plants).

    - **Megaspores** (or **macrospores**): meiospores that give rise to a female gametophyte, (in seed plants the gametophyte forms within the ovule).

- **Mitospores** (or **conidia**, **conidiospores**): spores produced by mitosis; they are characteristic of Ascomycetes. Fungi in which only mitospores are found are called "mitosporic fungi" or "anamorphic fungi", and are previously classified under the taxon Deuteromycota (See Teleomorph, anamorph and holomorph).

18.3.4   By mobility

Spores can be differentiated by whether they can move or not.

- **Zoospores**: mobile spores that move by means of one or more flagella, and can be found in some algae and fungi.

- **Aplanospores**: immobile spores that may nevertheless potentially grow flagella.

- **Autospores**: immobile spores that cannot develop flagella.

- **Ballistospores**: spores that are actively discharged from the body of the fungal fruiting body. Most basidiospores are also ballistospores, and another notable example is spores of *Pilobolus*.

- **Statismospores**: spores that are not actively discharged from the fungal fruiting body. Examples are puffballs.

18.4   Anatomy

Under high magnification, spores can be categorized as either **monolete spores** or **trilete spores**. In monolete spores, there is a single line on the spore indicating the axis on which the mother spore was split into four along a vertical axis. In trilete spores, all four spores share a common origin and are in contact with each other, so when they separate, each spore shows three lines radiating from a center pole.

18.4.1   Spore tetrads and trilete spores

Main article: Evolutionary history of plants

Envelope-enclosed spore tetrads are taken as the earliest evidence of plant life on land,[4] dating from the mid-Ordovician (early Llanvirn, ~470 million years ago), a period from which no macrofossils have yet been recovered.[5] Individual trilete spores resembling those of modern cryptogamic plants first appeared in the fossil record at the end of the Ordovician period.[6]

18.5   Dispersal

Spores being ejected by fungi.

In fungi, both asexual and sexual spores or sporangiospores of many fungal species are actively dispersed by forcible ejection from their reproductive structures. This ejection ensures exit of the spores from the reproductive structures as well as travelling through the air over long distances. Many fungi thereby possess specialized mechanical and physiological mechanisms as well as spore-surface structures, such as hydrophobins, for spore ejection. These mechanisms include, for example, forcible discharge of ascospores enabled by the structure of the ascus and accumulation of osmolytes in the fluids of the ascus that lead to explosive discharge of the ascospores into the air.[7] The forcible discharge of single spores termed *ballistospores* involves formation of a small drop of water (Buller's drop), which upon contact with the spore leads to its projectile release with an initial acceleration of more than 10,000 g.[8] Other fungi rely on alternative mechanisms for spore release, such as external mechanical forces, exemplified by puffballs. Attracting insects, such as flies, to fruiting structures, by virtue of their having lively colours and a putrid odour, for dispersal of fungal spores is yet another strategy, most prominently used by the stinkhorns.

In Common Smoothcap moss (*Atrichum undulatum*), the vibration of sporophyte has been shown to be an important mechanism for spore release.[9]

In the case of spore-shedding vascular plants such as ferns, wind distribution of very light spores provides great capacity for dispersal. Also, spores are less subject to animal predation than seeds because they contain almost no food reserve; however they are more subject to fungal and bacterial predation. Their chief advantage is that, of all forms of progeny, spores require the least energy and materials to produce.

In the spikemoss *Selaginella lepidophylla*, dispersal is achieved in part by an unusual type of diaspore, a tumbleweed.[10]

18.6   Gallery

- Microscopic view of dehisced fern sporangia (no spores are visible)

- Microscopic view of Equisetum spores

- Spore clusters, formed inside sporangia of *Reticularia olivacea* from pine forests of the Eastern Ukraine.

- Internal surface of the peridium of the *Tubifera dudkae* with spores

18.7   See also

- Alternation of generations

- Auxiliary cell

- Bioaerosol

- Cryptospores

- Endospore

- Evolutionary history of plants

- Fern

- Sporophyte

18.8   References

[1] Spore FAQ

[2] "Myxozoa." Tree of Life web project. Ivan Fiala 10 July 2008. Web. 14 Jan. 2014. <http://tolweb.org/Myxozoa/2460>

[3] Microsporidia (Protozoa): A Handbook of Biology and Research Techniques at the Wayback Machine (archived June 26, 2008). modares.ac.ir

[4] Gray, J.; Chaloner, W. G.; Westoll, T. S. (1985). "The Microfossil Record of Early Land Plants: Advances in Understanding of Early Terrestrialization, 1970–1984". *Philosophical Transactions of the Royal Society B* **309** (1138): 167–195. Bibcode:1985RSPTB.309..167G. doi:10.1098/rstb.1985.0077. JSTOR 2396358.

[5] Wellman, C.H., Gray, J. (2000). "The microfossil record of early land plants". *Philosophical Transactions of the Royal Society B* **355** (1398): 717–732. doi:10.1098/rstb.2000.0612. PMC 1692785. PMID 10905606.

[6] Steemans, P.; Herisse, A. L.; Melvin, J.; Miller, M. A.; Paris, F.; Verniers, J.; Wellman, C. H. (2009). "Origin and Radiation of the Earliest Vascular Land Plants". *Science* **324** (5925): 353–353. Bibcode:2009Sci...324..353S. doi:10.1126/science.1169659. ISSN 0036-8075. PMID 19372423.

[7] Trail F. (2007). "Fungal cannons: explosive spore discharge in the Ascomycota". *FEMS Microbiology Letters* **276** (1): 12–8. doi:10.1111/j.1574-6968.2007.00900.x. PMID 17784861.

[8] Pringle A, Patek SN, Fischer M, Stolze J, Money NP. (2005). "The captured launch of a ballistospore". *Mycologia* **97** (4): 866–71. doi:10.3852/mycologia.97.4.866. PMID 16457355.

[9] Johansson, Lönnell, Sundberg and Hylander (2014) Release thresholds for moss spores: the importance of turbulence and sporophyte length. Journal of Ecology. n/a-n/a.

[10] "False Rose of Jericho – Selaginella lepidophyllaFalse Rose of Jericho – Selaginella lepidophylla". *Plant- and Flower guide*. February 2009. Retrieved 1 February 2010.

Chapter 19

Morula

For the South African football player, see Lebohang Morula.

A **morula** (Latin, *morus*: mulberry) is an early stage embryo consisting of cells (called blastomeres) in a solid ball contained within the zona pellucida.[1][2]

A morula is distinct from a blastocyst in that a morula (3-4 days post fertilization) is an 16 cell mass in a spherical shape whereas a blastocyst (4-5 days post fertilization) has a cavity inside the zona pellucida along with an inner cell mass. A morula, if untouched and allowed to remain implanted, will eventually develop into a blastocyst.[3]

The morula is produced by a series of cleavage divisions of the early embryo, starting with the single-celled zygote. Once the embryo has divided into 16 cells, it begins to resemble a mulberry, hence the name *morula* (Latin, *morus*: mulberry).[4] Within a few days after fertilization, cells on the outer part of the morula become bound tightly together with the formation of desmosomes and gap junctions, becoming nearly indistinguishable. This process is known as compaction.[5][6] A cavity forms inside the morula, by the active transport of sodium ions from trophoblast cells and osmosis of water. This results in a hollow ball of cells known as the blastocyst.[7][8] The blastocyst's outer cells will become the first embryonic epithelium (the trophectoderm). Some cells, however, will remain trapped in the interior and will become the inner cell mass (ICM), and are pluripotent. In mammals (except monotremes), the ICM will ultimately form the "embryo proper", while the trophectoderm will form the placenta and extra-embryonic tissues. However, reptiles have a different ICM. The stages are prolonged and divided in 4 parts.[9][10][11][12]

19.1 See also

- Cleavage (embryo)

- Blastula

19.2 References

[1] Boklage, Charles E. (2009). *How New Humans Are Made: Cells and Embryos, Twins and Chimeras, Left and Right, Mind/Self/Soul, Sex, and Schizophrenia*. World Scientific. p. 217. ISBN 9789812835130.

[2] "The Early Embryology of the Chick". UNSW Embryology. Retrieved 2015-03-03.

[3] "The Morula and Blastocyst". the Endowment for Human Development. Retrieved 11 April 2015.

[4] Sherman, Lawrence S. et al., eds. (2001). *Human embryology* (3rd ed.). Elsevier Health Sciences. p. 20. ISBN 978-0-443-06583-5.

[5] Chard, Tim & Lilford, Richard (1995). *Basic sciences for obstetrics and gynaecology*. Springer. p. 18. ISBN 978-3-540-19903-8.

[6] Mercader, Amparo et al. (2008). "Human embryo culture". In Lanza, Robert & Klimanskaya, Irina. *Essential stem cell methods*. Academic Press. p. 343. ISBN 978-0-12-374741-9.

[7] Patestas, Maria Antoniou & Gartner, Leslie P. (2006). *A textbook of neuroanatomy*. Wiley-Blackwell. p. 11. ISBN 978-1-4051-0340-4.

[8] Geisert, R.D. & Malayer, J.R. (2000). "Implantation: Blastocyst formation". In Hafez, B. & Hafez, Elsayed S.E. *Reproduction in farm animals*. Wiley-Blackwell. p. 118. ISBN 978-0-683-30577-7.

[9] Morali, Olivier G. et al. (2005). "Epithelium-Mesenchyme Transitions are Crucial Morphogenetic Events Occurring During Early Development". In Savagner, Pierre. *Rise and fall of epithelial phenotype: concepts of epithelial-mesenchymal transition*. Springer. p. 16. ISBN 978-0-306-48239-7.

[10] Birchmeier, Carmen; et al. (1997). "Morphogenesis of epithelial cells". In Paul, Leendert C. & Issekutz, Thomas B. *Adhesion molecules in health and disease*. CRC Press. p. 208. ISBN 978-0-8247-9824-6.

[11] Nagy, András (2003). *Manipulating the mouse embryo: a laboratory manual*. CSHL Press. pp. 60–61. ISBN 978-0-87969-591-0.

[12] Connell, R.J. & Cutner, A. (2001). "Basic Embryology". In Cardozo, Linda & Staskin, David. *Textbook of female urology and urogynaecology*. Taylor & Francis. p. 92. ISBN 978-1-901865-05-9.

19.3   Further reading

- "Regulative development in mammals"

Chapter 20

Callus (cell biology)

Plant callus (plural *calluses* or *calli*) is a mass of unorganized parenchyma cells derived from plant tissue (explants) for use in biological research and biotechnology. In plant biology, callus cells are those cells that cover a plant wound.[1] Callus formation is induced from plant tissues after surface sterilization and plating onto *in vitro* tissue culture medium. Plant growth regulators, such as auxins, cytokinins, and gibberellins, are supplemented into the medium to initiate callus formation or somatic embryogenesis. Callus initiation has been described for all major groups of land plants.

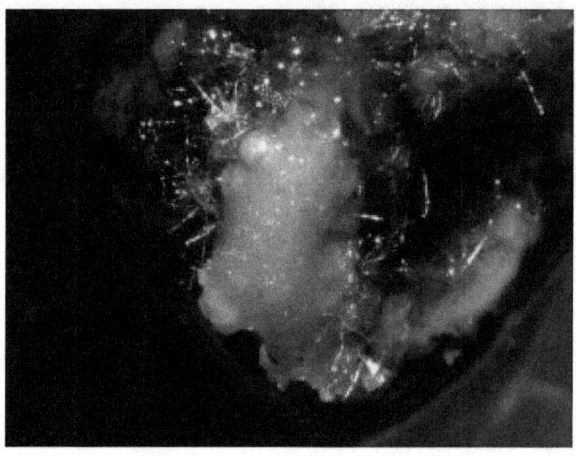

Callus cells forming during a process called "induction" in Pteris vittata

Callus Nicotiana tabacum

20.1 Callus induction and tissue culture

Plant species representing all major land plant groups have been shown to be capable of producing callus in tissue culture. [2][3][4][5][6][7][8][9][10][11][12] A callus cell culture is usually sustained on gel medium. Callus induction medium consists of agar and a mixture of macronutrients and micronutrients for the given cell type. There are several types of basal salt mixtures used in plant tissue culture, but most notably modified Murashige and Skoog medium,[13]

White's medium,[14] and woody plant medium.[15] Vitamins are also provided to enhance growth such as Gamborg B5 vitamins.[16] For plant cells, enrichment with nitrogen, phosphorus, and potassium is especially important. Plant callus is usually derived from somatic tissues. The tissues used to initiate callus formation depends on plant species and which tissues are available for explant culture. The cells that give rise to callus and somatic embryos usually undergo rapid division or are partially undifferentiated such as meristematic tissue. In alfalfa, *Medicago truncatula*, however callus and somatic embryos are derived from mesophyll cells that undergo dedifferentiation.[17] Plant hormones are used to initiate callus growth.

20.2 Morphology

Specific auxin to cytokinin ratios in plant tissue culture medium give rise to an unorganized growing and dividing mass of callus cells. Callus cultures are often broadly classified as being either compact or friable. Friable calluses fall apart easily, and can be used to generate cell suspension

151

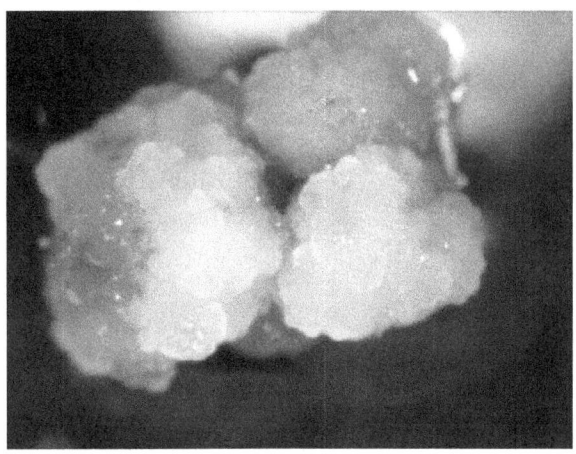

Callus induced from Pteris vittata *gametophytes*

cultures. Callus can directly undergo direct organogenesis and/or embryogenesis where the cells will form an entirely new plant.

20.3   Callus cells deaths

Callus can brown and die during culture, but the causes for callus browning are not well understood. In *Jatropha curcas* callus cells, small organized callus cells became disorganized and varied in size after browning occurred.[18] Browning has also been associated with oxidation and phenolic compounds in both explant tissues and explant secretions.[19] In rice, presumably, a condition which is favorable for scutellar callus induction induces necrosis too.[20]

20.4   Uses

Callus cells are not necessarily genetically homogeneous because a callus is often made from structural tissue, not individual cells. Nevertheless, callus cells are often considered similar enough for standard scientific analysis to be performed as if on a single subject. For example, an experiment may have half a callus undergo a treatment as the experimental group, while the other half undergoes a similar but non-active treatment as the control group.

Plant calli can differentiate into a whole plant, a process called regeneration, through addition of plant hormones in culture medium. This ability is known as totipotency. Regeneration of a whole plant from a single cell allows researchers to recover whole plants that have a copy of the transgene in every cell. Regeneration of a whole plant that has some genetically transformed cells and some untrans-

formed cells is called a chimera. In general, chimeras are not useful for genetic research or agricultural applications.

Genes can be inserted into callus cells using biolistic bombardment, also known as a gene gun, or *Agrobacterium tumefaciens*. Cells that receive the gene of interest can then be recovered into whole plants using a combination of plant hormones. The whole plants that are recovered can be used to experimentally determine gene function(s), or to enhance crop plant traits for modern agriculture.

Callus is of particular use in micropropagation where it can be used to grow genetically identical copies of plants with desirable characteristics.

20.5   History

Engraving of Henri-Louis Duhamel du Monceau by François-Hubert Drouais.   He is shown working on his Éléments d'architecture navale, *his most famous work.   Duhamel du Monceau was the first to describe callus formation that he observed growing over the wound of an elm tree.*

Henri-Louis Duhamel du Monceau investigated wound-healing responses in elm trees, and was the first to report formation of callus on live plants.[21]

In 1908, E. F. Simon was able to induce callus from poplar stems that also produced roots and buds.[22] The first reports of callus induction *in vitro* came from three independent researchers in 1939.[23] P. White induced callus derived from tumor-developing procambial tissues of hybrid *Nicotiana glauca* that did not require hormone supplementation.[14] Gautheret and Nobecourt were able to maintain callus cultures of carrot using auxin hormone additions.

20.6   See also

- Embryo Rescue

- Somatic Embryogenesis

- Chimera (plant)

- Hyperhydricity

20.7   References

[1] What is Plant Tissue Culture?

[2] Takeda, Reiji; Katoh, Kenji.   "Growth and sesquiterpenoid production by *Calypogeia granulata* inoue cells in suspension culture".   *Planta* **151** (6):   525–530. doi:10.1007/BF00387429.

[3] Peterson, M (2003).   "Cinnamic acid 4-hydroxylase from cell cultures of the hornwort *Anthoceros agrestis*".   *Planta* **217** (1):   96–101.   doi:10.1007/s00425-002-0960-9.   PMID 12721853.

[4] Beutelmann, P.; Bauer, L. (1 January 1977).   "Purification and identification of a cytokinin from moss callus cells".   *Planta* **133** (3):   215–217.   doi:10.1007/BF00380679.

[5] Atmane, N. "Histological analysis of indirect somatic embryogenesis in the Marsh clubmoss *Lycopodiella inundata* (L.) Holub (Pteridophytes)".   *Plant Science* **156** (2):   159–167.   doi:10.1016/S0168-9452(00)00244-2.

[6] Yang, Xuexi; Chen, Hui; Xu, Wenzhong; He, Zhenyan; Ma, Mi.   "Hyperaccumulation of arsenic by callus, sporophytes and gametophytes of *Pteris vittata* cultured *in vitro*".   *Plant Cell Reports* **26** (10):   1889–1897.   doi:10.1007/s00299-007-0388-6.

[7] Chavez, V. M.; Litz, R. E.; Monroy, M.; Moon, P. A.; Vovides, A. M.   "Regeneration of *Ceratozamia euryphyllidia* (Cycadales, Gymnospermae) plants from embryogenic leaf cultures derived from mature-phase trees".   *Plant Cell Reports* **17** (8):   612–616.   doi:10.1007/s002990050452.

[8] Jeon, MeeHee; Sung, SangHyun; Huh, Hoon; Kim, Young-Choong.   "Ginkgolide B production in cultured cells derived from *Ginkgo biloba* L. leaves".   *Plant Cell Reports* **14** (8). doi:10.1007/BF00232783.

[9] Finer, John J.; Kriebel, Howard B.; Becwar, Michael R. (1 January 1989).   "Initiation of embryogenic callus and suspension cultures of eastern white pine (*Pinus strobus* L.)".   *Plant Cell Reports* **8** (4):   203–206. doi:10.1007/BF00778532.

[10] O'Dowd, Niamh A.; McCauley, Patrick G.; Richardson, David H. S.; Wilson, Graham.   "Callus production, suspension culture and *in vitro* alkaloid yields of Ephedra".   *Plant Cell, Tissue and Organ Culture* **34** (2):   149–155. doi:10.1007/BF00036095.

[11] Chen, Ying-Chun; Chang, Chen; Chang, Wei-chin. "A reliable protocol for plant regeneration from callus culture of Phalaenopsis". *In Vitro Cellular & Developmental Biology – Plant* **36** (5): 420–423. doi:10.1007/s11627-000-0076-5.

[12] Burris, Jason N.; Mann, David G. J.; Joyce, Blake L.; Stewart, C. Neal (10 October 2009).   "An Improved Tissue Culture System for Embryogenic Callus Production and Plant Regeneration in Switchgrass (*Panicum virgatum* L.)".   *BioEnergy Research* **2** (4):   267–274.   doi:10.1007/s12155-009-9048-8.

[13] Murashige, Toshio; F. Skoog (July 1962).   "A Revised Medium for Rapid Growth and Bio Assays with Tobacco Tissue Cultures".   *Physiologia Plantarum* **15** (3):   473–497. doi:10.1111/j.1399-3054.1962.tb08052.x.

[14] White, P. R. (Feb 1939). "Potentially unlimited growth of excised plant callus in an artificial nutrient". *American Journal of Botany* **26** (2): 59–4. doi:10.2307/2436709. JSTOR 2436709.

[15] Lloyd, G; B McCown (1981).   "Commercially-feasible micropropagation of mountain laurel, Kalmia latifolia, by use of shoot-tip culture".   *Combined Proceedings, International Plant Propagators' Society* **30**: 421–427.

[16] Gamborg, OL; RA Miller; K Ojima (April 1968).   "Nutrient requirements of suspension cultures of soybean root cells".   *Experimental Cell Research* **50** (1):   151–158. doi:10.1016/0014-4827(68)90403-5. PMID 5650857.

[17] Wang, X.-D.; Nolan, K. E.; Irwanto, R. R.; Sheahan, M. B.; Rose, R. J. (10 January 2011).   "Ontogeny of embryogenic callus in Medicago truncatula: the fate of the pluripotent and totipotent stem cells".   *Annals of Botany* **107** (4):   599–609. doi:10.1093/aob/mcq269.

[18] He, Yang; Guo, Xiulian; Lu, Ran; Niu, Bei; Pasapula, Vijaya; Hou, Pei; Cai, Feng; Xu, Ying; Chen, Fang. "Changes in morphology and biochemical indices in browning callus derived from Jatropha curcas hypocotyls".   *Plant Cell, Tissue and Organ Culture (PCTOC)* **98** (1):   11–17. doi:10.1007/s11240-009-9533-y.

[19] Dan, Yinghui; Armstrong, Charles L.; Dong, Jimmy; Feng, Xiaorong; Fry, Joyce E.; Keithly, Greg E.; Martinell, Brian J.; Roberts, Gail A.; Smith, Lori A.; Tan, Lalaine J.; Duncan, David R. "Lipoic acid—an unique plant transformation enhancer". *In Vitro Cellular & Developmental Biology - Plant* **45** (6): 630–638. doi:10.1007/s11627-009-9227-5.

[20] Pazuki, Arman & Sohani, Mehdi (2013). "Phenotypic evaluation of scutellum-derived calluses in 'Indica' rice cultivars" (PDF). *Acta Agriculturae Slovenica* **101** (2): 239–247. doi:10.2478/acas-2013-0020. Retrieved February 2, 2014.

[21] Razdan, M. K. (2003). *Introduction to plant tissue culture* (2. ed.). Enfield, NH [u.a.]: oxford Publishers. ISBN 1-57808-237-4.

[22] Gautheret, Roger J. (1 December 1983). "Plant tissue culture: A history". *The Botanical Magazine Tokyo* **96** (4): 393–410. doi:10.1007/BF02488184.

[23] Chawla, H.S. (2002). *Introduction to plant biotechnology* (2nd ed.). Enfield, N.H.: Science Publishers. ISBN 1-57808-228-5.

Chapter 21

Endothelial stem cell

Endothelial stem cells (ESCs) are one of three types of stem cells found in bone marrow. They are multipotent, which describes the ability to give rise to many cell types, whereas a pluripotent stem cell can give rise to all types. ESCs have the characteristic properties of a stem cell: self-renewal and differentiation. These parent stem cells, ESCs, give rise to progenitor cells, which are intermediate stem cells that lose potency. Progenitor stem cells are committed to differentiating along a particular cell developmental pathway. ESCs will eventually produce endothelial cells (ECs), which create the thin-walled endothelium that lines the inner surface of blood vessels and lymphatic vessels.

21.1 Sources

ECs were first thought to arise from extraembryonic tissues because blood vessels were observed in the avian and mammalian embryos. However, after histological analysis, it was seen that ECs were in the embryo. This meant that blood vessels come from an intraembryonic source, the mesoderm.[1]

21.2 Properties

21.2.1 Self-renewal and differentiation

Stem cells have the unique ability make identical copies of themselves. This property maintains unspecialized and un-differentiated cells within the body. Differentiation is the process by which a cell becomes more specialized. For stem cells, this usually occurs through several stages, where a cell proliferates giving rise to daughter cells that are further specialized.[2] For example, an endothelial progenitor cell (EPC) is more specialized than an ESC, and an EC is more specialized than an EPC. The further specialized a cell is, the more differentiated it is and as a result it is considered to be more committed to a certain cellular lineage.[2]

21.2.2 Blood vessel formation

Blood vessels are made of a thin layer of ECs. As part of the circulatory system, blood vessels play a critical role in transporting blood throughout the body. Consequently, ECs have unique functions such as fluid filtration, homeostasis and hormone trafficking. ECs are the most differentiated form of an ESC. Formation of new blood vessels occurs by two different processes: vasculogenesis and angiogenesis.[3] The former requires differentiation of endothelial cells from hemangioblasts and then the further organization into a primary capillary network. The latter occurs when new vessels are built from preexisting blood vessels.[3]

21.2.3 Markers

The vascular system is made up of two parts: 1) Blood vasculature 2) Lymphatic vessels

Both parts consist of ECs that show differential expression of various genes. A study showed that ectopic expression of Prox-1 in blood vascular ECs (BECs) induced one-third of LEC specific gene expression. Prox-1is a homeobox transcription factor found in lymphatic ECs (LECs). For example, specific mRNAs such as VEGFR-3 and p57Kip2 were expressed by the BEC that was induced to express Prox-1.[4]

Lymphatic-specific vascular endothelial growth factors VEGF-C and VEGF-D function as ligands for the vascular endothelial growth factor receptor 3 (VEGFR-3). The ligand-receptor interaction is essential for normal development of lymphatic tissues.[5]

Tal1 gene is specifically found in the vascular endothelium and developing brain.[5] This gene encodes the basic helix-loop-helix structure and functions as a transcription factor. Embryos lacking *Tal1* fail to develop past embryonic day 9.5. However, the study found that *Tal1* is actually required for vascular remodeling of the capillary network, rather than early endothelial development itself.[5]

Fetal liver kinase-1 (Flk-1) is a cell surface receptor protein that is commonly used as a marker for ESCs and EPCs.[2]

CD34 is another marker that can be found on the surface of ESCs and EPCs. It is characteristic of hematopoietic stem cells, as well as muscle stem cells.[2]

21.3 Role in formation of vascular system

The two lineages arising from the EPC and the hematopoietic progenitor cell (HPC) form the blood circulatory system. Hematopoietic stem cells can of course undergo self-renewal, and are multipotent cells that give rise to erythrocytes (red blood cells), megakaryocytes/platelets, mast cells, T-lymphocytes, B-lymphocytes, dendritic cells, natural killer cells, monocyte/macrophage, and granulocytes.[6] A study found that in the beginning stages of mouse embryogenesis, commencing at embryonic day 7.5, HPCs are produced close to the emerging vascular system. In the yolk sac's blood islands, HPCs and EC lineages emerge from the extraembryonic mesoderm in near unison. This creates a formation in which early erythrocytes are enveloped by angioblasts, and together they give rise to mature ECs. This observation gave rise to the hypothesis that the two lineages come from the same precursor, termed hemangioblast.[5] Even though there is evidence that corroborates a hemangioblast, the isolation and exact location in the embryo has been difficult to pinpoint. Some researchers have found that cells with hemangioblast properties have been located in the posterior end of the primitive streak during gastrulation.[1]

In 1917, Florence Sabin first observed the development of blood vessels and red blood cells in the yolk sac of chick embryos occur in close proximity and time.[7] Then, in 1932, Murray detected the same event and created the term "hemangioblast" for what Sabin had seen.[8]

Further evidence to corroborate hemangioblasts come from the expression of various genes such as CD34 and Tie2 by both lineages. The fact that this expression was seen in both EC and HPC lineages led researchers to propose a common origin. However, endothelial markers like Flk1/VEGFR-2 are exclusive to ECs but stop HPCs from progressing into an EC. It is accepted that VEGFR-2+ cells are a common precursor for HPCs and ECs. If the *Vegfr3* gene is deleted then both HPC and EC differentiation comes to a halt in embryos. VEGF promotes angioblast differentiation; whereas, VEGFR-1 stops the hemangioblast from becoming an EC. In addition, basic fibroblast growth factor FGF-2 is also involved in promoting angioblasts from the mesoderm. After angioblasts commit to becoming an EC, the angioblasts gather and rearrange to assemble in a tube similar to a capillary. Angioblasts can travel during the formation of the circulatory system to configure the branches to allow for directional blood flow. Pericytes and smooth muscle cells encircle ECs when they are differentiating into arterial or venous arrangements. Surrounding the ECs creates a brace to help stabilize the vessels known as the pericellular basal lamina. It is suggested pericytes and smooth muscle cells come from neural crest cells and the surrounding mesenchyme.[5]

21.4 Role of insulin-like growth factors in endothelium differentiation

ECs derived from stem cells are the beginning of vasculogenesis.[9] Vasculogenesis is the new production of a vascular network from mesodermal progenitor cells. This can be distinguished from angiogenesis, which is the creation of new capillaries from vessels that already exist through the process of splitting or sprouting.[10] This can occur "in vitro" in embryoid bodies (EB) derived from embryonic stem cells; this process in EB is similar to "in vivo" vasculogenesis. Important signaling factors for vasculogenesis are TGF-β, BMP4, and VEGF, all of which promote pluripotent stem cells to differentiate into mesoderm, endothelial progenitor cells, and then into mature endothelium.[9]

It is well established that insulin-like growth factor (IGF) signaling is important for cell responses such as mitogenesis, cell growth, proliferation, angiogenesis, and differentiation. IGF1 and IGF2 increase the production of ECs in EB. A method that IGF employs to increase vasculogenesis is upregulation of VEGF. Not only is VEGF critical for mesoderm cells to become an EC, but also for EPCs to differentiate into mature endothelium. Understanding this process can lead to further research in vascular regeneration.[9]

21.5 Animal models of vasculogenesis

There are a number of models used to study vasculogenesis. Avian embryos, Xenopus laevis embryos, are both fair models. However, zebrafish and mouse embryos have widespread use for easily observed development of vascular systems, and the recognition of key parts of molecular regulation when ECs differentiate.[1]

21.6 Role in recovery

ESCs and EPCs eventually differentiate into ECs. The endothelium secretes soluble factors to regulate vasodilatation and to preserve homeostasis.[11] When there is any dysfunction in the endothelium, the body aims to repair the damage. Resident ESCs can generate mature ECs that replace the damaged ones.[12] However, the intermediate progenitor cell cannot always generate functional ECs. This is because some of the differentiated cells may just have angiogenic properties.[12]

Studies have shown that when vascular trauma occurs, EPCs and circulating endothelial progenitors (CEPs) are attracted to the site due to the release of specific chemokines.[13] CEPs are derived from EPCs within the bone marrow, and the bone marrow is a reservoir of stem and progenitor cells. These cell types accelerate the healing process and prevent further complications such as hypoxia by gathering the cellular materials to reconstruct the endothelium.[13]

Endothelium dysfunction is a prototypical characteristic of vascular disease, common in patients with autoimmune diseases such as systemic lupus erythematosus.[14] Further, there is an inverse relationship between age and levels of EPCs. With a decline in EPCs the body loses its ability to repair the endothelium.[12]

The use of stem cells for treatment has become a growing interest in the scientific community. Distinguishing between an ESC and its intermediate progenitor is nearly impossible,[2] so research is now being done broadly on EPCs. One study showed that brief exposure to sevoflurane promoted growth and proliferation of EPCs.[15] Sevoflurane is used in general anesthesia, but this finding shows the potential to induce endothelial progenitors. Using stem cells for cell replacement therapies is known as "regenerative medicine", which is a booming field that is now working on transplanting cells as opposed to bigger tissues or organs.[15]

21.7 Role in cancer

Understanding more about ESCs is important in cancer research. Tumours induce angiogenesis, which is the formation of new blood vessels. These cancerous cells do this by secreting factors such as VEGF and by reducing the amount of PGK, an anti-VEGF enzyme. The result is an uncontrolled production of beta-catenin, which regulates cell growth and cell mobility. With uncontrolled beta-catenin, the cell loses its adhesive properties. As ECs get packed together to create the lining for a new blood vessel, a single cancer cell is able to travel through the vessel to a distant site. If that cancer cell implants itself and begins forming a new tumour, the cancer has metastasized.[16]

21.8 Future efforts

Stem cells have always been a huge interest for scientists due to their unique properties that make them unlike any other cell in the body. Generally, the idea boils down to harnessing the power of plasticity and the ability to go from an unspecialized cell to a highly specialized differentiated cell. ESCs play an incredibly important role in establishing the vascular network that is vital for a functional circulatory system. Consequently, EPCs are under study to determine the potential for treatment of ischemic heart disease.[17] Scientists are still trying to find a way to definitely distinguish the stem cell from the progenitor. In the case of endothelial cells, it is even difficult to distinguish a mature EC from an EPC. However, because of the multipotency of the ESC, the discoveries made about EPCs will parallel or understate the powers of the ESC.[17]

21.9 See also

- endometrial regenerative cells, also known as endometrial stem cells, derived from mammalian uterus lining.

21.10 References

[1] Ferguson JW, Kelley RW, Patterson C. (2005). "Mechanisms of endothelial differentiation in embryonic vasculogenesis". *Journal of the American Heart Association* 25: 2246–2254. doi:10.1161/01.atv.0000183609.55154.44.

[2] Bethesda MD. (6 April 2009). "Stem Cell Basics". *In Stem Cell Information*. National Institutes of Health, U.S. Department of Health and Human Services. Retrieved 6 March 2012.

[3] Gehling U, Ergun S, Schumacher U, Wagener C, Pantel K, Otte M, Schuch G, Schafhausen P, Mende T, Kilic N, Kluge K, Schafer B, Hossfeld D, Fiedler W. (2000). "In vitro differentiation of endothelial cells from AC133-positive progenitor cells". *American Journal of Hematology* 95 (10): 3106–3112.

[4] Petrova TV, Makinen T, Makela TP, Saarela J, Virtanen I, Ferrell RE, Finegold DN, Kerjaschki D, Y;a-Herttuala S, Alitalo K. (2002). "Lymphatic endothelial reprogramming of vascular endothelial cells by the Prox-1 homeobox transcription factor". *EMBO Journal* 21 (17): 4593–4599. doi:10.1093/emboj/cdf470. PMC 125413. PMID 12198161.

[5] Kubo H. Alitalo K. (2003). "The bloody fate of endothelial stem cells". *Genes & Development* **17**: 322–329. doi:10.1101/gad.1071203.

[6] Seita J. Weissman IL. (2010). "Hematopoietic stem cell: self-renewal versus differentiation". *Systems Biology and Medicine* **2**: 640–653. doi:10.1002/wsbm.86.

[7] Sabin F. (1917). "Preliminary note on the differentiation of angioblasts and the method by which they produce blood-vessels, blood-plasma and red blood-cells as seen in the living chick". *The Anatomical Record* **13**: 199–204. doi:10.1002/ar.1090130403.

[8] Murray PDF. (1932). "The development in vitro of the blood of the early chick embryo". *Proceedings of the Royal Society of London. Series B, Containing Papers of a Biological Character* **111**: 497–521.

[9] Piecewicz SM. Pandey A. Roy B. Xiang SH. Zetter BR. Sengupta S. (2012). "Insulin-like growth factors promote vasculogenesis in embryonic stem cells". *PLoS ONE* **21** (17): e32191. doi:10.1371/journal.pone.0032191.

[10] Kovacic JC. Moore J. Herbert A. Ma D. Boehm M. Graham RM. (2008). "Endothelial Progenitor Cells, Angioblasts, and Angiogenesis- Old terms Reconsidered from a new current perspective". *Trends in Cardiovascular Medicine* **18**: 45–51. doi:10.1016/j.tcm.2007.12.002. PMID 18308194.

[11] Cheek D. Graulty R. Bryant S. (2002). "Meet the multitasking endothelium". *Nursing Made Incredibly Easy!* **6** (4): 18–25. doi:10.1097/01.nme.0000324934.19114.e0.

[12] Siddique A. Shantsila E. Lip G. Varma C. (2010). "Endothelial progenitor cells: what use for the cardiologist?". *Journal Angiogenesis Research* **2** (6). doi:10.1186/2040-2384-2-6.

[13] Rafil S. Lyden D. (2003). "Therapeutic stem and progenitor cell transplantation for organ vascularization and regeneration". *Nature Medicine* **9** (6): 702–12. doi:10.1038/nm0603-702. PMID 12778169.

[14] Deanfield J, Donald A, Ferri C, Giannattasio C, Halcox J, Halligan S, Lerman A, Mancia G, Oliver JJ, Pessina AC, Rizzoni D, Rossi GP, Salvetti A, Schiffrin EL, Taddei S, Webb DJ (2005). "Endothelial function and dysfunction. Part I: Methodological issues for assessment in the different vascular beds: a statement by the Working Group on Endothelin and Endothelial Factors of the European Society of Hypertension". *Journal of Hypertension* **23** (1): 7–17. doi:10.1097/00004872-200501000-00004. PMID 15643116.

[15] Lucchinetti E, Zeisberger SM, Baruscotti I, Wacker J, Feng J, Dubey R, Zisch AH, Zaugg M. (2009). "Stem cell-like human endothelial progenitors show enhanced colony-forming capacity after brief sevoflurane exposure: preconditioning of angiogenic cells by volatile anesthetics". *Anesthesia & Analgesia* **109** (4): 1117–26. doi:10.1213/ane.0b013e3181b5a277.

[16] Enzyme eliminated by cancer cells holds promise for cancer treatment

[17] Fan CL, Li Y, Gao PJ, Liu JJ, Zhang XJ, Zhu DL. (2003). "Differentiation of endothelial progenitor cells from human umbilical cord blood CD 34+ cells in vitro". *Acta Pharmacologica Sinica* **24** (3): 212–218.

Chapter 22

Hematopoietic stem cell

Hematopoietic stem cells (**HSCs**) or **hemocytoblasts** are the stem cells that give rise to all the other blood cells through the process of haematopoiesis. They are derived from mesoderm and located in the red bone marrow, which is contained in the core of most bones.

They give rise to both the myeloid and lymphoid lineages of blood cells. (Myeloid cells include monocytes, macrophages, neutrophils, basophils, eosinophils, erythrocytes, dendritic cells, and megakaryocytes or platelets. Lymphoid cells include T cells, B cells, and natural killer cells.) The definition of hematopoietic stem cells has changed in the last two decades. The hematopoietic tissue contains cells with long-term and short-term regeneration capacities and committed multipotent, oligopotent, and unipotent progenitors. HSCs constitute 1:10.000 of cells in myeloid tissue.

HSCs are a heterogeneous population. The third category consists of the balanced (Bala) HSC, whose L/M ratio is between 3 and 10. Only the myeloid-biased and -balanced HSCs have durable self-renewal properties. In addition, serial transplantation experiments have shown that each subtype preferentially re-creates its blood cell type distribution, suggesting an inherited epigenetic program for each subtype.

HSC studies through much of the past half century have led to a much deeper understanding. More recent advances have resulted in the use of HSC transplants in the treatment of cancers and other immune system disorders.[1]

22.1 Sources

HSCs are found in the bone marrow of adults, specially in the pelvis, femur, and sternum. They are also found in umbilical cord blood and, in small numbers, in peripheral blood.[2]

Stem and progenitor cells can be taken from the pelvis, at the iliac crest, using a needle and syringe.[3] The cells can

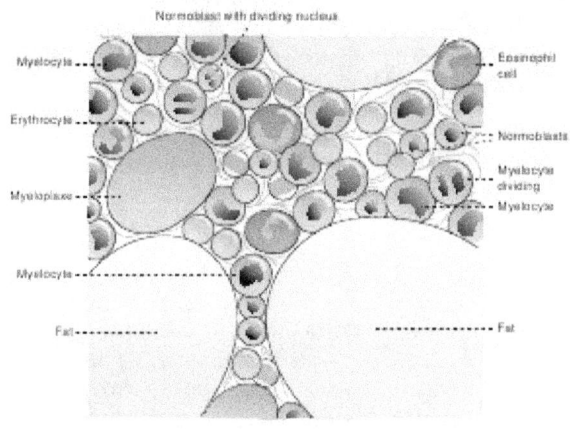

Sketch of bone marrow and its cells

be removed as liquid (to perform a smear to look at the cell morphology) or they can be removed via a core biopsy (to maintain the architecture or relationship of the cells to each other and to the bone).

In order to harvest stem cells from the circulating peripheral blood, blood donors are injected with a cytokine, such as granulocyte-colony stimulating factor (G-CSF), that induces cells to leave the bone marrow and circulate in the blood vessels.

In mammalian embryology, the first definitive HSCs are detected in the AGM (aorta-gonad-mesonephros), and then massively expanded in the fetal liver prior to colonising the bone marrow before birth.[4]

22.2 Functional characteristics

22.2.1 Multipotency and self-renewal

HSCs can replenish all blood cell types (i.e., are multipotent) and self-renew. A small number of HSCs can expand to generate a very large number of daughter HSCs.

This phenomenon is used in bone marrow transplantation, when a small number of HSCs reconstitute the hematopoietic system. This process indicates that, subsequent to bone marrow transplantation, symmetrical cell divisions into two daughter HSCs must occur.

Stem cell self-renewal is thought to occur in the stem cell niche in the bone marrow, and it is reasonable to assume that key signals present in this niche will be important in self-renewal. There is much interest in the environmental and molecular requirements for HSC self-renewal, as understanding the ability of HSC to replenish themselves will eventually allow the generation of expanded populations of HSC *in vitro* that can be used therapeutically.

22.2.2 Stem cell heterogeneity

It was originally believed that all HSCs were alike in their self-renewal and differentiation abilities. This view was first challenged by the 2002 discovery by the Muller-Sieburg group in San Diego, who illustrated that different stem cells can show distinct repopulation patterns that are epigenetically predetermined intrinsic properties of clonal Thy-1^{lo} Sca-1$^+$ lin$^-$ c-kit$^+$ HSC.[5][6][7] The results of these clonal studies led to the notion of **lineage bias**. Using the ratio $\rho = L/M$ of lymphoid (L) to myeloid (M) cells in blood as a quantitative marker, the stem cell compartment can be split into three categories of HSC. **Balanced (Bala) HSCs** repopulate peripheral white blood cells in the same ratio of myeloid to lymphoid cells as seen in unmanipulated mice (on average about 15% myeloid and 85% lymphoid cells, or $3 \leq \rho \leq 10$). **Myeloid-biased (My-bi) HSCs** give rise to very few lymphocytes resulting in ratios $0 < \rho < 3$, while **lymphoid-biased (Ly-bi) HSCs** generate very few myeloid cells, which results in lymphoid-to-myeloid ratios of $\rho > 10$. All three types are normal types of HSC, and they do not represent stages of differentiation. Rather, these are three classes of HSC, each with an epigenetically fixed differentiation program. These studies also showed that lineage bias is not stochastically regulated or dependent on differences in environmental influence. My-bi HSC self-renew longer than balanced or Ly-bi HSC. The myeloid bias results from reduced responsiveness to the lymphopoetin interleukin 7 (IL-7).[6]

Subsequently, other groups confirmed and highlighted the original findings.[8] For example, the Eaves group confirmed in 2007 that repopulation kinetics, long-term self-renewal capacity, and My-bi and Ly-bi are stably inherited intrinsic HSC properties.[9] In 2010, the Goodell group provided additional insights about the molecular basis of lineage bias in side population (SP) SCA-1$^+$ lin$^-$ c-kit$^+$ HSC.[10] As previously shown for IL-7 signaling, it was found that a member of the transforming growth factor family (TGF-beta) induces and inhibits the proliferation of My-bi and Ly-bi HSC, respectively.

22.2.3 Functions

A *cobblestone area-forming cell (CAFC)* assay is a cell culture-based empirical assay. When plated onto a confluent culture of stromal feeder layer, a fraction of HSCs creep between the gaps (even though the stromal cells are touching each other) and eventually settle between the stromal cells and the substratum (here the dish surface) or trapped in the cellular processes between the stromal cells. Emperipolesis is the *in vivo* phenomenon in which one cell is completely engulfed into another (e.g. thymocytes into thymic nurse cells); on the other hand, when *in vitro*, lymphoid lineage cells creep beneath nurse-like cells, the process is called pseudoemperipolesis. This similar phenomenon is more commonly known in the HSC field by the cell culture terminology *cobble stone area-forming cells (CAFC)*, which means areas or clusters of cells look dull cobblestone-like under phase contrast microscopy, compared to the other HSCs, which are refractile. This happens because the cells that are floating loosely on top of the stromal cells are spherical and thus refractile. However, the cells that creep beneath the stromal cells are flattened and, thus, not refractile. The mechanism of pseudoemperipolesis is only recently coming to light. It may be mediated by interaction through CXCR4 (CD184) the receptor for CXC Chemokines (e.g., SDF1) and $\alpha4\beta1$ integrins.[11]

22.2.4 Mobility

HSCs have a higher potential than other immature blood cells to pass the bone marrow barrier, and, thus, may travel in the blood from the bone marrow in one bone to another bone. If they settle in the thymus, they may develop into T cells. In the case of fetuses and other extramedullary hematopoiesis, HSCs may also settle in the liver or spleen and develop.

This ability is the reason why HSCs may be harvested directly from the blood.

22.3 Physical characteristics

With regard to morphology, hematopoietic stem cells resemble lymphocytes. They are non-adherent, and rounded, with a rounded nucleus and low cytoplasm-to-nucleus ratio. Since primitive hematopoietic stem cells (PHSCs) cannot be isolated as a pure population, it is not possible to identify them in a microscope. The above description is based on the morphological characteristics of a heterogeneous population, of which PHSCs are a component.

22.4 Markers

In reference to phenotype, hematopoeitic stem cells are identified by their small size, lack of lineage (lin) markers, low staining (side population) with vital dyes such as rhodamine 123 (rhodamine^{DULL}, also called rho^{lo}) or Hoechst 33342, and presence of various antigenic markers on their surface.

22.4.1 Cluster of differentiation and other markers

Many of these markers belong to the cluster of differentiation series, like: CD34, CD38, CD90, CD133, CD105, CD45, and also c-kit, - the receptor for stem cell factor. The hematopoietic stem cells are negative for the markers that are used for detection of lineage commitment, and are, thus, called Lin-; and, during their purification by FACS, a mixture of up to 14 different mature blood-lineage-marker antibodies are used to deplete the lin+ cells or late multipotent progenitors (MPP)s: e.g., CD13 & CD33 for myeloid, CD71 for erythroid, CD19 for B cells, CD61 for megakaryocytic, etc. for humans; and, B220 (murine CD45) for B cells, Mac-1 (CD11b/CD18) for monocytes, Gr-1 for Granulocytes, Ter119 for erythroid cells, Il7Ra, CD3, CD4, CD5, CD8 for T cells, etc. (for mice)

There are many differences between the human and mice hematopoietic cell markers for the commonly accepted type of hematopoietic stem cells.[12]

- **Mouse HSC** : CD34$^{lo/-}$, SCA-1$^+$, Thy1.1$^{+/lo}$, CD38$^+$, C-kit$^+$, lin$^-$

- **Human HSC** : CD34$^+$, CD59$^+$, Thy1/CD90$^+$, CD38$^{lo/-}$, C-kit/CD117$^+$, lin$^-$

However, not all stem cells are covered by these combinations that, nonetheless, have become popular. In fact, even in humans, there are hematopoietic stem cells that are CD34$^-$/CD38$^-$.[13][14] Also some later studies suggested that earliest stem cells may lack c-kit on the cell surface.[15] For human HSCs use of CD133 was one step ahead as both CD34$^+$ and CD34$^-$ HSCs were CD133$^+$.

Traditional purification method used to yield a reasonable purity level of mouse hematopoietic stem cells, in general, requires a large(~10-12) battery of markers, most of which were surrogate markers with little functional significance, and thus partial overlap with the stem cell populations and

sometimes other closely related cells that are not stem cells. Also, some of these markers (e.g., Thy1) are not conserved across mouse species, and use of markers like CD34$^-$ for HSC purification requires mice to be at least 8 weeks old.

22.4.2 SLAM code

Alternative methods that could give rise to a similar or better harvest of stem cells is an active area of research, and are presently emerging. One such method uses a signature of *SLAM* family of cell surface molecules. The SLAM (Signaling lymphocyte activation molecule) family is a group of more than 10 molecules whose genes are located mostly tandemly in a single locus on chromosome 1 (mouse), all belonging to a subset of the immunoglobulin gene superfamily, and originally thought to be involved in T-cell stimulation. This family includes CD48, CD150, CD244, etc., CD150 being the founding member, and, thus, also known as slamF1, i.e., SLAM family member 1.

The signature **SLAM codes** for the hemopoietic hierarchy are:

- **Hematopoietic stem cells (HSC)** : CD150$^+$CD48$^-$CD244$^-$

- **Multipotent progenitor cells (MPPs)** : CD150$^-$CD48$^-$CD244$^+$

- **Lineage-restricted progenitor cells (LRPs)** : CD150$^-$CD48$^+$CD244$^+$

- **Common myeloid progenitor (CMP)** : lin$^-$SCA-1$^-$c-kit$^+$CD34$^+$CD16/32^{mid}

- **Granulocyte-macrophage progenitor (GMP)** : lin$^-$SCA-1$^-$c-kit$^+$CD34$^+$CD16/32^{hi}

- **Megakaryocyte-erythroid progenitor (MEP)** : lin$^-$SCA-1$^-$c-kit$^+$CD34$^-$CD16/32^{low}

For HSCs, CD150$^+$CD48$^-$ was sufficient instead of CD150$^+$CD48$^-$CD244$^-$ because CD48 is a ligand for CD244, and both would be positive only in the activated lineage-restricted progenitors. It seems that this code was more efficient than the more tedious earlier set of the large number of markers, and are also conserved across the mouse strains; however, recent work has shown that this method excludes a large number of HSCs and includes an equally large number of non-stem cells.[16] [17] CD150$^+$CD48$^-$ gave stem cell purity comparable to Thy1^{lo}SCA-1$^+$lin$^-$c-kit$^+$ in mice.[18]

22.4.3 LT-HSC/ST-HSC/Early MPP/Late MPP

Irving Weissman's group at Stanford University was the first to isolate mouse hematopoietic stem cells in 1988 and was also the first to work out the markers to distinguish the mouse long-term (LT-HSC) and short-term (ST-HSC) hematopoietic stem cells (self-renew-capable), and the Multipotent progenitors (MPP, low or no self-renew capability — the later the developmental stage of MPP, the lesser the self-renewal ability and the more of some of the markers like CD4 and CD135):

- **LT-HSC** : $CD34^-$, $CD38^-$, $SCA-1^+$, $Thy1.1^{+/lo}$, C-kit^+, lin^-, $CD135^-$, $Slamf1/CD150^+$

- **ST-HSC** : $CD34^+$, $CD38^+$, $SCA-1^+$, $Thy1.1^{+/lo}$, C-kit^+, lin^-, $CD135^-$, $Slamf1/CD150^+$, Mac-1 $(CD11b)^{lo}$

- **Early MPP** : $CD34^+$, $SCA-1^+$, $Thy1.1^-$, C-kit^+, lin^-, $CD135^+$, $Slamf1/CD150^-$, Mac-1 $(CD11b)^{lo}$, $CD4^{lo}$

- **Late MPP** : $CD34^+$, $SCA-1^+$, $Thy1.1^-$, C-kit^+, lin^-, $CD135^{high}$, $Slamf1/CD150^-$, Mac-1 $(CD11b)^{lo}$, $CD4^{lo}$

22.5 Nomenclature of hematopoietic colonies and lineages

Between 1948 and 1950, the Committee for Clarification of the Nomenclature of Cells and Diseases of the Blood and Blood-forming Organs issued reports on the nomenclature of blood cells.[19][20] An overview of the terminology is shown below, from earliest to final stage of development:

- [root]blast

- pro[root]cyte

- [root]cyte

- meta[root]cyte

- mature cell name

The root for erythrocyte colony-forming units (CFU-E) is "rubri", for granulocyte-monocyte colony-forming units (CFU-GM) is "granulo" or "myelo" and "mono", for lympocyte colony-forming units (CFU-L) is "lympho" and for megakaryocyte colony-forming units (CFU-Meg) is "megakaryo". According to this terminology, the stages of red blood cell formation would be: rubriblast, prorubricyte, rubricyte, metarubricyte, and erythrocyte. However, the

following nomenclature seems to be, at present, the most prevalent:

Osteoclasts also arise from hemopoietic cells of the monocyte/neutrophil lineage, specifically CFU-GM.

22.6 Colony-forming units

In the context of hematopoietic stem cells, a colony-forming unit is a subtype of HSC. (This sense of the term is different from colony-forming units of microbes, which is a cell counting unit.) There are various kinds of HSC colony-forming units:

- Colony-forming unit lymphocyte (CFU-L)

- Colony-forming unit erythrocyte (CFU-E)

- Colony-forming unit granulo-monocyte (CFU-GM)

- Colony-forming unit megakaryocyte (CFU-Meg)

- Colony-forming unit Basophil (CFU-B)

- Colony-forming unit Eosinophil (CFU-Eo)

The above CFUs are based on the lineage. Another CFU, the *colony-forming unit–spleen* (**CFU–S**) was the basis of an *in vivo* clonal colony formation, which depends on the ability of infused bone marrow cells to give rise to clones of maturing hematopoietic cells in the spleens of irradiated mice after 8 to 12 days. It was used extensively in early studies, but is now considered to measure more mature progenitor or **Transit Amplifying Cells** rather than stem cells.

22.7 HSC repopulation kinetics

Hematopoietic stem cells (HSC) cannot be easily observed directly, and, therefore, their behaviors need to be inferred indirectly. Clonal studies are likely the closest technique for single cell in vivo studies of HSC. Here, sophisticated experimental and statistical methods are used to ascertain that, with a high probability, a single HSC is contained in a transplant administered to a lethally irradiated host. The clonal expansion of this stem cell can then be observed over time by monitoring the percent donor-type cells in blood as the host is reconstituted. The resulting time series is defined as the repopulation kinetic of the HSC.

The reconstitution kinetics are very heterogeneous. However, using symbolic dynamics, one can show that they fall into a limited number of classes.[21] To prove this, several hundred experimental repopulation kinetics from clonal $Thy-1^{lo}$ $SCA-1^+$ lin^- $c-kit^+$ HSC were translated into symbolic sequences by assigning the symbols "+", "-", "~" whenever two successive measurements of the percent donor-type cells have a positive, negative, or unchanged slope, respectively. By using the Hamming distance, the re-population patterns were subjected to cluster analysis yield-ing 16 distinct groups of kinetics. To finish the empirical proof, the Laplace add-one approach was used to determine that the probability of finding kinetics not contained in these 16 groups is very small. By corollary, this result shows that the hematopoietic stem cell compartment is also heteroge-neous by dynamical criteria.

22.8   DNA damage and aging

DNA strand breaks accumulate in long term HSCs during aging.[22] This accumulation is associated with a broad at-tenuation of DNA repair and response pathways that de-pends on HSC quiescence.[22] Non-homologous end join-ing (NHEJ) is a pathway that repairs double-strand breaks in DNA. NHEJ is referred to as "non-homologous" because the break ends are directly ligated without the need for a ho-mologous template. The NHEJ pathway depends on several proteins including ligase 4, DNA polymerase mu and NHEJ factor 1 (NHEJ1, also known as Cernunnos or XLF).

DNA ligase 4 (Lig4) has a highly specific role in the repair of double-strand breaks by NHEJ. Lig4 deficiency in the mouse causes a progressive loss of HSCs during aging.[23] Deficiency of lig4 in pluripotent stem cells results in ac-cumulation of DNA double-strand breaks and enhanced apoptosis.[24]

In polymerase mu mutant mice, hematopoietic cell devel-opment is defective in several peripheral and bone marrow cell populations with about a 40% decrease in bone marrow cell number that includes several hematopoietic lineages.[25] Expansion potential of hematopoietic progenitor cells is also reduced. These characteristics correlate with reduced ability to repair double-strand breaks in hematopoietic tis-sue.

Deficiency of NHEJ factor 1 in mice leads to premature aging of hematopoietic stem cells as indicated by several lines of evidence including evidence that long-term re-population is defective and worsens over time.[26] Using a human induced pluripotent stem cell model of NHEJ1 deficiency, it was shown that NHEJ1 has an important role in promoting survival of the primitive hematopoietic progenitors.[27] These NHEJ1 deficient cells possess a weak NHEJ1-mediated repair capacity that is apparently inca-pable of coping with DNA damages induced by physiolog-ical stress, normal metabolism, and ionizing radiation.[27]

The sensitivity of haematopoietic stem cells to Lig4, DNA polymerase mu and NHEJ1 deficiency suggests that NHEJ

is a key determinant of the ability of stem cells to maintain themselves against physiological stress over time.[23] Rossi et al.[28] found that endogenous DNA damage accumulates with age even in wild type HSCs, and suggested that DNA damage accrual may be an important physiological mecha-nism of stem cell aging.

22.9   See also

- Hematopoiesis

- Hematopoietic stem cell transplantation

22.10   References

[1] "5. Hematopoietic Stem Cells." Stem Cell Information. Na-tional Institutes of Health, U.S. Department of Health and Human Services. 17 Jun 2011. Web. 9 Nov 2013. <http: //stemcells.nih.gov/info/scireport/pages/chapter5.aspx >

[2] http://cordadvantage.com/cord-blood-101/ hematopoietic-stem-cell

[3] "Bone Marrow Transplant Process". *Mayo Clinic*. Retrieved 18 March 2015.

[4] Dzierzak & Speck, Of lineage and legacy: the development of mammalian hematopoietic stem cells, Nature Immunol-ogy, 2008

[5] Muller-Sieburg CE, Cho RH, Thoman M, Adkins B, Sieburg HB, "Deterministc regulation of haematopoietic stem cell self-renewal and differentiation" *Blood* 2002; 100; 1302-9

[6] Muller-Sieburg CE, Cho RH, Karlson L, Huang JF, Sieburg HB (2004). "Myeloid-biased hematopoietic stem cells have extensive self-renewal capacity but generate diminished progeny with impaired IL-7 responsiveness". *Blood* 103: 4111–8. doi:10.1182/blood-2003-10-3448.

[7] Sieburg HB, Cho RH, Dykstra B, Eaves CJ, Muller-Sieburg CE (2006). "The haematopoietic stem cell compartment consists of a limited number of discrete stem cell subsets". *Blood* 107: 2311–6. doi:10.1182/blood-2005-07-2970.

[8] Schroeder T (2010). "Haematopoietic Stem Cell Hetero-geneity: Subtypes, Not Unpredictable Behavior". *Cell Stem Cell* 6: 203–207. doi:10.1016/j.stem.2010.02.006.

[9] Dykstra B, et al. "Long-Term Propagation of Distinct Hematopoietic Differentiation Programs In Vivo". *Cell Stem Cell* 1 (2): 218–229. doi:10.1016/j.stem.2007.05.015.

[10] Challen G, Boles NC, Chambers SM, Goodell MA (2010). "Distinct Haematopoietic Stem Cell Subtypes Are Differen-tially Regulated by TGF-beta1". *Cell Stem Cell* 6: 265–278. doi:10.1016/j.stem.2010.02.002.

[11] Burger JA, Spoo A, Dwenger A, Burger M, Behringer D. CXCR4 chemokine receptors (CD184) and alpha4beta1 integrins mediate spontaneous migration of human CD34+ progenitors and acute myeloid leukaemia cells beneath marrow stromal cells (pseudoemperipolesis).

[12] http://stemcells.nih.gov/info/scireport/chapter5.asp

[13] Bhatia M., Bonnet D., Murdoch B., Gan O.I., Dick J.E. (1998). "A newly discovered class of human haematopoietic cells with SCID-repopulating activity". *Nature Medicine* **4** (9): 1038–1045. doi:10.1038/2023.

[14] "CD34- Hematopoietic Stem Cells: Current Concepts and Controversies". *Stem Cells* **21**: 15–20. 2003. doi:10.1634/stemcells.21-1-15.

[15] Doi H. et al. (1997). "Pluripotent hemopoietic stem cells are c-kit<low". *Proc. Natl. Acad. Sci. USA* **94**: 2513–2517. doi:10.1073/pnas.94.6.2513.

[16] David C Weksberg, Stuart M Chambers, Nathan C Boles, and Margaret A Goodell CD150 negative Side Population cells represent a functionally distinct population of long-term haematopoietic stem cells" *Blood* 2007 : blood-2007-09-115006v1

[17] Gary Van Zant Stem cell markers: less is more! Blood 107: 855-856.

[18] Kiel; et al. "SLAM Family Receptors Distinguish Hematopoietic Stem and Progenitor Cells and Reveal Endothelial Niches for Stem Cells". *Cell* **121**: 1109–1121. doi:10.1016/j.cell.2005.05.026.

[19] "First report of the Committee for Clarification of the Nomenclature of Cells and Diseases of the Blood and Blood-forming Organs". *Am J Clin Pathol* **18**: 443–450. 1948.

[20] "Third, fourth and fifth reports of the committee for clarification of the nomenclature of cells and diseases of the blood and blood-forming organs". *Am J Clin Pathol* **20** (6): 562–79. 1950. PMID 15432355.

[21] Sieburg HB, Muller-Sieburg CE (2004). "Classification of short kinetics by shape". *In Silico Biol.* **4** (2): 209–17.

[22] Beerman I, Seita J, Inlay MA, Weissman IL, Rossi DJ (2014). "Quiescent hematopoietic stem cells accumulate DNA damage during aging that is repaired upon entry into cell cycle". *Cell Stem Cell* **15** (1): 37–50. doi:10.1016/j.stem.2014.04.016. PMID 24813857.

[23] Nijnik A, Woodbine L, Marchetti C, Dawson S, Lambe T, Liu C, Rodrigues NP, Crockford TL, Cabuy E, Vindigni A, Enver T, Bell JI, Slijepcevic P, Goodnow CC, Jeggo PA, Cornall RJ (2007). "DNA repair is limiting for haematopoietic stem cells during ageing". *Nature* **447** (7145): 686–90. doi:10.1038/nature05875. PMID 17554302.

[24] Tilgner K, Neganova I, Moreno-Gimeno I, Al-Aama JY, Burks D, Yung S, Singhapol C, Saretzki G, Evans J, Gorbunova V, Gennery A, Przyborski S, Stojkovic M, Armstrong L, Jeggo P, Lako M (2013). "A human iPSC model of Ligase IV deficiency reveals an important role for NHEJ-mediated-DSB repair in the survival and genomic stability of induced pluripotent stem cells and emerging haematopoietic progenitors". *Cell Death Differ.* **20** (8): 1089–100. doi:10.1038/cdd.2013.44. PMC 3705601. PMID 23722522.

[25] Lucas D, Escudero B, Ligos JM, Segovia JC, Estrada JC, Terrados G, Blanco L, Samper E, Bernad A (2009). "Altered hematopoiesis in mice lacking DNA polymerase mu is due to inefficient double-strand break repair". *PLoS Genet.* **5** (2): e1000389. doi:10.1371/journal.pgen.1000389. PMC 2638008. PMID 19229323.

[26] Avagyan S, Churchill M, Yamamoto K, Crowe JL, Li C, Lee BJ, Zheng T, Mukherjee S, Zha S (2014). "Hematopoietic stem cell dysfunction underlies the progressive lymphocytopenia in XLF/Cernunnos deficiency". *Blood* **124** (10): 1622–5. doi:10.1182/blood-2014-05-574863. PMC 4155271. PMID 25075129.

[27] Tilgner K, Neganova I, Singhapol C, Saretzki G, Al-Aama JY, Evans J, Gorbunova V, Gennery A, Przyborski S, Stojkovic M, Armstrong L, Jeggo P, Lako M (2013). "Brief report: a human induced pluripotent stem cell model of cernunnos deficiency reveals an important role for XLF in the survival of the primitive hematopoietic progenitors". *Stem Cells* **31** (9): 2015–23. doi:10.1002/stem.1456. PMID 23818183.

[28] Rossi DJ, Bryder D, Seita J, Nussenzweig A, Hoeijmakers J, Weissman IL (2007). "Deficiencies in DNA damage repair limit the function of haematopoietic stem cells with age". *Nature* **447** (7145): 725–9. doi:10.1038/nature05862. PMID 17554309.

22.11   Additional images

- Hematopoiesis

22.12   External links

- Fact sheet about blood stem cells on EuroStemCell

- Hematopoietic stem cells at the US National Library of Medicine Medical Subject Headings (MeSH)

Chapter 23

Mesenchymal stem cell

Mesenchymal stem cells, or **MSCs**, are multipotent stromal cells that can differentiate into a variety of cell types,[1] including: osteoblasts (bone cells),[2] chondrocytes (cartilage cells),[3] myocytes (muscle cells)[4] and adipocytes (fat cells). This phenomenon has been documented in specific cells and tissues in living animals and their counterparts growing in tissue culture.

23.1 Definition

While the terms *mesenchymal stem cell* and *marrow stromal cell* have been used interchangeably, neither term is sufficiently descriptive:

- Mesenchyme is embryonic connective tissue that is derived from the mesoderm and that differentiates into hematopoietic and connective tissue, whereas MSCs do not differentiate into hematopoietic cells.[5]

- Stromal cells are connective tissue cells that form the supportive structure in which the functional cells of the tissue reside. While this is an accurate description for one function of MSCs, the term fails to convey the relatively recently discovered roles of MSCs in the repair of tissue.[6]

- Because the cells, called MSCs by many labs today, can encompass multipotent cells derived from other non-marrow tissues, such as placenta,[7] umbilical cord blood, adipose tissue, adult muscle, corneal stroma[8] or the dental pulp of deciduous baby teeth, yet do not have the capacity to reconstitute an entire organ, the term multipotent stromal cell has been proposed as a better replacement.

The youngest, most primitive MSCs can be obtained from the umbilical cord tissue, namely Wharton's jelly and the umbilical cord blood. However the MSCs are found in much higher concentration in the Wharton's jelly compared to the umbilical cord blood, which is a rich source of hematopoietic stem cells. The umbilical cord is easily obtained after the birth of the newborn, is normally thrown away, and poses no risk for collection. The umbilical cord MSCs have more primitive properties than other adult MSCs obtained later in life, which might make them a useful source of MSCs for clinical applications.

An extremely rich source for mesenchymal stem cells is the developing tooth bud of the mandibular third molar. While considered multipotent, they may prove to be pluripotent. The stem cells eventually form enamel, dentin, blood vessels, dental pulp, and nervous tissues, including a minimum of 29 different unique end organs. Because of extreme ease in collection at 8–10 years of age before calcification, and minimal to no morbidity, they will probably constitute a major source for personal banking, research, and multiple therapies. These stem cells have been shown capable of producing hepatocytes.

Additionally, amniotic fluid has been shown to be a rich source of stem cells. As many as 1 in 100 cells collected during amniocentesis has been shown to be a pluripotent mesenchymal stem cell.[9]

Adipose tissue is one of the richest sources of MSCs. There are more than 500 times more stem cells in 1 gram of fat than in 1 gram of aspirated bone marrow. Adipose stem cells are actively being researched in clinical trials for treatment of a variety of diseases.

The presence of MSCs in peripheral blood has been controversial. However, a few groups have successfully isolated MSCs from human peripheral blood and been able to expand them in culture.[10] Australian company Cynata also claims the ability to mass-produce MSCs from induced pluripotent stem cells obtained from blood cells using the method of K. Hu et al.[11][12]

23.2 Characteristics

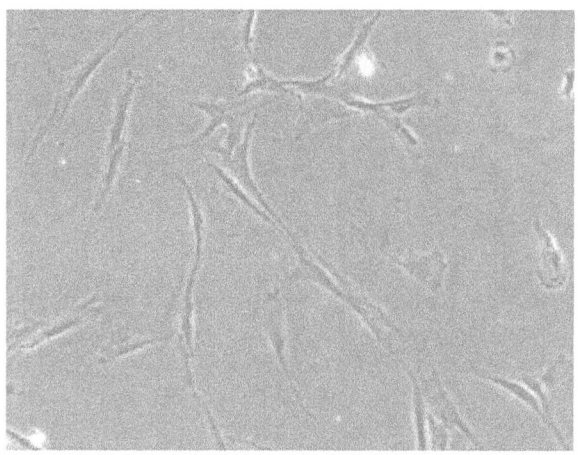

Human bone marrow derived Mesenchymal stem cell showing fibroblast like morphology seen under phase contrast microscope (carl zeiss axiovert 40 CFL) at 63 x magnification

23.2.1   Morphology

Mesenchymal stem cells are characterized morphologically by a small cell body with a few cell processes that are long and thin. The cell body contains a large, round nucleus with a prominent nucleolus, which is surrounded by finely dispersed chromatin particles, giving the nucleus a clear appearance. The remainder of the cell body contains a small amount of Golgi apparatus, rough endoplasmic reticulum, mitochondria, and polyribosomes. The cells, which are long and thin, are widely dispersed and the adjacent extracellular matrix is populated by a few reticular fibrils but is devoid of the other types of collagen fibrils.[13][14]

23.2.2   Detection

The International Society for Cellular Therapy (ISCT) has proposed a set of standards to define MSCs. A cell can be classified as an MSC if it shows plastic adherent properties under normal culture conditions and has a fibroblast-like morphology. In fact, some argue that MSCs and fibroblasts are functionally identical.[15] Furthermore, MSCs can undergo osteogenic, adipogenic and chondrogenic differentiation ex-vivo. The cultured MSCs also express on their surface CD73, CD90 and CD105, while lacking the expression of CD11b, CD14, CD19, CD34, CD45, CD79a and HLA-DR surface markers.[16]

23.2.3   Differentiation capacity

MSCs have a great capacity for self-renewal while maintaining their multipotency. Beyond that, there is little that can be definitively said. The standard test to confirm mul-

tipotency is differentiation of the cells into osteoblasts, adipocytes, and chondrocytes as well as myocytes and neurons. MSCs have been seen to even differentiate into neuron-like cells,[17] but there is lingering doubt whether the MSC-derived neurons are functional.[18] The degree to which the culture will differentiate varies among individuals and how differentiation is induced, e.g., chemical vs. mechanical;[19] and it is not clear whether this variation is due to a different amount of "true" progenitor cells in the culture or variable differentiation capacities of individuals' progenitors. The capacity of cells to proliferate and differentiate is known to decrease with the age of the donor, as well as the time in culture. Likewise, whether this is due to a decrease in the number of MSCs or a change to the existing MSCs is not known.

23.2.4   Immunomodulatory effects

Numerous studies have demonstrated that human MSCs avoid allorecognition, interfere with dendritic cell and T-cell function, and generate a local immunosuppressive microenvironment by secreting cytokines.[20] It has also been shown that the immunomodulatory function of human MSC is enhanced when the cells are exposed to an inflammatory environment characterised by the presence of elevated local interferon-gamma levels.[21] Other studies contradict some of these findings, reflecting both the highly heterogeneous nature of MSC isolates and the considerable differences between isolates generated by the many different methods under development.[22]

23.3   Culturing

The majority of modern culture techniques still take a colony-forming unit-fibroblasts (CFU-F) approach, where raw unpurified bone marrow or ficoll-purified bone marrow Mononuclear cell are plated directly into cell culture plates or flasks. Mesenchymal stem cells, but not red blood cells or haematopoetic progenitors, are adherent to tissue culture plastic within 24 to 48 hours. However, at least one publication has identified a population of non-adherent MSCs that are not obtained by the direct-plating technique.[23]

Other flow cytometry-based methods allow the sorting of bone marrow cells for specific surface markers, such as STRO-1.[24] STRO-1+ cells are generally more homogenous, and have higher rates of adherence and higher rates of proliferation, but the exact differences between STRO-1+ cells and MSCs are not clear.[25]

Methods of immunodepletion using such techniques as MACS have also been used in the negative selection of MSCs.[26]

The supplementation of basal media with fetal bovine serum or human platelet lysate is common in MSC culture. Prior the use of platelet lysates for MSC culture, the pathogen inactivation process is recommended to prevent pathogen transmission.[27]

23.4 Cancer

Mesenchymal stem cells have been shown to contribute to cancer progression in a number of different cancers, particularly the Hematological malignancies because they contact the transformed blood cells in the bone marrow.[28]

23.5 Medical use

Typical gross appearance of a tubular cartilaginous construct engineered from amniotic mesenchymal stem cells

The mesenchymal stem cells can be activated and mobilized if needed. However, the efficiency is very low. For instance, damage to muscles heals very slowly but further study into mechanisms of MSC action may provide avenues for increasing their capacity for tissue repair.[29]

Many of the early clinical successes using intravenous transplantation have come in systemic diseases like graft versus host disease and sepsis. However, it is becoming more accepted that diseases involving peripheral tissues, such as inflammatory bowel disease, may be better treated with methods that increase the local concentration of cells.[30] Direct injection or placement of cells into a site in need of repair may be the preferred method of treatment, as vascular delivery suffers from a "pulmonary first pass effect" where intravenous injected cells are sequestered in the lungs.[31] Clinical case reports in orthopedic applications have been published, though the number of patients treated is small

and these methods still lack rigorous study demonstrating effectiveness. Wakitani has published a small case series of nine defects in five knees involving surgical transplantation of mesenchymal stem cells with coverage of the treated chondral defects.[32]

In treating autoimmune disease

At least 218 clinical trials investigating the efficacy of mesenchymal stem cells in treating diseases have been initiated - many of which being autoimmune diseases. [33] Promising results have been shown in a variety of conditions, such as graft versus host disease, Crohn's disease, multiple sclerosis, systemic lupus erythematosus, and systemic sclerosis. [34] While their anti-inflammatory/immunomodulatory effects appear to greatly ameliorate autoimmune disease severity, the durability of these effects remain to be seen.

Clinical trials of cryopreserved MSCs

Scientists have reported that MSCs when transfused immediately within few hours post thawing may show reduced function or show decreased efficacy in treating diseases as compared to those MSCs which are in log phase of cell growth, so cryopreserved MSCs should be brought back into log phase of cell growth in *in vitro* culture before these are administered for clinical trials or experimental therapies, re-culturing of MSCs will help in recovering from the shock the cells get during freezing and thawing. Various clinical trials on MSCs have failed which used cryopreserved product immediately post thaw as compared to those clinical trials which used fresh MSCs.[35]

23.6 History

In 1924, Russian-born morphologist Alexander A. Maximow used extensive histological findings to identify a singular type of precursor cell within mesenchyme that develops into different types of blood cells.[36]

Scientists Ernest A. McCulloch and James E. Till first revealed the clonal nature of marrow cells in the 1960s.[37][38] An *ex vivo* assay for examining the clonogenic potential of multipotent marrow cells was later reported in the 1970s by Friedenstein and colleagues.[39][40] In this assay system, stromal cells were referred to as colony-forming unit-fibroblasts (CFU-f).

The first clinical trials of MSCs were completed in 1995 when a group of 15 patients were injected with cultured MSCs to test the safety of the treatment. Since then, over 200 clinical trials have been started. However, most are still in the safety stage of testing.[7]

Subsequent experimentation revealed the plasticity of marrow cells and how their fate could be determined by environmental cues. Culturing marrow stromal cells in the presence of osteogenic stimuli such as *ascorbic acid*, *inorganic phosphate*, and *dexamethasone* could promote their differentiation into osteoblasts. In contrast, the addition of *transforming growth factor-beta* (TGF-b) could induce chondrogenic markers.

23.7   Application in Therapy

Statistical-based analysis of MSC therapy for osteo-diseases inferred that most studies are still under investigation. There are different follow-up times that indicate we are still far from reaching the final conclusion. [41]

23.8   See also

- Bone marrow

- Intramembranous ossification

- Mesenchyme

- Multipotency

23.9   References

[1] Nardi, N. Beyer; da Silva Meirelles, L. (2006). "Mesenchymal Stem Cells: Isolation, In Vitro Expansion and Characterization". In Wobus, Anna M.; Boheler, Kenneth. *Stem Cells*. Handbook of experimental pharmacology **174**. pp. 249–82. doi:10.1007/3-540-31265-X_11. ISBN 978-3-540-77854-7.

[2] Oni OO (1992). "Early histological and ultrastructural changes in medullary fracture callus". *J Bone Joint Surg Am* **74** (4): 633–4. PMID 1583062.

[3] Brighton CT, Hunt RM (1997). "Early histologic and ultrastructural changes in microvessels of periosteal callus". *J Orthop Trauma* **11** (4): 244–53. doi:10.1097/00005131-199705000-00002. PMID 9258821.

[4] Pittenger (April 1999). "Multilineage potential of adult human mesenchymal stem cells". *Science* **284** (5411): 143–147. doi:10.1126/science.284.5411.143. PMID 10102814.

[5] Porcellini A (2009). "Regenerative medicine: a review". *Revista Brasileira de Hematologia e Hemoterapia* **31** (Suppl. 2). doi:10.1590/S1516-84842009000800017.

[6] Valero MC, Huntsman HD, Liu J, Zou K, Boppart MD (2012). "Eccentric exercise facilitates mesenchymal stem cell appearance in skeletal muscle". *PLoS ONE* **7** (1): e29760. doi:10.1371/journal.pone.0029760. PMC 3256189. PMID 22253772.

[7] Wang S, et al. (2012). "Clinical applications of mesenchymal stem cells". *JOURNAL OF HEMATOLOGY & ONCOLOGY* **5** (19). doi:10.1186/1756-8722-5-19.

[8] Branch MJ, Hashmani K, Dhillon P, Jones DR, Dua HS, Hopkinson A (2012). "Mesenchymal stem cells in the human corneal limbal stroma". *Invest. Ophthalmol. Vis. Sci.* **53** (9): 5109–16. doi:10.1167/iovs.11-8673. PMID 22736610.

[9] "What is Cord Tissue?". CordAdvantage.com.

[10] Chong PP, Selvaratnam L, Abbas AA, Kamarul T (2012). "Human peripheral blood derived mesenchymal stem cells demonstrate similar characteristics and chondrogenic differentiation potential to bone marrow derived mesenchymal stem cells". *J. Orthop. Res.* **30** (4): 634–42. doi:10.1002/jor.21556. PMID 21922534.

[11] Dylan Bushell-Embling (2015-02-19). "Cynata achieves world-first MSC breakthrough". LifeScientist Australia.

[12] Hu K et.al (2011). "Efficient generation of transgene-free induced pluripotent stem cells from normal and neoplastic bone marrow and cord blood mononuclear cells.". *Blood* **117** (14): e109–119. doi:10.1182/blood-2010-07-298331. PMID 21296996.

[13] Netter, Frank H. (1987). *Musculoskeletal system: anatomy, physiology, and metabolic disorders*. Summit, New Jersey: Ciba-Geigy Corporation. p. 134. ISBN 0-914168-88-6.

[14] Brighton CT, Hunt RM (1991). "Early histological and ultrastructural changes in medullary fracture callus". *The Journal of Bone and Joint Surgery* **73** (6): 832–47. PMID 2071617.

[15] Hematti P (2012). "Mesenchymal stromal cells and fibroblasts: a case of mistaken identity?". *Cytotherapy* **14** (5): 516–21. doi:10.3109/14653249.2012.677822. PMID 22458957.

[16] Dominici M, Le Blanc K, Mueller I, Slaper-Cortenbach I, Marini F, Krause D, Deans R, Keating A, Prockop Dj, Horwitz E (1 January 2006). "Minimal criteria for defining multipotent mesenchymal stromal cells. The International Society for Cellular Therapy position statement". *Cytotherapy* **8** (4): 315–317. doi:10.1080/14653240600855905. PMID 16923606.

[17] Jiang Y, Jahagirdar BN, Reinhardt RL, Schwartz RE, Keene CD, Ortiz-Gonzalez XR, Reyes M, Lenvik T, Lund T, Blackstad M, Du J, Aldrich S, Lisberg A, Low WC, Largaespada DA, Verfaillie CM (2002). "Pluripotency of mesenchymal stem cells derived from adult marrow". *Nature* **418** (6893): 41–49. doi:10.1038/nature00870. PMID 12077603.

[18] Franco Lambert AP, Fraga Zandonai A, Bonatto D, Cantarelli Machado D, Pêgas Henriques JA (2009). "Differentiation of human adipose-derived adult stem cells into neuronal tissue: Does it work?". *Differentiation* **77** (3): 221–8. doi:10.1016/j.diff.2008.10.016. PMID 19272520.

[19] Engler AJ, Sen S, Sweeney HL, Discher DE (2006). "Matrix Elasticity Directs Stem Cell Lineage Specification". *Cell* **126** (4): 677–89. doi:10.1016/j.cell.2006.06.044. PMID 16923388.

[20] Ryan JM, Barry FP, Murphy JM, Mahon BP (2005). "Mesenchymal stem cells avoid allogeneic rejection". *Journal of Inflammation* **2**: 8. doi:10.1186/1476-9255-2-8. PMC 1215510. PMID 16045800.

[21] Ryan JM, Barry F, Murphy JM, Mahon BP (2007). "Interferon-γ does not break, but promotes the immunosuppressive capacity of adult human mesenchymal stem cells". *Clinical & Experimental Immunology* **149** (2): 353–63. doi:10.1111/j.1365-2249.2007.03422.x. PMC 1941956. PMID 17521318.

[22] Phinney DG, Prockop DJ (2007). "Concise Review: Mesenchymal Stem/Multipotent Stromal Cells: The State of Transdifferentiation and Modes of Tissue Repair-Current Views". *Stem Cells* **25** (11): 2896–902. doi:10.1634/stemcells.2007-0637. PMID 17901396.

[23] Wan C, He Q, McCaigue M, Marsh D, Li G (2006). "Nonadherent cell population of human marrow culture is a complementary source of mesenchymal stem cells (MSCs)". *Journal of Orthopaedic Research* **24** (1): 21–8. doi:10.1002/jor.20023. PMID 16419965.

[24] Gronthos S, Graves SE, Ohta S, Simmons PJ (1994). "The STRO-1+ fraction of adult human bone marrow contains the osteogenic precursors". *Blood* **84** (12): 4164–73. PMID 7994030.

[25] Oyajobi BO, Lomri A, Hott M, Marie PJ (1999). "Isolation and Characterization of Human Clonogenic Osteoblast Progenitors Immunoselected from Fetal Bone Marrow Stroma Using STRO-1 Monoclonal Antibody". *Journal of Bone and Mineral Research* **14** (3): 351–61. doi:10.1359/jbmr.1999.14.3.351. PMID 10027900.

[26] Tondreau T, Lagneaux L, Dejeneffe M, Delforge A, Massy M, Mortier C, Bron D (1 January 2004). "Isolation of BM mesenchymal stem cells by plastic adhesion or negative selection: phenotype, proliferation kinetics and differentiation potential". *Cytotherapy* **6** (4): 372–379. doi:10.1080/14653240410004943. PMID 16146890.

[27] Iudicone P, Fioravanti D, Bonanno G, Miceli M, Lavorino C, Totta P, Frati L, Nuti M, Pierelli L (Jan 2014). "Pathogen-free, plasma-poor platelet lysate and expansion of human mesenchymal stem cells". *J Transl Med.* **12**: 28. doi:10.1186/1479-5876-12-28. PMC 3918216. PMID 24467837.

[28] Torsvik A, Bjerkvig R (2013). "Mesenchymal stem cell signaling in cancer progression". *Cancer Treat Rev.* 2 **39** (2): 180–8. doi:10.1016/j.ctrv.2012.03.005. PMID 22494966.

[29] Heirani-Tabasi A, Hassanzadeh M, Hemmati-Sadeghi S, Shahriyari M, Raeesolmohaddeseen M (2015). "Mesenchymal Stem Cells: Defining the Future of Regenerative Medicine". *Journal of Genes and Cells* **1** (2): 34–9. doi:10.15562/gnc.15.

[30] Manieri NA, Stappenbeck TS (2011). "Mesenchymal stem cell therapy of intestinal disease: are their effects systemic or localized?". *Current Opinion in Gastroenterology* **27** (2): 119–24. doi:10.1097/MOG.0b013e3283423f20. PMID 21150589.

[31] Fischer UM, Harting MT, Jimenez F, Monzon-Posadas WO, Xue H, Savitz SI, Laine GA, Cox CS (2009). "Pulmonary Passage is a Major Obstacle for Intravenous Stem Cell Delivery: The Pulmonary First-Pass Effect". *Stem Cells and Development* **18** (5): 683–92. doi:10.1089/scd.2008.0253. PMC 3190292. PMID 19099374.

[32] Wakitani S, Nawata M, Tensho K, Okabe T, Machida H, Ohgushi H (2007). "Repair of articular cartilage defects in the patello-femoral joint with autologous bone marrow mesenchymal cell transplantation: three case reports involving nine defects in five knees". *Journal of Tissue Engineering and Regenerative Medicine* **1** (1): 74–9. doi:10.1002/term.8. PMID 18038395.

[33] Sharma RR, Pollock K, Hubel A, McKenna D (2014). "Mesenchymal stem or stromal cells: a review of clinical applications and manufacturing practices.". *Transfusion* **54** (5): 1418–37. doi:10.1111/trf.12421. PMID 24898458.

[34] Figueroa FE, Carrión F, Villanueva S, Khoury M (2012). "Mesenchymal stem cell treatment for autoimmune diseases: a critical review.". *Biol Res.* **45** (3): 269–77. doi:10.4067/S0716-97602012000300008. PMID 23283436.

[35] François M, Copland IB, Yuan S, Romieu-Mourez R, Waller EK, Galipeau J (2012). "Cryopreserved mesenchymal stromal cells display impaired immunosuppressive properties as a result of heat-shock response and impaired interferon-γ licensing". *Cytotherapy* **14** (2): 147–52. doi:10.3109/14653249.2011.623691. PMC 3279133. PMID 22029655.

[36] Sell, Stewart (Stem cell handbook). Humana Press. p. 143. Check date values in: |date= (help); Missing or empty |title= (help)

[37] Becker AJ, McCULLOCH EA, Till JE (1963). "Cytological Demonstration of the Clonal Nature of Spleen Colonies Derived from Transplanted Mouse Marrow Cells". *Nature* **197** (4866): 452–4. doi:10.1038/197452a0. PMID 13970094.

[38] Siminovitch L, Mcculloch EA, Till JE (1963). "The distribution of colony-forming cells among spleen colonies". *Journal of Cellular and Comparative Physiology* **62** (3): 327–36. doi:10.1002/jcp.1030620313. PMID 14086156.

[39] Friedenstein AJ, Deriglasova UF, Kulagina NN, Panasuk AF, Rudakowa SF, Luriá EA, Ruadkow IA (1974). "Precursors for fibroblasts in different populations of hematopoietic cells as detected by the in vitro colony assay method". *Experimental hematology* **2** (2): 83–92. PMID 4455512.

[40] Friedenstein AJ, Gorskaja JF, Kulagina NN (1976). "Fibroblast precursors in normal and irradiated mouse hematopoietic organs". *Experimental hematology* **4** (5): 267–74. PMID 976387.

[41] Mohammad Mousaei Ghasroldasht, Muhammad Irfan-Maqsood, Maryam M. Matin, Hamid Reza Bidkhori, Hojjat Naderi-Meshkin, Ali Moradi and Ahmad Reza Bahrami (2014). "Mesenchymal stem cell based therapy for osteo-diseases". *Cell Biology International* **9999** (11): 1–5. doi:10.1002/cbin.10293.

23.10   External links

- Mesenchymal stem cells fact sheet, published June 2012, scientist-reviewed and not too technical

- Mesenchymal Stem Cell Research at Johns Hopkins University

- Murphy MB, Moncivais K, Caplan AI (2013). "Mesenchymal stem cells: environmentally responsive therapeutics for regenerative medicine". *Experimental & Molecular Medicine* **45** (11): e54. doi:10.1038/emm.2013.94. PMC 3849579. PMID 24232253.

Chapter 24

Neural stem cell

Neural stem cells (NSCs) are self-renewing, multipotent cells that generate the main phenotype of the nervous system. Stem cells are characterized by their capability to differentiate into multiple cell types via exogenous stimuli from their environment.[1] They undergo asymmetric cell division into two daughter cells, one non-specialized and one specialized. NSCs primarily differentiate into neurons, astrocytes, and oligodendrocytes.[2]

24.1 History

In 1989, Sally Temple described multipotent, self-renewing progenitor and stem cells in the subventricular zone (SVZ) of the mouse brain.[3] In 1992, Brent A. Reynolds and Samuel Weiss were the first to isolate neural progenitor and stem cells from the adult striatal tissue, including the SVZ — one of the neurogenic areas — of adult mice brain tissue.[4] In the same year the team of Constance Cepko and Evan Y. Snyder were the first to isolate multipotent cells from the mouse cerebellum and stably transfected them with the oncogene v-myc.[5] Interestingly, this molecule is one of the genes widely used now to reprogram adult non-stem cells into pluripotent stem cells. Since then, neural progenitor and stem cells have been isolated from various areas of the adult brain, including non-neurogenic areas, such as the spinal cord, and from various species including humans.[6][7]

24.2 Aging and development

24.2.1 *In vivo* origin

There are two basic types of stem cell: adult stem cells, which are limited in their ability to differentiate, and embryonic stem cells (ESCs), which are pluripotent. ESCs are not limited to a particular cell fate; rather they have the capability to differentiate into any cell type.[1] ESCs are derived from the inner cell mass of the blastocyst with the po-

tential to self-replicate.[2]

NSCs are considered adult stem cells because they are limited in their capability to differentiate. NSCs are generated throughout an adult's life via the process of neurogenesis.[8] Since neurons do not divide within the central nervous system (CNS), NSCs can be differentiated to replace lost or injured neurons or in many cases even glial cells.[2] NSCs are differentiated into new neurons within the SVZ of lateral ventricles, a remnant of the embryonic germinal neuroepithelium, as well as the dentate gyrus of the hippocampus.[8]

24.2.2 *In vitro* origin

Adult NSCs were first isolated from mouse striatum in the early 1990s. They are capable of forming multipotent neurospheres when cultured *in vitro*. Neurospheres can produce self-renewing and proliferating specialized cells. These neurospheres can differentiate to form the specified neurons, glial cells, and oligodendrocytes.[2][8] In previous studies, cultured neurospheres have been transplanted into the brains of immunodeficient neonatal mice and have shown engraftment, proliferation, and neural differentiation.[8]

24.2.3 NSC communication and migration

NSCs are stimulated to begin differentiation via exogenous cues from the microenvironment, or stem cell niche. This capability of the NSCs to replace lost or damaged neural cells is called neurogenesis.[2] Some neural cells are migrated from the SVZ along the rostral migratory stream which contains a marrow-like structure with ependymal cells and astrocytes when stimulated. The ependymal cells and astrocytes form glial tubes used by migrating neuroblasts. The astrocytes in the tubes provide support for the migrating cells as well as insulation from electrical and chemical signals released from surrounding cells. The astrocytes are the primary precursors for rapid cell amplification. The neuroblasts form tight chains and migrate towards the specified site of cell damage to repair or replace neural cells. One example is a neuroblast migrating towards the olfactory bulb to differentiate into periglomercular or granule neurons which have a radial migration pattern rather than a tangential one.[9]

On the other hand, the dentate gyrus neural stem cells produce excitatory granule neurons which are involved in learning and memory. One example of learning and memory is pattern separation, a cognitive process used to distinguish similar inputs.[2]

24.2.4   Aging

Neural stem cell proliferation declines as a consequence of aging.[10] Various approaches have been taken to counteract this age-related decline.[11] Because FOXO proteins regulate neural stem cell homeostasis,[12] FOXO proteins have been used to protect neural stem cells by inhibiting Wnt signaling.[13]

24.3   Functions of NSCs during differentiation and disease

Epidermal growth factor (EGF) and fibroblast growth factor (FGF) are mitogens that promote neural progenitor and stem cell growth *in vitro*, though other factors synthesized by the neural progenitor and stem cell populations are also required for optimal growth.[14] It is hypothesized that neurogenesis in the adult brain originates from NSCs. The origin and identity of NSCs in the adult brain remain to be defined.

24.3.1   Function of NSCs during differentiation

The most widely accepted model of an adult NSC is a radial, astrocytes-like, GFAP-positive cell. Quiescent stem cells are Type B that are able to remain in the quiescent state due to the renewable tissue provided by the specific niches composed of blood vessels, astrocytes, microglia, ependymal cells, and extracellular matrix present within the brain. These niches provide nourishment, structural support, and protection for the stem cells until they are activated by external stimuli. Once activated, the Type B cells develop into Type C cells, active proliferating intermediate cells, which then divide into neuroblasts consisting of Type A cells. The undifferentiated neuroblasts form chains that migrate and develop into mature neurons. In the olfactory bulb, they mature into GABAergic granule neurons, while in the hippocampus they mature into dentate granule cells.[15]

24.3.2   Function of NSCs during disease

NSCs have an important role during development producing the enormous diversity of neurons, astrocytes and oligodendrocytes in the developing CNS. They also have important role in adult animals, for instance in learning and hippocampal plasticity in the adult mice in addition to supplying neurons to the olfactory bulb in mice.[8]

Notably the role of NSCs during diseases is now being elucidated by several research groups around the world. The responses during stroke, multiple sclerosis, and Parkinson's disease in animal models and humans is part of the current investigation. The results of this ongoing investigation may have future applications to treat human neurological diseases.[8]

Neural stem cells have been shown to engage in migration and replacement of dying neurons in classical experiments performed by Sanjay Magavi and Jeffrey Macklis.[16] Using a laser-induced damage of cortical layers, Magavi showed that SVZ neural progenitors expressing Doublecortin, a critical molecule for migration of neuroblasts, migrated long distances to the area of damage and differentiated into mature neurons expressing NeuN marker. In addition Masato Nakafuku's group from Japan showed for the first time the role of hippocampal stem cells during stroke in mice.[17] These results demonstrated that NSCs can engage in the adult brain as a result of injury. Furthermore, in 2004 Evan Y. Snyder's group showed that NSCs migrate to brain tumors in a directed fashion. Jaime Imitola, M.D and colleagues from Harvard demonstrated for the first time, a molecular mechanism for the responses of NSCs to injury. They showed that chemokines released during injury such as SDF-1a were responsible for the directed migration of human and mouse NSCs to areas of injury in mice.[18] Since then other molecules have been found to participate in the responses of NSCs to injury. All these results have been widely reproduced and expanded by other investigators joining the classical work of Richard L. Sidman in Autoradiography to visualize neurogenesis during development, and neurogenesis in the adult by Joseph Altman in 1960's, as evidence of the responses of adult NSCs activities and neurogenesis during homeostasis and injury.

The search for additional mechanisms that operate in the injury environment and how they influence the responses of NSCs during acute and chronic disease is matter of intense research.[19]

24.4   Potential clinical applications

24.4.1 Regenerative therapy of the CNS

Cell death is a characteristic of acute CNS disorders as well as neurodegenerative disease. The loss of cells is amplified by the lack of regenerative abilities for cell replacement and repair in the CNS. One way to circumvent this is to use cell replacement therapy via regenerative NSCs. NSCs can be cultured *in vitro* as neurospheres. These neurospheres are composed of neural stem cells and progenitors (NSPCs) with growth factors such as EGF and FGF. The withdrawal of these growth factors activate differentiation into neurons, astrocytes, or oligodendrocytes which can be transplanted within the brain at the site of injury. The benefits of this therapeutic approach have been examined in Parkinson's disease, Huntington's disease, and multiple sclerosis. NSPCs induce neural repair via intrinsic properties of neuroprotection and immunomodulation. Some possible routes of transplantation include intracerebral transplantation and xenotransplantation.[20][21]

An alternative therapeutic approach to the transplantation of NSPCs is the pharmacological activation of endogenous NSPCs (eNSPCs). Activated eNSPCs produce neurotrophic factors,several treatments that activate a pathway that involves the phosphorylation of STAT3 on the serine residue and subsequent elevation of Hes3 expression (STAT3-Ser/Hes3 Signaling Axis) oppose neuronal death and disease progression in models of neurological disorder.[22][23]

24.5 Basic laboratory studies

24.5.1 Generation of 3D *in vitro* models of the human CNS

Human midbrain-derived neural progenitor cells (hmNPCs) have the ability to differentiate down multiple neural cell lineages that lead to neurospheres as well as multiple neural phenotypes. The hmNPC can be used to develop a 3D *in vitro* model of the human CNS. There are two ways to culture the hmNPCs, the adherent monolayer and the neurosphere culture systems. The neurosphere culture system has previously been used to isolate and expand CNS stem cells by its ability to aggregate and proliferate hmNPCs under serum-free media conditions as well as with the presence of epidermal growth factor (EGF) and fibroblast growth factor-2 (FGF2). Initially, the hmNPCs were isolated and expanded before performing a 2D differentiation which was used to produce a single-cell suspension. This single-cell suspension helped achieve a homogenous 3D structure of uniform aggregate size. The 3D aggregation formed neurospheres which was used to form an *in vitro* 3D CNS model.[24]

24.5.2 Neural stem cells and bioactive scaffolds as traumatic brain injury treatment

Traumatic Brain Injury (TBI) can deform the brain tissue, leading to necrosis primary damage which can then cascade and activate secondary damage such as excitotoxicity, inflammation, ischemia, and the breakdown of the blood-brain-barrier. Damage can escalate and eventually lead to apoptosis or cell death. Current treatments focus on preventing further damage by stabilizing bleeding, decreasing intracranial pressure and inflammation, and inhibiting pro-apoptoic cascades. In order to repair TBI damage, an upcoming therapeutic option involves the use of NSCs derived from the embryonic peri-ventricular region. Stem cells can be cultured in a favorable 3-dimensional, low cytotoxic environment, a hydrogel, that will increase NSC survival when injected into TBI patients. The intracerebrally injected, primed NSCs were seen to migrate to damaged tissue and differentiate into oligodendrocytes or neuronal cells that secreted neuroprotective factors.[25][26]

24.5.3 Galectin-1 in neural stem cells

Galectin-1 is expressed in adult NSCs and has been shown to have a physiological role in the treatment of neurological disorders in animal models. There are two approaches to using NSCs as a therapeutic treatment: (1) stimulate intrinsic NSCs to promote proliferation in order to replace injured tissue, and (2) transplant NSCs into the damaged brain area in order to allow the NSCs to restore the tissue. Lentivirus vectors were used to infect human NSCs (hNSCs) with Galectin-1 which were later transplanted into the damaged tissue. The hGal-1-hNSCs induced better and faster brain recovery of the injured tissue as well as a reduction in motor and sensory deficits as compared to only hNSC transplantation.[9]

24.6 Assays

Neural stem cells are routinely studied *in vitro* using a method referred to as the Neurosphere Assay (or Neurosphere culture system), first developed by Reynolds and Weiss.[4] Neurospheres are intrinsically heterogeneous cellular entities almost entirely formed by a small fraction (1 to 5%) of slowly dividing neural stem cells and by their progeny, a population of fast-dividing nestin-positive progenitor cells.[4][27][28] The total number of these progenitors determines the size of a neurosphere and, as a result, disparities in sphere size within different neurosphere populations may reflect alterations in the proliferation, survival and/or differentiation status of their neural progenitors. Indeed, it has been reported that loss of β1-integrin in a neurosphere culture does not significantly affect the capacity of β1-integrin deficient stem cells to form new neurospheres, but it influences the size of the neurosphere: β1-integrin deficient neurospheres were overall smaller due to increased cell death and reduced proliferation.[29]

While the Neurosphere Assay has been the method of choice for isolation, expansion and even the enumeration of neural stem and progenitor cells, several recent publications have highlighted some of the limitations of the neurosphere culture system as a method for determining neural stem cell frequencies.[30] In collaboration with Reynolds, STEMCELL Technologies has developed a collagen-based assay, called the Neural Colony-Forming Cell (NCFC) Assay, for the quantification of neural stem cells. Importantly, this assay allows discrimination between neural stem and progenitor cells.[31]

24.7   Neural Stem Cell Institute

The damaged CNS tissue has very limited regenerative and repair capacity so that loss of neurological function is often chronic and progressive. Cell replacement from stem cells is being actively pursued as a therapeutic option. In 2009, a research institute dedicated solely to translating neural stem research into therapies for patients was created outside of Albany, New York, The Neural Stem Cell Institute.

24.8   See also

- Astrocyte
- Induced progenitor stem cells
- Neurogenesis
- Neuron
- Oligodendrocyte
- Stem cell

24.9   References

- J Cancer Res Ther. 2011 Jan-Mar;7(1):58-63. doi: 10.4103/0973-1482.80463

Intensity-modulated radiation to spare neural stem cells in brain tumors: a computational platform for evaluation of physical and biological dose metrics. Jaganathan A, Tiwari M, Phansekar R, Panta R, Huilgol N.

[1] Clarke, D.; Johansson, C; Wilbertz, J; Veress, B; Nilsson, E; Karlstrom, H; Lendahl, U; Frisen, J (2000). "Generalized Potential of Adult Neural Stem Cells.". Science 288 (5471): 1660–63. Bibcode:2000Sci...288.1660C. doi:10.1126/science.288.5471.1660. PMID 10834848.

[2] Alenzi, F; Bahkali, A (2011). "Stem cells: Biology and clinical potential". African Journal of Biotechnology 10 (86): 19929–40. doi:10.5897/ajbx11.046.

[3] Temple, S (1989). "Division and differentiation of isolated CNS blast cells in microculture". Nature 340 (6233): 471–73. Bibcode:1989Natur.340..471T. doi:10.1038/340471a0.

[4] Reynolds, B.; Weiss, S (1992). "Generation of neurons and astrocytes from isolated cells of the adult mammalian central nervous system". Science 255 (5052): 1707–10. Bibcode:1992Sci...255.1707R. doi:10.1126/science.1553558. PMID 1553558.

[5] Snyder, Evan Y.; Deitcher, David L.; Walsh, Christopher; Arnold-Aldea, Susan; Hartwieg, Erika A.; Cepko, Constance L. (1992). "Multipotent neural cell lines can engraft and participate in development of mouse cerebellum". Cell 68 (1): 33–51. doi:10.1016/0092-8674(92)90204-P. PMID 1732063.

[6] Zigova, Tanja; Sanberg, Paul R.; Sanchez-Ramos, Juan Raymond, eds. (2002). Neural stem cells: methods and protocols. Humana Press. ISBN 978-0-89603-964-3. Retrieved 18 April 2010.

[7] Taupin, Philippe; Gage, Fred H. (2002). "Adult neurogenesis and neural stem cells of the central nervous system in mammals". Journal of Neuroscience Research 69 (6): 745–9. doi:10.1002/jnr.10378. PMID 12205667.

[8] Paspala, S; Murthy, T; Mahaboob, V; Habeeb, M (2011). "Pluripotent stem cells – A review of the current status in neural regeneration". Neurology India 59 (4): 558–65. doi:10.4103/0028-3886.84338. PMID 21891934.

[9] Sakaguchi, M; Okano, H (2012). "Neural stem cells, adult neurogenesis, and galectin-1: From bench to bedside". Developmental Neurobiology 72 (7): 1059–67. doi:10.1002/dneu.22023. PMID 22488739.

[10] Kuhn HG, Dickinson-Anson H, Gage FH; Dickinson-Anson; Gage (1996). "Neurogenesis in the dentate gyrus of the adult rat: age-related decrease of neuronal progenitor proliferation" (PDF). Journal of Neuroscience 16 (6): 2027–2033. PMID 8604047.

[11] Artegiani B, Calegari F; Calegari (2012). "Age-related cognitive decline: can neural stem cells help us?". Aging 4 (3): 176–186. PMC 3348478. PMID 22466406.

[12] Renault VM, Rafalski VA, Morgan AA, Salih DA, Brett JO, Webb AE, Villeda SA, Thekkat PU, Guillerey C, Denko NC, Palmer TD, Butte AJ, Brunet A; Rafalski; Morgan; Salih; Brett; Webb; Villeda; Thekkat; Guillerey; Denko;

Palmer; Butte; Brunet (2009). "FoxO3 regulates neural stem cell homeostasis". *CELL: Stem Cell* **5** (5): 527–539. doi:10.1016/j.stem.2009.09.014. PMC 2775802. PMID 19896443.

[13] Paik JH, Ding Z, Narurkar R, Ramkissoon S, Muller F, Kamoun WS, Chae SS, Zheng H, Ying H, Mahoney J, Hiller D, Jiang S, Protopopov A, Wong WH, Chin L, Ligon KL, DePinho RA; Ding; Narurkar; Ramkissoon; Muller; Kamoun; Chae; Zheng; Ying; Mahoney; Hiller; Jiang; Protopopov; Wong; Chin; Ligon; Depinho (2009). "FoxOs cooperatively regulate diverse pathways governing neural stem cell homeostasis". *CELL: Stem Cell* **5** (5): 540–553. doi:10.1016/j.stem.2009.09.013. PMC 3285492. PMID 19896444.

[14] Taupin, Philippe; Ray, Jasodhara; Fischer, Wolfgang H; Suhr, Steven T; Hakansson, Katarina; Grubb, Anders; Gage, Fred H (2000). "FGF-2-Responsive Neural Stem Cell Proliferation Requires CCg, a Novel Autocrine/Paracrine Cofactor". *Neuron* **28** (2): 385–97. doi:10.1016/S0896-6273(00)00119-7. PMID 11144350.

[15] Bergstrom, T; Forsbery-Nilsson, K (2012). "Neural stem cells: Brain building blocks and beyond". *Upsala Journal of Medical Sciences* **117** (2): 132–42. doi:10.3109/03009734.2012.665096. PMC 3339545. PMID 22512245.

[16] MacKlis, Jeffrey D.; Magavi, Sanjay S.; Leavitt, Blair R. (2000). "Induction of neurogenesis in the neocortex of adult mice". *Nature* **405** (6789): 951–5. doi:10.1038/35016083. PMID 10879536.

[17] Nakatomi, Hirofumi; Kuriu, Toshihiko; Okabe, Shigeo; Yamamoto, Shin-Ichi; Hatano, Osamu; Kawahara, Nobutaka; Tamura, Akira; Kirino, Takaaki; Nakafuku, Masato (2002). "Regeneration of Hippocampal Pyramidal Neurons after Ischemic Brain Injury by Recruitment of Endogenous Neural Progenitors". *Cell* **110** (4): 429–41. doi:10.1016/S0092-8674(02)00862-0. PMID 12202033.

[18] Imitola, Jaime; Raddassi K, Park KI, Mueller FJ, Nieto M, Teng YD, Frenkel D, Li J, Sidman RL, Walsh California, Snyder EY, Khoury SJ.; In Park, Kook; Mueller, Franz-Josef; Nieto, Marta; Teng, Yang D.; Frenkel, Dan; Li, Jianxue; Sidman, Richard L.; Walsh, Christopher A.; Snyder, Evan Y.; Khoury, Samia J. (December 28, 2004). "Directed migration of neural stem cells to sites of CNS injury by the stromal cell-derived factor 1alpha/CXC chemokine receptor 4 pathway". *PNAS* **101** (52): 18117–22. Bibcode:2004PNAS..10118117I. doi:10.1073/pnas.0408258102. PMC 536055. PMID 15608062.

[19] Sohur US, US.; Emsley JG; Mitchell BD; Macklis JD. (September 29, 2006). "Adult neurogenesis and cellular brain repair with neural progenitors, precursors and stem cells". *Philos Trans R Soc Lond B Biol Sci.* **361** (1473): 1477–97. doi:10.1098/rstb.2006.1887. PMC 1664671. PMID 16939970.

[20] Bonnamain, V; Neveu, I; Naveilhan, P (2012). "Neural stem/progenitor cells as promising candidates for regenerative therapy of the central nervous system". *Frontiers in Cellular Neuroscience* **6**.

[21] Xu, X; Warrington, A; Bieber, A; Rodriguez, M (2012). "Enhancing Central Nervous System Repair-The Challenges". *CNS Drugs* **25** (7): 555–73. doi:10.2165/11587830 (inactive 2015-02-01). PMC 3140701. PMID 21699269.

[22] Androutsellis-Theotokis A, Leker RR, Soldner F; et al. (August 2006). "Notch signalling regulates stem cell numbers in vitro and in vivo". *Nature* **442** (7104): 823–6. Bibcode:2006Natur.442..823A. doi:10.1038/nature04940. PMID 16799564.

[23] Androutsellis-Theotokis A, Rueger MA, Park DM; et al. (August 2009). "Targeting neural precursors in the adult brain rescues injured dopamine neurons". *Proc. Natl. Acad. Sci. U.S.A.* **106** (32): 13570–5. Bibcode:2009PNAS..10613570A. doi:10.1073/pnas.0905125106. PMC 2714762. PMID 19628689.

[24] Brito, C; Simao, D; Costa, I; Malpique, R; Pereira, C; Fernandes, P; Serra, M; Schwarz, S; Schwarz, J; Kremer, E; Alves, P (2012). "Generation and genetic modification of 3D cultures of human dopaminergic neurons derived from neural progenitor cells". <*Methods* **56** (3): 452–60. doi:10.1016/j.ymeth.2012.03.005.

[25] Stabenfeldt, S; Irons, H; LaPlace, M (2011). "Stem Cells and Bioactive Scaffolds as a Treatment for Traumatic Brain Injury". *Current Stem Cell Research & Therapy* **6** (3): 208–20. doi:10.2174/157488811796575396.

[26] Ratajczak, J; Zuba-Surma, E; Paczkowska, K; Kucia, M; Nowacki, P; Ratajczak, MZ (2011). "Stem cells for neural regeneration--a potential application of very small embryonic-like stem cells". *J Physiol Pharmacol.* **62** (1): 3–12. PMID 21451204.

[27] Campos, L. S.; Leone, DP; Relvas, JB; Brakebusch, C; Fässler, R; Suter, U; Ffrench-Constant, C (2004). "β1 integrins activate a MAPK signalling pathway in neural stem cells that contributes to their maintenance". *Development* **131** (14): 3433–44. doi:10.1242/dev.01199. PMID 15226259.

[28] Lobo, M. V. T.; Alonso, F. J. M.; Redondo, C.; Lopez-Toledano, M. A.; Caso, E.; Herranz, A. S.; Paino, C. L.; Reimers, D.; Bazan, E. (2003). "Cellular Characterization of Epidermal Growth Factor-expanded Free-floating Neurospheres". *Journal of Histochemistry & Cytochemistry* **51** (1): 89–103. doi:10.1177/002215540305100111. PMID 12502758.

[29] Leone, D. P.; Relvas, JB; Campos, LS; Hemmi, S; Brakebusch, C; Fässler, R; Ffrench-Constant, C; Suter, U (2005). "Regulation of neural progenitor proliferation and survival by β1 integrins". *Journal of Cell Science* **118** (12): 2589–99. doi:10.1242/jcs.02396. PMID 15928047.

[30] Singec, Ilyas; Knoth, Rolf; Meyer, Ralf P; Maciaczyk, Jaroslaw; Volk, Benedikt; Nikkhah, Guido; Frotscher, Michael; Snyder, Evan Y (2006). "Defining the actual sensitivity and specificity of the neurosphere assay in stem cell biology". *Nature Methods* **3** (10): 801–6. doi:10.1038/nmeth926. PMID 16990812.

[31] Louis, Sharon A.; Rietze, Rodney L.; Deleyrolle, Loic; Wagey, Ravenska E.; Thomas, Terry E.; Eaves, Allen C.; Reynolds, Brent A. (2008). "Enumeration of Neural Stem and Progenitor Cells in the Neural Colony-Forming Cell Assay". *Stem Cells* **26** (4): 988–96. doi:10.1634/stemcells.2007-0867. PMID 18218818.

Chapter 25

Precursor cell

In cytology, a **precursor cell**, also called a **blast cell** or simply **blast**, is a type of partially differentiated, usually unipotent cell that has lost most or all of the stem cell multipotency.

Usually a precursor cell is a stem cell which has the capacity to differentiate into only one cell types. Sometimes *precursor cell* is used as an alternative term for unipotent stem cells.

A blastoma is a type of cancer created by malignancies of precursor cells.

In embryology, precursor cells are a group of cells that differentiate later into one organ.

25.1 Cytological types

- Oligodendrocyte precursor cell
- Myeloblast
- Thymocyte
- Meiocyte
- Megakaryoblast
- Promegakaryocyte
- Melanoblast
- Lymphoblast
- Bone marrow precursor cells
- Normoblast
- Angioblast (endothelial precursor cells)
- Myeloid precursor cells

25.2 References

Precursor cell – Britannica Online Encyclopedia

25.3 External links

- NIF Search - Precursor Cell via the Neuroscience Information Framework

Chapter 26

Blastocyst

For the non-species specific developmental stage, see Blastula. For the single-celled parasite, see Blastocystis.

The **blastocyst** is a structure formed in the early development of mammals. It possesses an inner cell mass (ICM) which subsequently forms the embryo. The outer layer of the blastocyst consists of cells collectively called the trophoblast. This layer surrounds the inner cell mass and a fluid-filled cavity known as the blastocoele. The trophoblast gives rise to the placenta. The name "blastocyst" arises from the Greek βλαστός blastos ("a sprout") and κύστις kystis ("bladder, capsule").

In humans, blastocyst formation begins about 5 days after fertilization, when a fluid-filled cavity opens up in the morula, a ball consisting of a few dozen cells. The blastocyst has a diameter of about 0.1-0.2 mm and comprises 200-300 cells following rapid cleavage (cell division). After about 1 day (5–6 days post-fertilization), which is the time usually required to reach the uterus, the blastocyst begins to embed itself into the endometrium of the uterine wall where it will undergo later developmental processes, including gastrulation. Embedding of the blastocyst into the endometrium requires that it hatches from the zona pellucida, which prevents it from adhering to the oviduct as it makes its way to the uterus. The blastocyst is completely embedded in the endometrium only 11–12 days after fertilization.

The use of blastocysts in in-vitro fertilization (IVF) involves culturing a fertilized egg for five days before implanting it into the uterus. It can be a more viable method of fertility treatment than traditional IVF. The inner cell mass of blastocysts is also a source of embryonic stem cells.

26.1 Development cycle

During human embryogenesis, the blastocyst arises from the morula in the uterus, 5 days after fertilization. The early embryo undergoes cell differentiation and structural changes to become the blastocyst. It is then prepared for implantation into the uterine wall 6 days after fertilization. Implantation marks the end of the germinal stage of embryogenesis.[1]

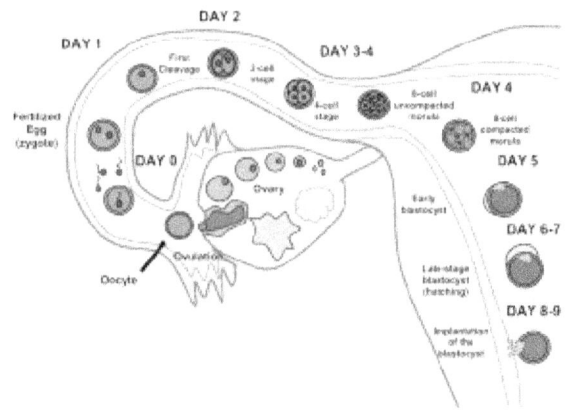

Early development of the embryo from ovulation through implantation in humans. The blastocyst stage occurs between 5 and 8-9 days following conception.

26.1.1 Blastocyst formation

The morula, which precedes the blastocyst, is an early embryo composed of 16 undifferentiated cells. Shortly following the morula's entry into the uterus from the Fallopian tube, the morula becomes the blastocyst through cellular differentiation and cavitation. The morula's cells differentiate into two types: an inner cell mass growing on the interior of the blastocoel and trophoblast cells growing on the exterior.[2] The animal pole refers to the side of the blastocyst where the ICM resides, while the vegetal pole is on the opposite side. Cavitation is the process by which a fluid cavity forms inside the embryo. The trophoblast cells pump sodium ions into the center of the embryo, which causes water to enter through osmosis. This forms an internal fluid-filled cavity called the blastocoel. This distinguishable blastocoel cavity in addition to cellular specification are both hallmark identities of the blastocyst.[3]

26.1.2 Implantation

Implantation is critical to the survival and development of the early embryo. It establishes a connection between the mother and the early embryo which will continue through the remainder of the pregnancy. Implantation is made possible through structural changes in both the blastocyst and endometrial wall.[4] The zona pellucida surrounding the blastocyst breaches, referred to as hatching. This removes the constraint on the physical size of the embryonic mass and exposes the outer cells of the blastocyst to the interior of the uterus. Furthermore, hormonal changes in the mother, specifically a peak in luteinizing hormone (LH) prepares the endometrium to receive the blastocyst and envelope it. Once bound to the extracellular matrix of the endometrium, trophoblast cells secrete enzymes and other factors to embed the blastocyst into the uterine wall. The enzymes released degrade the endometrial lining, while autocrine growth factors such as human chorionic gonadotropin (hCG) and insulin-like growth factor (IGF) allow the blastocyst to further invade the endometrium.[5]

Implantation in the uterine wall allows for the next step in embryogenesis, gastrulation, which includes formation of the placenta from trophoblastic cells and differentiation of the ICM into the amniotic sac and epiblast.

26.2 Structure

The blastocyst is made up of cells from the inner cell mass and the blastocoel.

There are two types of blastomere cells:[6]

- The inner cell mass, also known as the embryoblast, gives rise to the primitive endoderm and the epiblast.

  - The primitive endoderm develops into the amniotic sac which forms the fluid-filled cavity that the embryo resides in during pregnancy.[7]

  - The epiblast gives rise to the three germ layers of the developing embryo during gastrulation (endoderm, mesoderm, and ectoderm).

- The trophoblast is a layer of cells forming the outer ring of the blastocyst that combines with the maternal endometrium to form the placenta. Trophoblast cells also secrete factors to make the blastocoel.[8]

  - Cytotrophoblast is the inner layer of the trophoblast, composed of stem cells which give rise

to cells comprising the chorionic villi, placenta, and syncytiotrophoblast.

- Syncytiotrophoblast is the outermost layer of the trophoblast. These cells secrete proteolytic enzymes to break down the endometrial extracellular matrix to allow for implantation of the blastocyst in the uterine wall.[9]

The blastocoel fluid cavity contains amino acids, growth factors, and other necessary molecules for cellular differentiation.[10]

26.2.1 Cell specification

Multiple processes control cell lineage specification in the blastocyst to produce the trophoblast, epiblast, and primitive endoderm. These processes include: gene expression, cell signaling, cell-cell contact and positional relationships, and epigenetics.

Once the ICM has been established within the blastocyst, this cell mass prepares for further specification into the epiblast and primitive endoderm. This process of specification is determined in part by fibroblast growth factor (FGF) signaling which generates a MAP kinase pathway to alter cellular genomes.[11] Further segregation of blastomeres into the trophoblast and inner cell mass are regulated by the homeodomain protein, Cdx2. This transcription factor represses the expression of Oct4 and Nanog transcription factors in the trophectoderm.[12] These genomic alterations allow for the progressive specification of both epiblast and primitive endoderm lineages at the end of the blastocyst phase of development preceding gastrulation.

Trophoblasts express integrin on their cell surfaces which allow for adhesion to the extracellular matrix of the uterine wall. This interaction allows for implantation and also triggers further specification into the three different cell types, preparing the blastocyst for gastrulation.[13]

26.3 Clinical implications

26.3.1 Pregnancy tests

Levels of human chorionic gonadotropin secreted by the blastocyst during implantation is the factor measured in a pregnancy test. HCG can be measured in both the blood and urine to determine if a woman is pregnant. More hCG is secreted in a multiple pregnancy. Blood tests of hCG can also be used to check for abnormal pregnancies.[14]

26.3.2 *In vitro* fertilization

In vitro fertilization is an alternative to traditional *in vivo* fertilization for fertilizing an egg with sperm and implanting that embryo into a female's womb. For many years the embryo was inserted into the fallopian tube two to three days after fertilization. However at this stage of development it is very difficult to predict which embryos will develop best, and several embryos were typically implanted. Several implanted embryos helped to guarantee that there would be a developing fetus but it also led to the development of multiple fetuses. This was a major problem and drawback for using embryos to IVF.

A recent breakthrough for in vitro fertilization is the use of blastocysts. A blastocyst would be implanted five to six days after the eggs had been fertilized.[15] After five or six days it is much easier to determine which embryos will result in healthy live births. Knowing which embryos will succeed allows just one blastocyst to be implanted, cutting down dramatically on the health risk and expense of multiple births. Now that the nutrient sources for embryonic and blastocyst development has been determined, it is much easier to give embryos the correct nutrients in order to sustain them into the blastocyst phase. Blastocyst implantation through in vitro fertilization is a painless procedure in which a catheter is inserted into the vagina, guided through the cervix via ultrasound, into the uterus where the blastocysts are inserted into the womb.

Blastocysts also offer an advantage because they can be used to genetically test the cells to check for genomic problems. There are enough cells in a blastocyst that a few trophectoderm cells are able to be removed without disturbing the developing blastocyst. These cells can be tested for chromosome aneuploidy using preimplantation genetic screening (PGS).

26.4 See also

- Developmental biology

- Human embryogenesis

26.5 References

This article incorporates text in the public domain from the 20th edition of Gray's Anatomy (1918)

[1] Sherk, Stephanie Dionne (2006). "Prenatal Development". *Gale Encyclopedia of Children's Health*. Retrieved 2013-12-07.

[2] Clinic, Mayo (2012). "Fetal development: The first trimester". *Mayo Foundation for Medical Education*. Retrieved 2013-12-07.

[3] Gilbert SF. Developmental Biology. 6th edition. Sunderland (MA): Sinauer Associates; 2000. Early Mammalian Development. Available from: http://www.ncbi.nlm.nih.gov/books/NBK10052/

[4] Zhang, Shuang; Lin, Haiyan; Kong, Shuangbo; Wang, Shumin; Wang, Hongmei; Wang, Haibin; Armant, D. Randall (2013). "Physiological and molecular determinants of embryo implantation". *Molecular Aspects of Medicine* **34** (5): 939–80. doi:10.1016/j.mam.2012.12.011. PMID 23290997.

[5] Srisuparp, Santha; Strakova, Zuzana; Fazleabas, Asgerally T (2001). "The Role of Chorionic Gonadotropin (CG) in Blastocyst Implantation". *Archives of Medical Research* **32** (6): 627–34. doi:10.1016/S0188-4409(01)00330-7. PMID 11750740.

[6] Scott F. Gilbert (15 July 2013). *Developmental Biology*. Sinauer Associates, Incorporated. ISBN 978-1-60535-173-5.

[7] Schoenwolf, Gary C., and William J. Larsen. *Larsen's Human Embryology*. 4th ed. Philadelphia: Churchill Livingstone/Elsevier, 2009. Print.

[8] James, J. L; Stone, PR; Chamley, LW (2005). "Cytotrophoblast differentiation in the first trimester of pregnancy: Evidence for separate progenitors of extravillous trophoblasts and syncytiotrophoblast". *Reproduction* **130** (1): 95–103. doi:10.1530/rep.1.00723. PMID 15985635.

[9] Vićovac, L; Aplin, JD (1996). "Epithelial-mesenchymal transition during trophoblast differentiation". *Acta anatomica* **156** (3): 202–16. doi:10.1159/000147847. PMID 9124037.

[10] Gasperowicz, M.; Natale, D. R. C. (2010). "Establishing Three Blastocyst Lineages--Then What?". *Biology of Reproduction* **84** (4): 621–30. doi:10.1095/biolreprod.110.085209. PMID 21123814.

[11] Yamanaka, Y.; Lanner, F.; Rossant, J. (2010). "FGF signal-dependent segregation of primitive endoderm and epiblast in the mouse blastocyst". *Development* **137** (5): 715–24. doi:10.1242/dev.043471. PMID 20147376.

[12] Strumpf, D.; Mao, CA; Yamanaka, Y; Ralston, A; Chawengsaksophak, K; Beck, F; Rossant, J (2005). "Cdx2 is required for correct cell fate specification and differentiation of trophectoderm in the mouse blastocyst". *Development* **132** (9): 2093–102. doi:10.1242/dev.01801. PMID 15788452.

[13] C.H. Damsky; Librach, C; Lim, KH; Fitzgerald, ML; McMaster, MT; Janatpour, M; Zhou, Y; Logan, SK; Fisher, SJ (1994-12-01). "Integrin switching regulates normal trophoblast invasion". *Development* **120** (12): 3657–66. PMID 7529679.

[14] "Human Chorionic Gonadotropin (hCG)". *WebMD*. 2010. Retrieved 2013-12-07.

[15] Fong, C. Y.; Bongso, A.; Ng, S. C.; Anandakumar, C.; Trounson, A.; Ratnam, S. (1997). "Ongoing normal pregnancy after transfer of zona-free blastocysts: Implications for embryo transfer in the human". *Human Reproduction* **12** (3): 557–60. doi:10.1093/humrep/12.3.557. PMID 9130759.

26.6 External links

- Blastocyst transfer and fertility treatment

- Risks of blastocyst transfer

- Blastocyst photos at different stages of development

- Diagram at weber.edu

- Blastocyst Differentiation Diagram

Chapter 27

Fibroblast

A **fibroblast** is a type of cell that synthesizes the extracellular matrix and collagen,[1] the structural framework (stroma) for animal tissues, and plays a critical role in wound healing. Fibroblasts are the most common cells of connective tissue in animals.

27.1 Background information

Fibroblasts and fibrocytes are two states of the same cells, the former being the activated state, the latter the less active state, concerned with maintenance and tissue metabolism. Currently, there is a tendency to call both forms fibroblasts. The suffix "blast" is used in cellular biology to denote a stem cell or a cell in an activated state of metabolism.

Fibroblasts are morphologically heterogeneous with diverse appearances depending on their location and activity. Though morphologically inconspicuous, ectopically transplanted fibroblasts can often retain positional memory of the location and tissue context where they had previously resided, at least over a few generations. This remarkable behavior may lead to discomfort in the rare event that they stagnate there excessively.

27.2 Embryologic origin

The main function of fibroblasts is to maintain the structural integrity of connective tissues by continuously secreting precursors of the extracellular matrix. Fibroblasts secrete the precursors of all the components of the extracellular matrix, primarily the ground substance and a variety of fibers. The composition of the extracellular matrix determines the physical properties of connective tissues.

Like other cells of connective tissue, fibroblasts are derived from primitive mesenchyme. Thus they express the intermediate filament protein vimentin, a feature used as a marker to distinguish their mesodermal origin.[2] However, this test is not specific as epithelial cells cultured in vitro on adherent substratum may also express vimentin after some time.

In certain situations epithelial cells can give rise to fibroblasts, a process called epithelial-mesenchymal transition (EMT).

Conversely, fibroblasts in some situations may give rise to epithelia by undergoing a mesenchymal to epithelial transition (MET) and organizing into a condensed, polarized, laterally connected true epithelial sheet. This process is seen in many developmental situations (e.g. nephron and notocord development), as well as in wound healing and tumorigenesis.

27.3 Structure and function

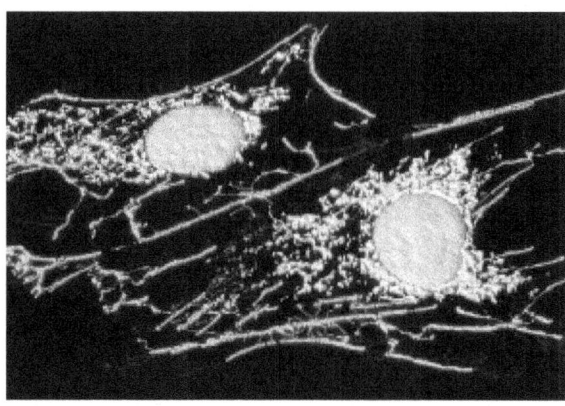

Microfilaments, mitochondria, and nuclei in fibroblast cells

Fibroblasts have a branched cytoplasm surrounding an elliptical, speckled nucleus having two or more nucleoli. Active fibroblasts can be recognized by their abundant rough endoplasmic reticulum. Inactive fibrocytes are smaller, spindle shaped, and have a reduced rough endoplasmic reticulum. Although disjointed and scattered when they have to cover a large space, fibroblasts, when crowded, often locally align in parallel clusters.

Fibroblasts make collagens, glycosaminoglycans, reticular and elastic fibers, glycoproteins found in the extracellular matrix, and cytokine TSLP. Growing individuals' fibroblasts are dividing and synthesizing ground substance. Tissue damage stimulates fibrocytes and induces the mitosis of fibroblasts.

Unlike the epithelial cells lining the body structures, fibroblasts do not form flat monolayers and are not restricted by a polarizing attachment to a basal lamina on one side, although they may contribute to basal lamina components in some situations (e.g. subepithelial myofibroblasts in intestine may secrete the $\alpha-2$ chain carrying component of the laminin which is absent only in regions of follicle-associated epithelia which lack the myofibroblast lining). Fibroblasts can also migrate slowly over substratum as individual cells, again in contrast to epithelial cells. While epithelial cells form the lining of body structures, it is fibroblasts and related connective tissues which sculpt the "bulk" of an organism.

The life span of a fibroblast, as measured in chick embryos, is 57 ± 3 days.[3]

27.4 Secondary actions

Mouse embryonic fibroblasts (MEFs) are often used as "feeder cells" in human embryonic stem cell research. However, many researchers are gradually phasing out MEFs in favor of culture media with precisely defined ingredients of exclusively human derivation. Further, the difficulty of exclusively using human derivation for media supplements is most often solved by the use of "defined media" where the supplements are synthetic and achieve the primary goal of eliminating the chance of contamination from derivative sources.

27.5 See also

- 3T3 cells
- Fibrocartilage callus
- Fibrous connective tissue

27.6 References

[1] "Fibroblast". *Genetics Home Reference*. U.S. National Library of Medicine. 2014-05-05. Retrieved 2014-05-10.

[2] Dave, Jui; Bayless, Kayla (May 1, 2014). "Vimentin as an Integral Regulator of Cell Adhesion and Endothelial Sprouting". *Microcirculation* 21 (4): 333–344.

[3] Weissmanshomer, P.; Fry, M. (1975). "Chick embryo fibroblasts senescence in vitro: Pattern of cell division and life span as a function of cell density". *Mechanisms of Ageing and Development* 4 (2): 159–166. doi:10.1016/0047-6374(75)90017-2. PMID 1152547.

27.7 External links

- UIUC Histology Subject *240*

- MedEd at Loyola *Histo/practical/ctproper/hp3-15.html*

Chapter 28

Trophoblast

Trophoblasts (from Greek *trephein*: to feed, and *blastos*: germinator) are cells forming the outer layer of a blastocyst, which provide nutrients to the embryo and develop into a large part of the placenta. They are formed during the first stage of pregnancy and are the first cells to differentiate from the fertilized egg. This layer of trophoblasts is also collectively referred to as "the trophoblast",[1] or, after gastrulation,[2] the **trophectoderm**, as it is then contiguous with the ectoderm of the embryo.

28.1 Structure

The trophoblast proliferates and differentiates into 2 cell layers at approximately 6 days after fertilization for humans:

28.2 Function

Trophoblasts are specialized cells of the placenta that play an important role in embryo implantation and interaction with the decidualised maternal uterus. The core of placental villi contain mesenchymal cells and placental blood vessels that are directly connected to the fetal circulation via the umbilical cord. This core is surrounded by two layers of trophoblast; a single layer of mononuclear cytotrophoblast that fuse together to form the overlying multinucleated syncytiotrophoblast layer that covers the entire surface of the placenta. It is this syncytiotrophoblast that is in direct contact with the maternal blood that reaches the placental surface, and thus facilitates the exchange of nutrients, wastes and gases between the maternal and fetal systems.

In addition, cytotrophoblast in the tips of villi can differentiate into another type of trophoblast called the extravillous trophoblast. Extravillous trophoblast grow out from the placenta and penetrate into the decidualised uterus. This process is essential not only for physically attaching the placenta to the mother, but also for altering the vasculature in the uterus to allow it to provide an adequate blood supply to the growing fetus as pregnancy progresses. Some of these trophoblast even replace the endothelial cells in the uterine spiral arteries as they remodel these vessels into wide bore conduits that are independent of maternal vasoconstriction. This ensures the fetus receives a steady supply of blood, and the placenta is not subjected to fluctuations in oxygen that could cause it damage.

28.3 Clinical significance

The invasion of a specific type of trophoblast (extravillous trophoblast) into the maternal uterus is a vital stage in the establishment of pregnancy. Failure of the trophoblast to invade sufficiently is important in the development of some cases of pre-eclampsia. Too firm an attachment may lead to placenta accreta.

Gestational trophoblastic disease represents a form of proliferation.

28.4 Additional images

- Blastodermic vesicle of Vespertilio murinus.
- Section through embryonic disk of Vespertilio murinus.
- Transverse section of a chorionic villus.
- Scheme of placental circulation.
- The initial stages of human embryogenesis

28.5 See also

- Syncytiotrophoblast
- Hydatidiform mole

28.6 References

[1] http://www.britannica.com/EBchecked/topic/606502/trophoblast

[2] Merriam-Webster's Medical Dictionary > trophectoderm Retrieved August 2010

28.7 External links

- Swiss embryology (from UL, UB, and UF) *iperiodembry/carnegie02*

Chapter 29

Stem cell controversy

The **stem cell controversy** is the consideration of the ethics of research involving the development, usage, and destruction of human embryos. Most commonly, this controversy focuses on embryonic stem cells. Not all stem cell research involves the creation, usage and destruction of human embryos. For example, adult stem cells, amniotic stem cells and induced pluripotent stem cells do not involve creating, using or destroying human embryos and thus are minimally, if at all, controversial. Many less controversial sources of acquiring stem cells include using cells from the umbilical cord, breast milk, and bone marrow.

29.1 Background

Main article: Stem cell

The use of stem cells has been happening for decades. In 1998, scientists discovered how to extract stem cells from human embryos. This discovery led to moral ethics questions concerning research involving embryo cells, such as what restrictions should be made on studies using these types of cells? At what point does one consider life to begin? Is it just to destroy an embryo cell if it has the potential to cure countless numbers of patients? Political leaders are debating how to regulate and fund research studies that involve the techniques used to remove the embryo cells. No clear consensus has emerged. Other recent discoveries may extinguish the need for embryonic stem cells.[1]

Since stem cells have the ability to differentiate into any type of cell, they offer something in the development of medical treatments for a wide range of conditions. Treatments that have been proposed include treatment for physical trauma, degenerative conditions, and genetic diseases (in combination with gene therapy). Yet further treatments using stem cells could potentially be developed thanks to their ability to repair extensive tissue damage.[2]

Great levels of success and potential have been shown from research using adult stem cells. In early 2009, the FDA ap-

proved the first human clinical trials using embryonic stem cells. Embryonic stem cells can become all cell types of the body which is called totipotent. Adult stem cells are generally limited to differentiating into different cell types of their tissue of origin. However, some evidence suggests that adult stem cell plasticity may exist, increasing the number of cell types a given adult stem cell can become. In addition, embryonic stem cells are considered more useful for nervous system therapies, because researchers have struggled to identify and isolate neural progenitors from adult tissues. Embryonic stem cells, however, might be rejected by the immune system - a problem which wouldn't occur if the patient received his or her own stem cells.

Some stem cell researchers are working to develop techniques of isolating stem cells that are as potent as embryonic stem cells, but do not require a human embryo.

Some believe that human skin cells can be coaxed to "dedifferentiate" and revert to an embryonic state. Researchers at Harvard University, led by Kevin Eggan and Savitri Marajh, have transferred the nucleus of a somatic cell into an existing embryonic stem cell, thus creating a new stem cell line.[3] Another study published in August 2006 also indicates that differentiated cells can be reprogrammed to an embryonic-like state by introducing four specific factors, resulting in induced pluripotent stem cells.[4]

Researchers at Advanced Cell Technology, led by Robert Lanza and Travis Wahl, reported the successful derivation of a stem cell line using a process similar to preimplantation genetic diagnosis, in which a single blastomere is extracted from a blastocyst.[5] At the 2007 meeting of the International Society for Stem Cell Research (ISSCR),[6] Lanza announced that his team had succeeded in producing three new stem cell lines without destroying the parent embryos. "These are the first human embryonic cell lines in existence that didn't result from the destruction of an embryo." Lanza is currently in discussions with the National Institutes of Health (NIH) to determine whether the new technique sidesteps U.S. restrictions on federal funding for ES cell research.[7]

Anthony Atala of Wake Forest University says that the fluid surrounding the fetus has been found to contain stem cells that, when utilized correctly, "can be differentiated towards cell types such as fat, bone, muscle, blood vessel, nerve and liver cells". The extraction of this fluid is not thought to harm the fetus in any way. He hopes "that these cells will provide a valuable resource for tissue repair and for engineered organs as well".[8]

29.2 Viewpoints

The status of the human embryo and human embryonic stem cell research is a controversial issue as, with the present state of technology, the creation of a human embryonic stem cell line requires the destruction of a human embryo. Stem cell debates have motivated and reinvigorated the pro-life movement, whose members are concerned with the rights and status of the embryo as an early-aged human life. They believe that embryonic stem cell research instrumentalizes and violates the sanctity of life and is tantamount to murder.[9] The fundamental assertion of those who oppose embryonic stem cell research is the belief that human life is inviolable, combined with the belief that human life begins when a sperm cell fertilizes an egg cell to form a single cell.

A portion of stem cell researchers use embryos that were created but not used in in vitro fertility treatments to derive new stem cell lines. Most of these embryos are to be destroyed, or stored for long periods of time, long past their viable storage life. In the United States alone, there have been estimates of at least 400,000 such embryos.[10] This has led some opponents of abortion, such as Senator Orrin Hatch, to support human embryonic stem cell research.[11] See Also Embryo donation.

Medical researchers widely submit that stem cell research has the potential to dramatically alter approaches to understanding and treating diseases, and to alleviate suffering. In the future, most medical researchers anticipate being able to use technologies derived from stem cell research to treat a variety of diseases and impairments. Spinal cord injuries and Parkinson's disease are two examples that have been championed by high-profile media personalities (for instance, Christopher Reeve and Michael J. Fox, who have lived with these conditions, respectively). The anticipated medical benefits of stem cell research add urgency to the debates, which has been appealed to by proponents of embryonic stem cell research.

In August 2000, The U.S. National Institutes of Health's Guidelines stated:

> "...research involving human pluripotent

stem cells...promises new treatments and possible cures for many debilitating diseases and injuries, including Parkinson's disease, diabetes, heart disease, multiple sclerosis, burns and spinal cord injuries. The NIH believes the potential medical benefits of human pluripotent stem cell technology are compelling and worthy of pursuit in accordance with appropriate ethical standards."

[12]

In 2006, researchers at Advanced Cell Technology of Worcester, Massachusetts, succeeded in obtaining stem cells from mouse embryos without destroying the embryos.[13] If this technique and its reliability are improved, it would alleviate some of the ethical concerns related to embryonic stem cell research.

Another technique announced in 2007 may also defuse the longstanding debate and controversy. Research teams in the United States and Japan have developed a simple and cost effective method of reprogramming human skin cells to function much like embryonic stem cells by introducing artificial viruses. While extracting and cloning stem cells is complex and extremely expensive, the newly discovered method of reprogramming cells is much cheaper. However, the technique may disrupt the DNA in the new stem cells, resulting in damaged and cancerous tissue. More research will be required before non-cancerous stem cells can be created.[14][15][16][17]

Update article to include 2009/2010 current stem cell usages in clinical trials.[18][19] The planned treatment trials will focus on the effects of oral lithium on neurological function in people with chronic spinal cord injury and those that have received umbilical cord blood mononuclear cell transplants to the spinal cord. The interest in these two treatments derives from recent reports indicating that **umbilical cord blood stem cells** may be beneficial for spinal cord injury and that lithium may promote regeneration and recovery of function after spinal cord injury. Both lithium and umbilical cord blood are widely available therapies that have long been used to treat diseases in humans.

29.2.1 Endorsement

- Embryonic stem cells have the potential to grow indefinitely in a laboratory environment and can differentiate into almost all types of bodily tissue. This makes embryonic stem cells a prospect for cellular therapies to treat a wide range of diseases.[20]

Human potential and humanity

This argument often goes hand-in-hand with the utilitarian argument, and can be presented in several forms:

- Embryos are not equivalent to human life while they are still incapable of surviving outside the womb (i.e. they only have the potential for life).

- More than a third of zygotes do not implant after conception.[21][22] Thus, far more embryos are lost due to chance than are proposed to be used for embryonic stem cell research or treatments.

- Blastocysts are a cluster of human cells that have not differentiated into distinct organ tissue; making cells of the inner cell mass no more "human" than a skin cell.[20]

- Some parties contend that embryos are not humans, believing that the life of *Homo sapiens* only begins when the heartbeat develops, which is during the 5th week of pregnancy,[23] or when the brain begins developing activity, which has been detected at 54 days after conception.[24]

Efficiency

- In vitro fertilization (IVF) generates large numbers of unused embryos (e.g. 70,000 in Australia alone).[20] Many of these thousands of IVF embryos are slated for destruction. Using them for scientific research uses a resource that would otherwise be wasted.[20]

- While the destruction of human embryos is required to establish a stem cell line, no new embryos have to be destroyed to work with existing stem cell lines. It would be wasteful not to continue to make use of these cell lines as a resource.[20]

Superiority

This is usually presented as a counter-argument to using adult stem cells as an alternative that doesn't involve embryonic destruction.

- Embryonic stem cells make up a significant proportion of a developing embryo, while adult stem cells exist as minor populations within a mature individual (e.g. in every 1,000 cells of the bone marrow, only 1 will be a usable stem cell). Thus, embryonic stem cells are likely to be easier to isolate and grow ex vivo than adult stem cells.[20]

- Embryonic stem cells divide more rapidly than adult stem cells, potentially making it easier to generate large numbers of cells for therapeutic means. In contrast, adult stem cell might not divide fast enough to offer immediate treatment.[20]

- Embryonic stem cells have greater plasticity, potentially allowing them to treat a wider range of diseases.[20]

- Adult stem cells from the patient's own body might not be effective in treatment of genetic disorders. Allogeneic embryonic stem cell transplantation (i.e. from a healthy donor) may be more practical in these cases than gene therapy of a patient's own cell.[20]

- DNA abnormalities found in adult stem cells that are caused by toxins and sunlight may make them poorly suited for treatment.[20]

- Embryonic stem cells have been shown to be effective in treating heart damage in mice.[20]

- Embryonic stem cells have the potential to cure chronic and degenerative diseases which current medicine has been unable to effectively treat.

Individuality

- Before the primitive streak is formed when the embryo attaches to the uterus at approximately 14 days after fertilization,two fertilized eggs can combine by fusing together and develop into one person (a tetragametic chimera). Since a fertilized egg has the potential to be two individuals or half of one, some believe it can only be considered a *potential* person, not an actual one. Those who subscribe to this belief then hold that destroying a blastocyst for embryonic stem cells is ethical.[25]

Viability

- Viability is another standard under which embryos and fetuses have been regarded as human lives. In the United States, the 1973 Supreme Court case of *Roe v. Wade* concluded that viability determined the permissibility of abortions performed for reasons other than the protection of the woman's health, defining *viability* as the point at which a fetus is "potentially able to live outside the mother's womb, albeit with artificial aid."[26] The point of viability was 24 to 28 weeks when the case was decided and has since moved to about 22 weeks due to advancement in medical technology. Embryos used in medical research for stem cells are well below development that would enable viability.

29.2.2 Objection

Alternatives

This argument is used by opponents of embryonic destruction as well as researchers specializing in adult stem cell research.

Pro-life supporters often claim that the use of adult stem cells from sources such as umbilical cord blood has consistently produced more promising results than the use of embryonic stem cells.[27] Furthermore, adult stem cell research may be able to make greater advances if less money and resources were channeled into embryonic stem cell research.[28]

In the past it has been a necessity to research embryonic stem cells and in doing so destroy them in order for research to progress.[29] As a result of the research done with both embryonic and adult stem cells new techniques may make the necessity for embryonic cell research obsolete. Because many of the restrictions placed on stem cell research have been based on moral dilemmas surrounding the use of embryonic cells there will likely be rapid advancement in the field as the techniques that created those issues are becoming less of a necessity.[30] Many funding and research restrictions on embryonic cell research will not impact research on IPSCs (induced pluripotent stem cells) allowing for a promising portion of the field of research to continue relatively unhindered by the ethical issues of embryonic research.[31]

Adult stem cells have provided many different therapies for illnesses such as Parkinson's Disease, leukemia, multiple sclerosis, lupus, sickle-cell anemia, and heart damage,[32] (to date, embryonic stem cells have also been used in treatment)[33] Moreover, there have been many advances in adult stem cell research, including a recent study where pluripotent adult stem cells were manufactured from differentiated fibroblast by the addition of specific transcription factors.[34] Newly created stem cells were developed into an embryo and were integrated into newborn mouse tissues, analogous to the properties of embryonic stem cells.

29.3 Stated views of groups

Main article: Stem cell laws

29.3.1 Government policy stances

Europe

Austria, Denmark, France, Germany, and Ireland do not allow the production of embryonic stem cell lines,[35] but the creation of embryonic stem cell lines is permitted in Finland, Greece, the Netherlands, Sweden, and the United Kingdom.[35]

United States

Main article: Stem cell laws and policy in the United States

Origins   In 1973, Roe v. Wade legalized abortion in the United States. Five years later, the first successful human *in vitro* fertilization resulted in the birth of Louise Brown in England. These developments prompted the federal government to create regulations barring the use of federal funds for research that experimented on human embryos. In 1995, the NIH Human Embryo Research Panel advised the administration of President Bill Clinton to permit federal funding for research on embryos left over from *in vitro* fertility treatments and also recommended federal funding of research on embryos specifically created for experimentation. In response to the panel's recommendations, the Clinton administration, citing moral and ethical concerns, declined to fund research on embryos created solely for research purposes,[36] but did agree to fund research on leftover embryos created by *in vitro* fertility treatments. At this point, the Congress intervened and passed the Dickey Amendment in 1995 (the final bill, which included the Dickey Amendment, was signed into law by Bill Clinton) which prohibited any federal funding for the Department of Health and Human Services be used for research that resulted in the destruction of an embryo regardless of the source of that embryo.

In 1998, privately funded research led to the breakthrough discovery of Human Embryonic Stem Cells (hESC). This prompted the Clinton Administration to re-examine guidelines for federal funding of embryonic research. In 1999, the president's National Bioethics Advisory Commission recommended that hESC harvested from embryos discarded after *in vitro* fertility treatments, but not from embryos created expressly for experimentation, be eligible for federal funding. Though embryo destruction had been inevitable in the process of harvesting hESC in the past (this is no longer the case[37][38][39][40]), the Clinton Administration had decided that it would be permissible under the Dickey Amendment to fund hESC research as long as such research did not itself directly cause the destruction of an embryo. Therefore, HHS issued its proposed regulation concerning hESC funding in 2001. Enactment of the new guidelines was delayed by the incoming George W. Bush administration which decided to reconsider the issue.

President Bush announced, on August 9, 2001 that federal funds, for the first time, would be made available for hESC research on currently existing embryonic stem cell lines. President Bush authorized research on existing human embryonic stem cell lines, not on human embryos under a specific, unrealistic timeline in which the stem cell lines must have been developed. However, the Bush Administration chose not to permit taxpayer funding for research on hESC cell lines not currently in existence, thus limiting federal funding to research in which "the life-and-death decision has already been made".[41] The Bush Administration's guidelines differ from the Clinton Administration guidelines which did not distinguish between currently existing and not-yet-existing hESC. Both the Bush and Clinton guidelines agree that the federal government should not fund hESC research that directly destroys embryos.

Neither Congress nor any administration has ever prohibited private funding of embryonic research. Public and private funding of research on adult and cord blood stem cells is unrestricted.

U.S. Congressional response   In April 2004, 206 members of Congress signed a letter urging President Bush to expand federal funding of embryonic stem cell research beyond what Bush had already supported.

In May 2005, the House of Representatives voted 238-194 to loosen the limitations on federally funded embryonic stem-cell research — by allowing government-funded research on surplus frozen embryos from in vitro fertilization clinics to be used for stem cell research with the permission of donors — despite Bush's promise to veto the bill if passed.[42] On July 29, 2005, Senate Majority Leader William H. Frist (R-TN), announced that he too favored loosening restrictions on federal funding of embryonic stem cell research.[43] On July 18, 2006, the Senate passed three different bills concerning stem cell research. The Senate passed the first bill (Stem Cell Research Enhancement Act), 63-37, which would have made it legal for the Federal government to spend Federal money on embryonic stem cell research that uses embryos left over from *in vitro* fertilization procedures.[44] On July 19, 2006 President Bush vetoed this bill. The second bill makes it illegal to create, grow, and abort fetuses for research purposes. The third bill would encourage research that would isolate pluripotent, i.e., embryonic-like, stem cells without the destruction of human embryos.

In 2005 and 2007, Congressman Ron Paul introduced the Cures Can Be Found Act,[45] with 10 cosponsors. With an income tax credit, the bill favors research upon non em-

bryonic stem cells obtained from placentas, umbilical cord blood, amniotic fluid, humans after birth, or unborn human offspring who died of natural causes; the bill was referred to committee. Paul argued that hESC research is outside of federal jurisdiction either to ban or to subsidize.[46]

Bush vetoed another bill, the Stem Cell Research Enhancement Act of 2007,[47] which would have amended the Public Health Service Act to provide for human embryonic stem cell research. The bill passed the Senate on April 11 by a vote of 63-34, then passed the House on June 7 by a vote of 247-176. President Bush vetoed the bill on July 19, 2007.[48]

On March 9, 2009, President Obama removed the restriction on federal funding for newer stem cell lines.   [49] Two days after Obama removed the restriction, the President then signed the Omnibus Appropriations Act of 2009, which still contained the long-standing Dickey-Wicker provision which bans federal funding of "research in which a human embryo or embryos are destroyed, discarded, or knowingly subjected to risk of injury or death;"[50] the Congressional provision effectively prevents federal funding being used to create new stem cell lines by many of the known methods. So, while scientists might not be free to create new lines with federal funding, President Obama's policy allows the potential of applying for such funding into research involving the hundreds of existing stem cell lines as well as any further lines created using private funds or state-level funding. The ability to apply for federal funding for stem cell lines created in the private sector is a significant expansion of options over the limits imposed by President Bush, who restricted funding to the 21 viable stem cell lines that were created before he announced his decision in 2001.[51] The ethical concerns raised during Clinton's time in office continue to restrict hESC research and dozens of stem cell lines have been excluded from funding, now by judgment of an administrative office rather than Presidential or legislative discretion.[52]

Funding   In 2005 the NIH funded $607 million worth of stem cell research, of which $39 million was specifically used for hESC.[53] Sigrid Fry-Revere has argued that private organizations, not the federal government, should provide funding for stem-cell research, so that shifts in public opinion and government policy would not bring valuable scientific research to a grinding halt[54]

In 2005 the State of California took out 3 billion dollars in bond loans to fund embryonic stem cell research in that state.[55]

Asia

China has one of the most permissive human embryonic

stem cell policies in the world. In the absence of a public controversy, human embryo stem cell research is supported by policies that allow the use of human embryos and therapeutic cloning.[56]

29.3.2 Religious views

Jewish view

According to Rabbi Levi Yitzchak Halperin of the Institute for Science and Jewish Law in Jerusalem, embryonic stem cell research is permitted so long as it has not been implanted in the womb. Not only is it permitted, but research is encouraged, rather than wasting it.

Similarly, the sole Jewish majority state, Israel permits research on embryonic stem cells.

Catholicism

The Catholic Church opposes human embryonic stem cell research calling it "an absolutely unacceptable act." The Church supports research that involves stem cells from adult tissues and the umbilical cord, as it "involves no harm to human beings at any state of development."[57]

Baptists

The Southern Baptist Convention opposes human embryonic stem cell research on the grounds that "Bible teaches that human beings are made in the image and likeness of God (Gen. 1:27; 9:6) and protectable human life begins at fertilization."[58] However, it supports adult stem cell research as it does "not require the destruction of embryos."[58]

Methodism

The United Methodist Church opposes human embryonic stem cell research, saying, "a human embryo, even at its earliest stages, commands our reverence."[59] However, it supports adult stem cell research, stating that there are "few moral questions" raised by this issue.[59]

Pentecostalism

The Assemblies of God opposes human embryonic stem cell research, saying, it "perpetuates the evil of abortion and should be prohibited."[60]

Islam

The religion of Islam favors the stance that scientific research and development in terms of stem cell research is allowed as long as it benefits society while using the least amount of harm to the subjects. Stem cell research is one of the most controversial topics of our time period and has raised many religious and ethical questions regarding the research being done. With there being no true guidelines set forth in the Qur'an against the study of biomedical testing, Muslims have adopted any new studies as long as the studies do not contradict another teaching in the Qur'an. One of the teachings of the Qur'an states that "Whosoever saves the life of one, it shall be if he saves the life of humankind" (5:32), it is this teaching that makes stem cell research acceptable in the Muslim faith because of its promise of potential medical breakthrough.[61]

The Church of Jesus Christ of Latter-day Saints

The First Presidency of The Church of Jesus Christ of Latter-day Saints "has not taken a position regarding the use of embryonic stem cells for research purposes. The absence of a position should not be interpreted as support for or opposition to any other statement made by Church members, whether they are for or against embryonic stem cell research."[62]

29.4 See also

- Stem cell laws
- Dickey-Wicker Amendment
- Medical ethics
- Stem Cell Research Enhancement Act
- Stem cell research policy
- Fetal tissue implant

29.5 References

[1] Genetic Science Learning Center. "The Stem Cell Debate: Is It Over?". *Learn Genetics*. Retrieved 15 January 2015.

[2] *Stem Cells for Tissue Regeneration and Joint Repair Science Daily* 29th March 2006.

[3] Cowan CA, Atienza J, Melton DA, Eggan K (August 2005). "Nuclear reprogramming of somatic cells after fusion with human embryonic stem cells". *Science* **309** (5739): 1369–73. doi:10.1126/science.1116447. PMID 16123299.

[4] Takahashi K and Yamanaka S (2006). "Induction of Pluripotent Stem Cells from Mouse Embryonic and Adult Fibroblast Cultures by Defined Factors". *Cell* **126** (4): 663–676. doi:10.1016/j.cell.2006.07.024. PMID 16904174.

[5] Klimanskaya I, Chung Y, Becker S, Lu SJ, Lanza R (November 2006). "Human embryonic stem cell lines derived from single blastomeres". *Nature* **444** (7118): 481–5. doi:10.1038/nature05142. PMID 16929302.

[6] ABC News: ABC News

[7] ScienceNOW - REUTERS: U.S. Company Says Grows Embryo-Safe Stem Cells

[8] Clout, Laura; Agencies (2007-09-01). "Scientists report alternative stem cell source". *The Daily Telegraph* (London). Retrieved 2007-09-20.

[9] "The stated reason for President Bush's objection to embryonic stem cell research is that 'murder is wrong'" (BBC)

[10] Weiss, Rick. (May 8, 2003) "400,000 Human Embryos Frozen in U.S.," *Washington Post.* Retrieved August 24, 2006.

[11] Connolly, Ceci. (July 30, 2005) "Frist Breaks With Bush On Stem Cell Research." *Washington Post.* Retrieved August 24, 2006.

[12] "NIH Publishes Final Guidelines for Stem Cell Research". National Institutes of Health. 2000. Retrieved 2007-04-29.

[13] "Deriving Stem Cells Without Killing Embryo". Medical News Today. 2006. Retrieved 2007-12-26.

[14] "New stem cell breakthrough". inthenews.co.uk. 2007. Retrieved 2007-12-26.

[15] Wade, Nicholas (June 6, 2007). "Biologists Make Skin Cells Work Like Stem Cells". The New York Times. Retrieved 2007-12-26.

[16] Weiss, Rick (June 7, 2007). "Scientists Use Skin To Create Stem Cells". The Washington Post. Retrieved 2007-12-26.

[17] "Scientists Convert Mouse Skin Cells to Stem Cells". Public Broadcasting Service. 2007. Retrieved 2007-12-26.

[18] http://www.clinicaltrials.gov/ct2/show/NCT01162915?term=stem+cells+covington&rcv_d=14

[19] http://clinicaltrials.gov/ct2/show/NCT01046786?term=China+Spinal+Cord+Injury+Network&rank=4

[20] "Arguments For Stem cell Research". Spinneypress. 2006. Archived from the original on 2008-02-01. Retrieved 2007-12-26.

[21] Raymond J. Devettere. Practical Decision Making in Health Care Ethics

[22] Kathleen Stassen Berger. The Developing Person Through the Life Span

[23] Greenfield, Marjorie. "Dr. Spock.com". Retrieved 2007-01-20.

[24] Singer, Peter. *Rethinking life & death: the collapse of our traditional ethics,* page 104 (St. Martins Press 1996). Retrieved 2007-03-04.

[25] West, Michael D.(2005) The Ethics of Genetic Engineering (At Issue Series). (pp 100-107) USA: Thomson Gale

[26] *Roe v. Wade,* 410 U.S. 113 (1973). Findlaw.com. Retrieved 2007-05-15

[27] Prentice, David. (October 17, 2005) "Live Patients & Dead Mice". *Christianity Today.* Retrieved on August 24, 2006.

[28] The Coalition of Americans for Research Ethics. "The "Political Science" of Stem Cells". Retrieved on July 16, 2006.

[29] BAERTSCHI, BERNARD, and ALEXANDRE MAURON. "Moral Status Revisited: The Challenge Of Reversed Potency." Bioethics 24.2 (2010): 96-103. Retrieved. 19 Apr. 2015.

[30] "Stem Cell Basics: Introduction" Bethesda, MD: National Institutes of Health (NIH), U.S. Department of Health and Human Services, 2009. Retrieved. 19 Apr. 2015

[31] Robertson, John A. "Embryo Stem Cell Research: Ten Years Of Controversy." Journal Of Law, Medicine & Ethics 38.2 (2010): 191-203. Retrieved. 19 Apr. 2015.

[32] Wislet-Gendebien, S., Laudet, E., Neirinckx, V., & Rogister, B. (2012). Adult Bone Marrow:Which Stem Cells for Cellular Therapy Protocols in Neurodegenerative Disorders?. Journal Of Biomedicine & Biotechnology, 20121-10. doi:10.1155/2012/601560

[33] Accessed: 11/1/10 | First person treated in milestone stem cell trial | 17:59 11 October 2010 by Andy Coghlan | http://www.newscientist.com/article/dn19570-first-person-treated-in-milestone-stem-cell-trial.html

[34] Cyranoski D (June 2007). "Simple switch turns cells embryonic". *Nature* **447** (7145): 618–9. doi:10.1038/447618a. PMID 17554270.

[35] Wiedemann PM, Simon J, Schicktanz S, Tannert C (October 2004). "The future of stem-cell research in Germany". *EMBO Rep.* **5** (10): 927–31. doi:10.1038/sj.embor.7400266. PMC 1299161. PMID 15459742. As noted before, the production of hESC lines is currently illegal in Germany; the 1990 Embryo Protection Act prohibits any utilization of the embryo that does not serve its preservation. ... Ireland, Austria, Denmark and France prohibit any production of hESC lines...Finland, Greece, the Netherlands, Sweden and the UK allow the production of hESC lines from surplus IVF embryos.

[36] "President Clinton's Comments on NIH and Human Embryo Research". *U.S. National Archives.* December 2, 1994. Retrieved 2006-07-19.

[37] http://www.nature.com/nature/journal/v444/n7118/full/nature05142.html

[38] Kao, CF; Chuang, CY; Chen, CH; Kuo, HC (2008). "Human pluripotent stem cells: Current status and future perspectives". *The Chinese journal of physiology* **51** (4): 214–25. PMID 19112879.

[39] http://www.advancedcell.com/news-and-media/cellular-therapy-press-releases/nih-approves-advanced-cell-technologys-stem-cell-line-for-federal-funding/

[40] http://www.advancedcell.com/documents/0000/0254/Press_Release_MA135_approval_-_FINAL.pdf

[41] The White House, Press Release, August 9, 2001

[42] "A Step Closer to Stem-Cell Heaven". Wired. 2005. Retrieved 2008-02-28.

[43] Connolly, Ceci (July 30, 2005). "Despite Bush Veto, Stem Cell Research Abounds". *Washington Post*. pp. A01. Retrieved 2006-07-21.

[44] Kellman, Laurie (July 18, 2006). "Senate Approves Embryo Stem Cell Bill". *Associated Press*. Retrieved 2006-07-18.

[45] H.R. 457, 110th Congress

[46] "Rights of Taxpayers is Missing Element in Stem Cell Debate". The Ron Paul Library.

[47] S. 5

[48] "Senate Approves Embryonic Stem Cell Bill", David Espo, Associated Press, April 12, 2007

[49] "Obama overturns Bush policy on stem cells" CNN, March 9, 2009

[50] Obama's Stem Cell Policy Hasn't Reversed Legislative Restrictions, Fox News, March 14, 2009

[51] Aldhous, Peter (March 2009). "Obama lifts research restrictions on embryonic stem cells". *New Scientist (online)* (Reed Business Information Ltd). Retrieved 2009-03-16. This frees biologists to work with a wide range of human ESCs - including cell lines created with state and private funding. But researchers are not expected to be able to use federal grants to create new cell lines. This is because of a 1996 law called the Dickey-Wicker amendment...

[52] "Statement from the Director on the Addition of New Lines to the Human Embryonic Stem Cell Registry" NIH, June 21, 2010

[53] "Estimates of Funding for Various Diseases, Conditions, Research Areas". National Institutes of Health. 2007. Retrieved 2008-01-21.

[54] "Best Hope Lies in Private Stem-Cell Funding". Retrieved 2008-01-31.

[55] http://cirm.ca.gov/

[56] " "Kerstin Klein, Illiberal Biopolitics, Human Embryos and the Stem Cell Controversy in China, 2010"] LSE

[57] "On Embryonic Stem Cell Research" (PDF). USCCB Publishing. Retrieved 2007-06-24.

[58] "Resolution On Human Embryonic And Stem Cell Research". Southern Baptist Convention. Retrieved 2007-06-24.

[59] "Ethics of Embryonic Stem Cell Research". The United Methodist Church. Retrieved 2007-06-24.

[60] "Sanctity of Human Life Including Abortion and Euthanasia" (PDF). Assemblies of God. Retrieved 2007-06-24.

[61] Agha, Fatima; Hayani, Al (2008). "Muslim perspectives on stem cell research and cloning". *Zygon* **4** (43): 783–795.

[62] "Embryonic Stem-cell Research". The Church of Jesus Christ of Latter-day Saints. Retrieved 2011-10-20.

29.6 External links

- Video: The Stem Cell Controversy January 18, 2006, Woodrow Wilson Center event featuring Robin Cook (novelist), William B. Hurlbut, and Michael D. West

- The Hinxton Group: An International Consortium on Stem Cells, Ethics & Law

- A Scientific-Industrial Complex? By Sigrid Fry-Revere

- On the Personhood of Pre-implantation Embryos

- Videos about Stem Cell Ethics and Controversy

29.7   Text and image sources, contributors, and licenses

29.7.1   Text

- **Stem cell** *Source:* https://en.wikipedia.org/wiki/Stem_cell?oldid=710955220 *Contributors:* AxelBoldt, Tobias Hoevekamp, Magnus Manske, Derek Ross, Sodium, Eloquence, Mav, Bryan Derksen, The Anome, Taw, Ed Poor, Alex.tan, Andre Engels, Youssefsan, Danny, SJK, Ellmist, AdamRetchless, Graft, Modemac, KF, Hfastedge, Michael Hardy, Lexor, Gabbe, Ixfd64, Nina, Tregoweth, Ronabop, Extro, JWSchmidt, Habj, Kimiko, Netsnipe, Big iron, Llull, Andres, Evercat, Goododa, Quizkajer, Mpt, Timwi, Nohat, Wikiborg, Glimz~enwiki, Wik, Steinsky, AHands, Haukurth, Tpbradbury, Grendelkhan, Saltine, Tero~enwiki, Tlotoxl, Samsara, Raul654, Frazzydee, Shafei, Owen, Seriv, Robbot, Paranoid, Astronautics~enwiki, Moriori, Fredrik, Chris 73, RedWolf, Jmabel, Goethean, ZimZalaBim, Altenmann, Pingveno, Henrygb, TMLutas, Oji-giri~enwiki, Matty j, Moink, Hadal, Johnstone, Aetheling, Seth Ilys, Alan Liefting, Jsan, Robert Happelberg, Decumanus, Centrx, JamesMLane, DocWatson42, Christopher Parham, DavidCary, Kim Bruning, Wolfkeeper, Tom harrison, Lupin, Brian Kendig, Bradeos Graphon, Everyking, Fleminra, Maha ts, Snap Davies, Jfdwolff, Sundar, Siroxo, BesigedB, Raekwon, CRUCIFYKEVIN, AdamJacobMuller, Bobblewik, Tagishsi-mon, RobM~enwiki, Peter Ellis, Gyrofrog, Wmahan, Espetkov, Adenosine, Chowbok, Gadfium, Pgan002, Toytoy, Jpkoester1, Vanished user svinet8j3ogifm98wjfgoi3tjosfg, Slowking Man, Antandrus, Beland, PDH, Jossi, Rdsmith4, Langston~enwiki, PFHLai, FrozenUmbrella, Yos-sarian, Vogon77, Gscshoyru, Neutrality, Joyous!, Jh51681, Grm wnr, Naus, Adashiel, Iwilcox, Grunt, Gazpacho, MikeF, Mike Rosoft, Oskar Sigvardsson, Mernen, Rfl, DanielCD, Mindspillage, Discospinster, Solitude, Rich Farmbrough, Rhobite, Vague Rant, NeuronExMachina, Melt-Banana, Paul August, Stbalbach, Bender235, ESkog, Richard Taylor, Austinbond, Violetriga, Fenice, Tooto, Brian0918, Howdy, Mwanner, Shanes, Freeze42084, RoyBoy, LordRM, Moilleadóir, Omhafeieio, Spoon!, MPS, Gyll, Bobo192, Spalding, Circeus, Key45, Hurricane111, Smalljim, Nectarflowed, Reinyday, Adraeus, Sasha Kopf, Brim, AllyUnion, Phidauex, Cohesion, Adrian~enwiki, Geoff.green, Arcadian, Jag123, Scapermoya, I9Q79oL78KiL0QTFHgyc, RMann, Sriram sh, Tomgally, Boredzo, Pschemp, Vanished user 19794758563875, Andrewbadr, Sr-lasky, Oarih, Obradovic Goran, Holdek, Sam Korn, Hahani hanuka, PochWiki, Benbread, Martyman, Truejim, Melah Hashamaim, Jblatt, Mark Lewis, Zachlipton, Defunkt, Alansohn, JYolkowski, ChrisGlew, Tpikonen, Free Bear, Arthena, Borisblue, Mu5ti, PatrickFisher, Pale-orthid, Riana, Nicholas Cimini, Axl, Seans Potato Business, Malo, Idont Havaname, Snowolf, Velella, ClockworkSoul, Keetoowah, Helixblue, Tycho, Cburnett, Evil Monkey, RJII, Tony Sidaway, AngryParsley, TenOfAllTrades, H2g2bob, Someoneinmyheadbutit'snotme, MIT Trekkie, Capecodeph, Johntex, Bookandcoffee, Dan100, Ceyockey, Tintin1107, Mathprog777, RyanGerbil10, Gatewaycat, ChipsAhoya, Gmaxwell, Marasmusine, Boothy443, Simetrical, Mel Etitis, Bushytails, OwenX, Woohookitty, Mindmatrix, Yansa, Jersyko, CyrilleDunant, Uncle G, Poly-paradigm, Ruud Koot, JeremyA, Ztrawhes, MONGO, Rtdrury, Tabletop, Schzmo, Grika, Terence, Brendanconway, Wayward, Gimboid13, Arcus~enwiki, Aeon221, GSlicer, Mandarax, Tslocum, Graham87, Alienus, Magister Mathematicae, MikeSar, Teflon Don, Olbrich~enwiki, Kbdank71, FreplySpang, BorgHunter, Rkevins, Canderson7, Sjakkalle, Rjwilmsi, Tizio, Nightscream, Strangethingintheland, Jivecat, Astro-naut, PinchasC, Quiddity, Tangotango, Gogglecollector, MZMcBride, Tawker, HandyAndy, Oblivious, DonSiano, Crazynas, ElKevbo, Lud-dite, Bob Wiyadabebe-lytsaboi, Sango123, Darqcyde, Yamamoto Ichiro, Ravidreams, Ravenswood, Ground Zero, Nihiltres, Crazycomputers, TheMidnighters, Who, Nivix, RexNL, Gurch, Brendan Moody, Atrix20, Stevenfruitsmaak, Alphachimp, Dayed, MOF, Butros, WouterBot, Ryddragyn, Youssefa, Chobot, Karch, DVdm, Stephen Compall, Bgwhite, Wjfox2005, Badanagram, Bisyork, The Rambling Man, YurikBot, Wavelength, TexasAndroid, Angus Lepper, Vedranf, RobotE, Systemfolder, RussBot, AVM, Hede2000, Chris Capoccia, RadioFan, Hydrar-gyrum, Stephenb, Bill52270, CambridgeBayWeather, Cryptic, Wimt, RadioKirk, Draeco, NawlinWiki, Theglockner, Wiki alf, Bachrach44, Snek01, Yegorka, Cquan, Howcheng, Davfoster88, Irishguy, Javeryt, Anetode, D. Wu, Brian Crawford, Aldux, Moe Epsilon, Aaron charles, Misza13, Bucketsofg, Aaron Schulz, DeadEyeArrow, Private Butcher, Oliverdl, Ilmaisin, Werdna, Jinkbl0t, Dark.ranger, User27091, Nick123, Searchme, Jkelly, FF2010, Herb West, Encephalon, Huangcjz, Closedmouth, Arthur Rubin, Fang Aili, Bucaro1975, Fstorino~enwiki, Esprit15d, Sean Whitton, BorgQueen, GraemeL, Natgoo, Skittle, Curpsbot-unicodify, Garion96, MagneticFlux, RunOrDie, Kungfuadam, Selkem, TLSuda, Jonathan.s.kt, Tom Morris, NetRolller 3D, Luk, Glisteringwaters, Frankie, SmackBot, Manogamez, Moeron, Ksargent, Zanter, KnowledgeOf-Self, TestPilot, Royalguard11, SearedIce, Pgk, Bomac, Lankenau, Kfor, Thunderboltz, Clpo13, Wolf ODonnell, Nickst, Anastrophe, Delldot, Michaelll, Brossow, Warren.cheung, Edgar181, Flux.books, Xaosflux, Yamaguchi⬚⬚, Cool3, Gilliam, Phizzy, DividedByNegativeZero, Bfig-gis, Betacommand, Boots bklyn, ERcheck, Viewtiful Zoidberg, Rmosler2100, Alenshrew, Keegan, RDBrown, Jprg1966, Master of Puppets, Skomae, EncMstr, Miquonranger03, Isaacsurh, Mattythewhite, Parkyk219, Wykis, JONJONAUG, CMacMillan, MDChanderson~enwiki, Cas-sivs, Toughpigs, Konstable, Antonrojo, VirtualSteve, Zsinj, Trekphiler, Can't sleep, clown will eat me, Snowmanradio, OOODDD, Avb, Ko-rinkami, Blankfaction.tk, EvelinaB, Rrburke, Huon, Cpt.squeaky, Dharmabum420, Brainhell, PiMaster3, Wordwhiz, Nakon, TheLimbicOne, Savidan, Funky Monkey, Blake-, Bigal78, Drunken Whale, MBCF, So cool~enwiki, TheAxeGrinder, Drphilharmonic, Andcarne, Freshy-ill, A.V.~enwiki, DMacks, Zeamays, Brainfood, Caelarch, Kuzaar, Nishkid64, Sil3nt4ss4sin, Niels, Alakey2010, Dbtfz, Kuru, AmiDaniel, Provider uk, Cholerashot, Tedwardo2, Perfectblue97, Chodorkovskiy, KillZOne, NYCJosh, Moop stick, Sigma77, 041744, Kyphe, Slakr, Hikoto, Shangrilaista, Tasc, Martinp23, Mr Stephen, Soccerkid1212, Arkrishna, Vir, Kamoesai, Impact red, Big Smooth, MTSbot~enwiki, Ryanjunk, Seemers, Autonova, Ginkgo100, Sonic3KMaster, The7thmagus, Jwalte04, Pegasus1138, Octane, Courcelles, Gilabrand, Tawker-bot2, Bioinformin, Daniel5127, Nobleeagle, George100, Altonbr, Crazydog115, Userdce, Krohon, The Haunted Angel, Switchercat, SkyWalker, JForget, Will314159, Alexei Kouprianov, InvisibleK, CmdrObot, The Librarian, Mirceat, Agathman, Scohoust, Iced Kola, Arkantosstevius, Jgwlaw, Nouseforaname312, Leevanjackson, Nadyes, Lmcelhiney, MiamiDolphins3, Argon233, Casper2k3, Sonare, Cydebot, Mblumber, CB-Fan, Lbuckler, RenamedUser2, Snowbord60, Eubanks718, Michaelas10, RelentlessRecusant, Gogo Dodo, Anthonyhcole, JFreeman, Lazydaisy, Dr.enh, Karafias, Naudefj, DumbBOT, Chrislk02, Dimitmant, Narayanese, Meghaljani, Sailer247, Incitatus4, Shrinkshooter, Malleus Fatuorum, Epbr123, Dr Aaron, MHGinTN, Bartimaeus666, Qwyrxian, Ucanlookitup, N5iln, Jdm64, Headbomb, Mishmash8, Marek69, Peter Znamen-skiy, Tapir Terrific, Brown boi5, Corporate.legal, Jonny-mt, K8hjelow, Masonian, Sean William, Dantheman531, M0s6p, Shibby970, Anti-VandalBot, Luna Santin, Seaphoto, Sziegler~enwiki, Fic-in, TimVickers, Byornski, Silver seren, Gdo01, Ellwyn, Megalaser, David Shankbone, BballJones, Arx Fortis, Myanw, JAnDbot, Mjmurphyjr, Husond, Lilblackme, Davewho2, K955301, Epeefleche, Svm2, Eurobas, Getaway, Jam-intime, Hut 8.5, Savant13, Lawilkin, Joecool94, Jkp44, Dave Nelson, Lulurascal, Acroterion, I80and, Magioladitis, Bongwarrior, VoABot II, Ishikawa Minoru, MastCell, Plain jack, Pugetbill, Jim Douglas, MatthewJS, Rootxploit, LeaHazel, Prestonmcconkie, ThomWatson, Choco-latelana, Hdynes, Hamiltonstone, Robert M. Hunt, Roadghost, Lethaniol, Parijata, Pflueger, DerHexer, Philg88, Wi-king, WLU, Heyhombre, DGG, James301294, Otvaltak, Pvosta, Yobol, MartinBot, PhuryPrime, Agricolae, Nadejla, Mister Congeniality, Theredbanana, Ricardogpn, Mike6271, Juansidious, Nashpaul, Pakraw, R'n'B, Averross, Ralphxcrunner, AlexiusHoratius, Blazersguy, Runninganok, Etparle, Cyrus And-iron, EdBever, Mediaworker, J.delanoy, Trusilver, AstroHurricane001, James cudahy, Deepakpurang, Numbo3, Peter Chastain, Richiekim,

Boghog, Uncle Dick, Debatesensei, Mbm124verizon, Prince wiki thai, Nanpae-enwiki, Keesiewonder, Gwilym84, Drumman91, Jv9mmm, Flyingdream, Rod57, Holyone2, JavaJawaUK, Dr d12, Notreallydavid, Max boettcher, Mikael Häggström, Skier Dude, Visual Basic 5.0, Magicnxnja, NewEnglandYankee, Harrisd5917, Cobi, Buhuzu, Flatterworld, Id711, Mullagatawny, Vikkidly, Vanished user 39948282, Snoop a la boop, Logiboy123, Pdcook, Useight, Xiahou, ThePointblank, Feverinlove, Xnuala, Xenonice, Bigminisachin1231, Milnivlek, Vinegarfod, Walik, Jeff G., Mocirne, Choach, Independovirus, Stable attractor, Philip Trueman, SamMichaels, Berlickshane, Lukey jones, Cosmic Latte, Baughnie, Anonymous Dissident, CoJaBo, Hank Ramsey, Sankalpdravid, Aymatth2, Oxfordwang, EdJ343, Xcrunner29, Corvus cornix, Leafyplant, Jvbishop, Drbohlav-enwiki, Bubbles3, Hockey87, Thestonesfieldmonster, T-woo22, Synthebot, Lova Falk, David Peacham, Falcon8765, Dragostanasie, Levtchenkov, OnlyWayne, C45207, Monty845, Doc James, Logan, SylviaStanley, Redfiona99, 1 c linguist, SieBot, Demong, Nonerds2, Jeanettephair, Maurauth, Dusti, Calliopejen1, Scarian, Jauerback, Caltas, Yintan, Blkpowernig2, LeadSongDog, Greenbough, Chimera16, Allmightyduck, Oxymoron83, Colin marks, Harry-enwiki, Lightmouse, Tombomp, Sunrise, 04spicerc, OKBot, Dlh-stablelights, Svick, Speimer, G.-M. Cupertino, StaticGull, Kagome 85, Sean.hoyland, Realm of Shadows, Jim hoerner, Hirenism, D3monicVeng3ance, Dabomb87, Florentino floro, TubularWorld, Regener, Naturespace, Forluvoft, Vanished user qkqknjitkcse45u3, ClueBot, Artichoker, Healthwise, The Thing That Should Not Be, Getwood, Egandrews, ArneLH, Plastikspork, Jan1nad, Meekywiki, Abbaroodle, Drmies, Lelestopme, LizardJr8, Yadmat, Duckzilla2000, Deselliers, Excirial, Jusdafax, Astoman, SpikeToronto, Alanliddell, NuclearWarfare, Aurora2698, Iohannes Animosus, InaMaka, Hendog456, Usbdriver, Change93, Mlaffs, Michael3180, Banman1004, 9Nak, Moosepers, Ted4d7, Stemcel2, Versus22, Shorthoped again, MelonBot, DumZiBoT, Badarkace, XLinkBot, Mjharrison, Jytdog, Ost316, Avoided, Blunub, Facts707, Nicolae Coman, SilvonenBot, Josh.K.Poole, Tobymanchee, RyanCross, Chrisdale92, Addbot, Darena mipt, Some jerk on the Internet, DOI bot, Toyokuni3, 旧字新字-enwiki, Ronhjones, Ashishlohorung, Jncraton, Fluffernutter, Diptanshu.D, MrOllie, Chamal N, Glane23, Ruineye, Favonian, Marcos987, ChenzwBot, 5 albert square, Chrismeehan, AgadaUrbanit, Numbo3-bot, Tide rolls, OlEnglish, Guyonthesubway, Gail, Angrysockhop, आशीष भटनागर, Adam.douggie, Luckas-bot, Yobot, HRED, Mirabellen, Les boys, Mmxx, THEN WHO WAS PHONE?, AnomieBOT, Ormers, SMC23, Noq, Danandlollie, Dwayne, Accuruss, JackieBot, Todderk, Glagolev, Kingpin13, Fatalityduff, KinStrategy, Ncrm, Peyton20, Deving, Materialscientist, Inthend9, The High Fin Sperm Whale, Citation bot, ArthurBot, LilHelpa, Xqbot, Poetnk, Nasnema, Prowler08, Mlpearc, Jezhotwells, Peterish, Amaury, Wiki emma johnson, Gregbrak, Dougofborg, Captain-n00dle, Fingerz, FrescoBot, Paine Ellsworth, Tobby72, Abbasbeyli, Vin9-enwiki, Topmec, Dobett, HJ Mitchell, Marcogreguzzoni, Citation bot 1, Pinethicket, Jonesey95, TheFourFreedoms, Acgator09, RedBot, Meaghan, Bgpaulus, Silicon Beach Comber, FoxBot, Jordgette, Fama Clamosa, Vrenator, Vikrantt, Allen4names, Bobbyus, Seahorseruler, Everyone Dies In the End, Weedwhacker128, Tbhotch, Minimac, DARTH SIDIOUS 2, Mean as custard, RjwilmsiBot, Bento00, Walkinxyz, Rogen123, Aircorn, Prasadmalladi, Androstachys, Dr.shosho, Wojcz, DASHBot, EmausBot, John of Reading, Davejohnsan, Jaysonlam, Immunize, Alchewizzard, Stemscientist, Nichefinder, RA0808, YELKENN, RenamedUser01302013, Vanished user zq46pw21, Obesegypsi, Solarra, Tommy2010, Wikipelli, K6ka, Green88en, Vindicata, Alextill95, John Cline, Potionism, Thunder365, Tish18, Everard Proudfoot, Elektrik Shoos, Dbsteve100, A930913, H3llBot, Wayne Slam, Ajournaleditor, Tolly4bolly, Larba200, Ohio34895, TyA, GeorgeBarnick, Ipscellmate, L Kensington, Bethelizebeth, Donner60, Cellmedsociety, Smackerlacker, Orange Suede Sofa, ChuispastonBot, UBERxL33T, Munna.alerts, Sunshine4921, Blonde chik819, Famedog, Xanchester, Gwen-chan, ClueBot NG, Scottttttt, Gareth Griffith-Jones, Galilsnap, Wikitavanti, MelbourneStar, Isgsimeng, Wimpus-enwiki, Cntras, Braincricket, O.Koslowski, Rezabot, Iknows, Widr, Korovin2, Exceedingly Rare, Europa284, Yeah nit, Drianmcniece, Helpful Pixie Bot, Thethird33, HMSSolent, Kjveeley, Lowercase sigmabot, BG19bot, Alasdair.g1998, Stemcellhub, Todobo, Spicywiener123, UConn Multimedia, Jimmybass, MusikAnimal, Stocx, Mark Arsten, Pramod Vora, Silvrous, Cabodoc, BreeRobinson, Snow Blizzard, MrBill3, Leo181, Catlith, Andrewdekamp, Aberdam, Garywest1111, Zeeaali, Irishboy5, Zachfdog, Heybbyy, BattyBot, ChrisGualtieri, JYBot, APerson, Webclient101, Ruha9, Syzygy2048, TwoTwoHello, Lugia2453, Isarra (HG), JakobSteenberg, Dmitry Dzhagarov, Wywin, PC-XT, Epicgenius, The colledge teacher, I am One of Many, Kamker, Seth brady, Iztwoz, Gavin300, Poothebob, Yolo12348, Stemcellmafia, Triscomn1, CensoredScribe, Mfoti610, MSperbeck, Lgirard, Zurc25, BigRigsFan3, Deleter9000, Stemcells90, Heimo79, Samatict, Ryan.gaylard, Yny501, Troller66, Dentinho98br, Div2005, Mbryantsa, Muse2000, Almeidaxavier.miguel, Joaunthy3111, Chaya5260, Askilli, Jla9715, Giancarlobasile, Lolfailpoop, Monkbot, Neurologicalinstitute, Saifu3727, Griff618, Wagetawny34, Kensta10, MelnaisDzons, Salamuddin.Shaikh89, 123lazyg, KasparBot, Han-Jun Cho, Debennett2 and Anonymous: 1540

- **Stem-cell therapy** *Source:* https://en.wikipedia.org/wiki/Stem-cell_therapy?oldid=713109814 *Contributors:* William Avery, D, Ixfd64, Andrewman327, Jondel, Jfdwolff, Beland, PDH, Demeter-enwiki, Gscshoyru, Discospinster, Rich Farmbrough, DJP, Adambro, Alansohn, Wouterstomp, Nicholas Cimini, SlimVirgin, Velella, Greg Kuperberg, H2g2bob, Ceyockey, Woohookitty, RHaworth, Benbest, Sjö, Rjwilmsi, Jivecat, Quiddity, MZMcBride, Naraht, Alphachimp, The Rambling Man, Wavelength, SLATE, Porturology, Madkayaker, Stephenb, Rsrikanth05, Cquan, Aaron charles, Tony1, PanchoS, Elkman, Kkmurray, FF2010, Elfalem, Mais oui!, Rathfelder, SmackBot, KnowledgeOfSelf, Ariedartin, Delldot, Mifren, Ohnoitsjamie, Chris the speller, Bluebot, RDBrown, Isaacsurh, Colonies Chris, Anabus, Can't sleep, clown will eat me, OOODDD, GVnayR, Kim Bach, Andcarne, DavidJ710, Mion, Kopf1988, Dicklyon, NaturalBornKiller, Alessandro57, Markbassett, Bioinformin, Kooperfan, CmdrObot, Dycedarg, Haridan, Cydebot, Anthonyhcole, Swakeman, Odie5533, Narayanese, Thijs!bot, Dr Aaron, N5iln, Headbomb, Marek69, Peter Znamenskiy, Alphius, Opelio, TimVickers, Mdotley, Jaredroberts, Kirrages, Lulurascal, VoABot II, MastCell, JamesBWatson, Drthiruna, Nposs, Millstone, David Eppstein, Desirsar, Hbent, Wikianon, Flowanda, Hdt83, MartinBot, DougCube, CliffC, Nono64, Runningamok, Lilac Soul, Svetovid, Awdavidoff, Rod57, Mikael Häggström, AntiSpamBot, Jemoro, Id711, Xiahou, Orphic, TravellerDMT-07, Jeff G., Larry R. Holmgren, QuackGuru, Philip Trueman, Scilit, Tkslaney, BotKung, Arcticdawg, Wolfrock, Cmcnicoll, Levtchenkov, Doc James, DivaNtrainin, SylviaStanley, Tresiden, Stemsales, Dawn Bard, Kotabatubara, Tiptoety, Sunrise, Speimer, Anchor Link Bot, Maralia, WikiLaurent, Parrotz1461, Denisarona, ClueBot, Neil bhullar, Foxj, The Thing That Should Not Be, Abbaroodle, Mheidara, Meeva, Brewcrewer, Jusdafax, Eeekster, Mlaffs, Stemcellguy, Zaledin, Versus22, DumZiBoT, XLinkBot, Jytdog, Skarebo, Jeorava, Esen O, Fluffernutter, Diptanshu.D, Jamieortiz, Shoushan, Vishwas samant, Cblack2000, Favonian, Chrismeehan, Scodinzolare, Tassedethe, Lolinfoman, Totorotroll, Gail, Yobot, Themfromspace, Agoldstand, AnomieBOT, Noq, Cutisderma, Citation bot, Tintenza, Addihockey10, Prowler08, Gap9551, RoodleDoodle, Lilk8emama, Jcjjhrsns, Bowenap, Zer02325, Kylelovesyou, Wiki emma johnson, Sematena, LincolnSt, FrescoBot, Naseem abi shaheen, Marina03, Meishern, Citation bot 1, Machn, Jonesey95, Neurosojourn, Stemcells21, Jdoenosh, Full-date unlinking bot, Leong Lau, Rebeccanichols09, Orenburg1, Vaishnav2010, Trappist the monk, LilyKitty, Mccartjt, DARTH SIDIOUS 2, RjwilmsiBot, Bento00, EmausBot, Archit894, Gfoley4, Müdigkeit, Siddharthnmishra, Racerx11, RA0808, EME44, Passionless, Tommy2010, Deirovic, JPtheK9, Cheungy97, Bvwaghmare, Hells Outkast, Manicjedi, Coasterlover1994, L Kensington, Cellmedsociety, Wakebrdkid, Cellmedsoc, ClueBot NG, Gareth Griffith-Jones, Morgankevinj huggle, Pthaivanich, Tideflat, Andrey Makhinya, Widr, Distshore78, Harsimaja, Proofofthis, Vigi.limi, BG19bot, Stemcellcures11, Tammie6123, Mysupertopup, Cabodoc, Abby26, MrBill3, Leo181, Zedshort, Ginger Maine Coon, Anbu121, Alisifer, FLOPROWEG, Samwalton9, MeanMotherJr, BattyBot, Cyberbot II, ChrisGualtieri, Biolprof, Khazar2, BrightStarSky, Dexbot, Hmainsbot1, Ruha9, Mogism, Everything Is Numbers, Dmitry Dzhagarov, Ettalynn7, Tcifelli, Copulative, Epicgenius, AndreasKyttaro, BradYard, Stemcellnow47, CensoredScribe, Exesop, Steven Levy MD, Mshams2, Pervaks, Feel Happy, Cbrillaz, Lung123!, Stashrules, Monkbot, Saifu3727, Ijnijnokm, Santanupsingh, Clasters, Typicaldiamonds, Carr 81, Sagewigan, Legenderfox, Rosea999, Erinreynolds7, Slipvoid, Nøkkenbuer, Happyhohaho, Duncan Thomson, Sweta Oza, Valdomri and Anonymous: 362

- **Cell (biology)** *Source:* https://en.wikipedia.org/wiki/Cell_(biology)?oldid=712854546 *Contributors:* AxelBoldt, Magnus Manske, Joao, Kpjas, Brion VIBBER, Mav, Bryan Derksen, Taw, Malcolm Farmer, RK, BenBaker, Andre Engels, LA2, Josh Grosse, Christian List, William Avery, Anthere, AdamRetchless, Heron, Icarus-enwiki, Patrick, Olrick, D., Michael Hardy, Zocky, Lexor, Gabbe, Ixfd64, Lquilter, GTBacchus, Dori, Kosebamse, 168..., Ahoerstemeier, Theresa knott, JWSchmidt, Julesd, Glenn, Andres, Evercat, Cherkash, Rob Hooft, Mxn, Smack, Jengod, Emperorbma, Ec5618, RodC, Timwi, Wikiborg, Selket, Steinsky, Tpbradbury, Marshman, Morwen, Itai, Omegatron, Ed g2s, Thue, Quoth-22, Topbanana, Bjarki S, Renato Caniatti-enwiki, Raul654, L-Bit, Doug swisher, DougS, Bcorr, Camerong, Francs2000, Shantavira, Rogper-enwiki, Donarreiskoffer, Robbot, Josh Cherry, Sander123, Vyasa, Romanm, Caknuck, Moink, Hadal, Fuelbottle, Pengo, Dina, Adam78, Mdmeginn, Alan Liefting, Fabiform, Giftlite, Christopher Parham, Crimson30, Nmg20, Wikilibrarian, Kim Bruning, Ævar Arnfjörð Bjarmason, Netoholic, Bradeos Graphon, Average Earthman, Everyking, Michael Devore, Bensaccount, Gareth Wyn, Hugh2414, Sietse, Siroxo, Kandar, Wmahan, Adenosine, Gadfium, Andycjp, Gazibara, Pearbonn, Antandrus, GeneMosher, Williamb, Onco p53, OverlordQ, PDH, Sean Heron, Vina, MacGyverMagic, Brian Brondel, Rdsmith4, Sam Hocevar, Vogon77, Gseshoyru, Creidieki, Kevin Rector, Deglr6328, Grunt, SYSS Mouse, Possession, ClockworkTroll, Discospinster, Zaheen, Rich Farmbrough, KillerChihuahua, Guanabot, Lejean2000, Kdammers, Vsmith, Xezbeth, CrisDias, 1pezguy, Mani1, Dmr2, Yersinia-enwiki, Bender235, ESkog, Kbh3rd, Kaisershatner, Violetriga, Pusher, Eric Forste, Pofkezas, Syp, Kaszeta, CanisRufus, Charm, Illumynite, Anphanax, Lycurgus, Skeppy, Koenige, Shanes, Art LaPella, RoyBoy, CDN99, Gylₑ, Bobo192, Whosyourjudas, ..Ajvol.., Mytildebang, Arcadian, Jag123, Dennis Valeev, Joe Jarvis, Man vyi, MPerel, Hagerman, EricAir, Orangemarlin, Ranveig, Jumbuck, Danski14, Bob rulz, Alansohn, Mo0, Jared81, Tek022, Wouterstomp, Scarecroe, Lightdarkness, Fawcett5, Bart133, Snowolf, ClockworkSoul, Tycho, Cburnett, TenOfAllTrades, Sciurinæ, Charlie123, Memenen, Justinus, Oleg Alexandrov, Rorschach, 2004-12-29T22:45Z, Rocastelo, Mark K. Jensen, Yuubinbako, WadeSimMiser, JeremyA, MONGO, CiTrusD, Kmg90, LadyofHats, Marco-Tolo, Palica, Tslocum, RichardWeiss, Justin Bailey, BD2412, Chun-hian, FreplySpang, Vvuppala, DePiep, Icey, BorgHunter, Edison, Drbogdan, Sjakkalle, Rjwilmsi, Harry491, SMC, Mitul0520, Nneonneo, Kalogeropoulos, ScottJ, Oliverkeenan, Matt Deres, Sango123, Yamamoto Ichiro, Ravidreams, Titoxd, Nihiltres, Crazycomputers, RexNL, Gurch, Jrtayloriv, R Lee E, TeaDrinker, Terrx, Nabarry, NetAddict, BradBeattie, Chris is me, Chobot, WriterHound, Whosasking, Kjlewis, YurikBot, Wavelength, TexasAndroid, Pile0nades, Icedemon, Phantomsteve, RussBot, Sputnikcccp, Reo On, Hydrargyrum, Stephenb, Emmanuelm, Gaius Cornelius, Yyy, Rsrikanth05, Pseudomonas, Wimt, NawlinWiki, Misos, Wiki alf, Astral, Jaxl, Justin Eiler, Chunky Rice, Joelr31, Irishguy, PeepP, Froth, Hv, Misza13, Chichui, Zwobot, Samir, Mysid, Jhinman, FF2010, 2over0, Zzuuzz, Lt-wiki-bot, Uartseieu, Imaninjapirate, Warfreak, Squeedlyspooch, Smoggyrob, Theda, Closedmouth, Jwissick, Jonaan6, Pb30, Kris33, BorgQueen, Amren, Anclation-enwiki, RenamedUser jaskldjslak904, Cjfsyntropy, Katieh5584, NeilN, GrinBot-enwiki, DVD R W, Dposse, Snalwibma, Hughitt1, Crystallina, KnightRider-enwiki, SmackBot, MattieTK, Moeron, Arti Sahajpal, TestPilot, Unyoyega, Blue520, KocjoBot-enwiki, Stepa, Vanished user 3dk2049pot4, Timotheus Canens, CuriousOliver, Zephyris, Pathless, Gilliam, Rickpearce, Ohnoitsjamie, Betacommand, Bluebot, Persian Poet Gal, NCurse, MK8, Sirex98, Master of Puppets, Oli Filth, IanBailey, Miquonranger03, SchfiftyThree, Akanemoto, Adamstevenson, Onkelschark, Darth Panda, Gracenotes, Nick Bond, Can't sleep, clown will eat me, TheRaven7, Shalom Yechiel, Snowmanradio, JonHarder, Andy120290, Threeafterthree, Dharmabum420, Krich, Flyguy649, Nakon, Jiddisch-enwiki, Funky Monkey, Dacoutts, Weregerbil, Iridescence, Zzorse, Andrew4010, Pilotguy, FelisLeo, Wikier.ko, Nmnogueira, The undertow, SashatoBot, EMan32x, Nishkid64, GiollaUidir, ArglebargleIV, Mouse Nightshirt, Zahid Abdassabur, Kuru, Scientizzle, Kipala, Epingchris, Sir Nicholas de Mimsy-Porpington, James.S. Chodorkovskiy, JorisvS, Accurizer, Goodnightmush, Mr. Lefty, Cielomobile, Extremophile, Jaredhelfer, 041744, Ckatz, Grumpyyoungman01, Slakr, Avs5221, The Missing Hour, Brazucs, Childzy, Spook', Ryulong, Citicat, Elb2000, Nbhatla, LaMenta3, Darry2385, DabMachine, Natpal, BranStark, WahreJakob, Nehrams2020, Mario2000, Delta x, Imad marie, Matzman, Tawkerbot2, Timrem, Chris55, The Haunted Angel, JForget, CmdrObot, Ale jrb, Scohoust, SupaStarGirl, Ninetyone, Nunquam Dormio, Kylu, N2e, Istrancis, Yarnalgo, Fletcher, Bbasen, Peripitus, Reywas92, Gogo Dodo, Was a bee, JFreeman, ST47, Julian Mendez, Tawkerbot4, DumbBOT, Narayanese, Kozuch, JayW, Greeneto, Woland37, Rjm656s, Thijs!bot, Epbr123, Aftenfour, Opabinia regalis, N5iln, Anupam, Headbomb, Marek69, Peter Znamenskiy, John254, A3RO, PaperTruths, James086, Zé da Silva, MesserWoland, Dgies, Thedarkestshadow, Dawnseeker2000, Hempfel, Escarbot, Ju6613r, David D., AntiVandalBot, Yomamais, Konman72, Luna Santin, Akradecki, Quintote, Jesse dudexz, TimVickers, Smartse, PhilipWest, Farosdaughter, Alvarogonzalezsotillo, Figma, Phil153, Res2216firestar, DCincarnate, JAnDbot, AOB, Leuko, Husond, MER-C, Plantsurfer, The Transhumanist, WikipedianProlific, Hello32020, Db099221, Owenozier, Chizeng, Andonic, Noobeditor, Hut 8.5, BrotherE, Bearly541, Jarkeld, .anacondabot, Acroterion, Bencherlite, FaerieInGrey, Magioladitis, Canjth, Bongwarrior, VoABot II, Brunoman1990, AuburnPilot, Tatwell, Wikidudeman, Jnb, CTF83!, Prestonmcconkie, Avicennasis, Catgut, Animum, Shim'on, Allstarecho, Spellmaster, DerHexer, JaGa, Linuxcity, Squidonius, SquidSK, Darthtire, Rustyfence, Hdt83, MartinBot, Gasheadsteve, Dietzel65, Arjun01, Robin63, Tuganax-enwiki, Rettetast, Tholly, R'n'B, Kateshortforbob, CommonsDelinker, AlexiusHoratius, LedgendGamer, J.delanoy, Pharaoh of the Wizards, CFCF, Bogey97, Nbauman, Hans Dunkelberg, Uncle Dick, A:f6, Cewoo, Eliz81, Extransit, Jerry, OttoMäkelä, Phatdan46, Keesiewonder, Acalamari, Katalaveno, Kangie, BMBTHC, CountZepplin, TomasBat, NewEnglandYankee, Raichu Trainer, Charmander trainer, 83d40m, Cackerman, FJPB, 2help, Juliancolton, Cometstyles, SlightlyMad, Erick.Antezana, Khargas, Ja 62, Ministry of truth 02, Useight, Magoscope, Idioma-bot, Kay for kate, Wikieditor06, Deor, VolkovBot, Lordmontu, Spancakes2, Flyingidiot, Adapter-enwiki, Jeff G., Lia Todua, HJ32, Philip Trueman, Drunkenmonkey, Jhon montes24, TXiKiBoT, Gato340, Susan Walton, Miranda, Ogger181, Wxyz999a, Anonymous Dissident, Qxz, Littlealien182, Anna Lincoln, Walteryoo93, Leafyplant, Sanfranman59, Abdullais4u, LeaveSleaves, Sc0ttkclark, Hobbit13, Earthdirt, SpecMode, Brainmuncher, Adam.J.W.C., Richwil, Synthebot, Strangerer, Duckttape17, Enviroboy, RaseaC, Brianga, Northfox, AlleborgoBot, Kehrbykid, NorbertR, Readthesign, NHRHS2010, Ivanivanovich, Demmy, GirasoleDE, SieBot, Funkamatic, ThomasTenCate, Restre419, Graham Beards, Danimsturr, BotMultichill, Jauerback, Gerakibot, Westerrer, Dawn Bard, Caltas, Jasujas0, Xymmax, TheScotsman1987, Out slide, Oxymoron83, Byrialbot, Xelaw, Rpgch, Lightmouse, Mercenario97, BenoniBot-enwiki, Paully111, Sunrise, Rewdewa, Robert W. A., Stfg, CryoSagittarius, Jmameren, Anchor Link Bot, Crazydoctor, Bhartta, Hamiltondaniel, Ascidian, Lloydpick, Micheb, Troy 07, WikipedianMarlith, WakingLili, ClueBot, PipepBot, The Thing That Should Not Be, Zfc89, Cheapshots, Blogeswar, Cryptographic hash, Uncle Milty, SuperHamster, Boing! said Zebedee, CounterVandalismBot, Woodardc, Blanchardb, Jordin986, Dylan620, Fatsamsgrandslam, Riccomario96, Aua, DragonBot, Jusdafax, Calimo, Nudve, KC109, Chance Jeong, Shinkolobwe, Jdmirfer, MacedonianBoy, Tyler, NuclearWarfare, Lunchscale, Sbfw, Feeuoo, Razorflame, NoriMori, Muro Bot, Cloning jedi, Aitias, Versus22, Timshiels, Johnuniq, SoxBot III, Psymier, DumZiBoT, XLinkBot, Soccerstar132, Pichpich, PseudoOne, Rror, Koolokamba, Dylandugan, Crowbarthe1337h4x0r, Nepenthes, Little Mountain 5, Treesoulja, NellieBly, Mifter, Ianbecerro, Poopybutts, JinJian, ZooFari, Kittymew1, Mowgali, Bookgeek10, Thatguyflint, MACHINAENIX, Addbot, Xp54321, Salwateama2008, DOI bot, Thecorduroysuit, PaterMcFly, GSMR, Ronhjones, Fielddaysunday, CanadianLinuxUser, Tedmund, Looie496, MrOllie, Download, LaaknorBot, Duckdude, Bernstein0275, Debresser, Dr. Universe, LinkFA-Bot, West.andrew.g, Numbo3-bot, Zachary Murray, Carbon arka, Bradht2710, Erutuon, Tide rolls, Krano, Ramsis II, Gail, Xenobot, Jarble, Kelseyking, Ettrig, Jim, Luckas-bot, Biji123, Yobot, 2D, Ptbotgourou, Les boys, Il MusLiM HyBRiD II, Testaccount rrrr, Medical geneticist, Ajh16, THEN WHO WAS PHONE?, Nallimbot, KamikazeBot, AnakngAraw, Eric-Wester, Magog the Ogre, Bility, Backslash Forwardslash, AnomieBOT, Galoubet, Savedhewadgh, Piano non troppo, AdjustShift, Kingpin13, EryZ, Bodiejnr, Materialscientist, Dracopacoboy, Socrates321, The High Fin Sperm Whale, Citation bot, Hatesschool, ArthurBot, Szxd, Xqbot, SciGuy013, Timir2, Jeffrey Mall, Gensanders, Wyklety, Tennispro427, Trivelt, GrouchoBot, DerryTaylor, Shirik, Mark Schierbecker, Gott wisst, Doulos Christos, IShadowed, Natural Cut, Shadowjams, Methcub, Vanished user giiw8u4ikmfw823foinc2, Erik9, Swimmrfrend97, Dougofborg, Mikesta2, Alisonthegreat, Sugarrush100, Sterimmer, Clearrise, GT5162, Bookwormlady1100, FrescoBot, Yssha, LucienBOT, Pepper, Domdomegg, Wikipe-tan, Galorr, Goodbye Galaxy, D'ohBot, Cargoking, Alxeedo, HJ Mitchell, Jcampsall, DivineAlpha, Stephen Morley, HamburgerRadio, Sarpelia, Pinethicket, HRoestBot, PrincessofLlyr, LinDrug, Hamtechperson, A8UDI, Zaluzar, MastiBot, Pertusaria, IP69.226.103.13, SpaceFlight89, Σ, Meaghan, Merlion444, KnowledgeRequire, Jeangabin, Gamewizard71, FoxBot, TobeBot, Trappist the monk, Hippy Sorix, DragonofFire, January, WeiAMS426, Gaby97123, Toejam1101, Wsander, Najzeko, JoshHoward77, Specs112, ShyHinaHyuga, Tbhotch, Reach Out to the Truth, Flatulent tube, Blackdeath15, DARTH SIDIOUS 2, Difu Wu, ScienceFreakGeek, Andrea105, Itinyboo, Jacyanda9, TjBot, Popular97, Kezari, NerdyScienceDude, Nd State Assosiation, Mandolinface, Leroylevine, DASHBot, EmausBot, Acather96, WikitanvirBot, Ajraddatz, Heracles31, Ndkl, Suh004757, ME IS NINJA, I be awsome, Wright496, Bruceplayzwow, NoisyJinx, The Last Kilroy, Jake, MrTranscript, Usama042, Abbypettis, Wak 999, Ren799, Madmaxz, Acer al2017, Werieth, ZéroBot, Josve05a, Aeonx, Sky380, Demiurge1000, L Kensington, Lawstubes, ChuispastonBot, Juanjosegreen2, Snehalshekatkar, Pooja raveendran, Шиманський Василь, Frietjes, Ww12, Helpful Pixie Bot, Tholme, BG19bot, Mohamed CJ, KingMunch, Cadiomals, Ahora, NotWith, Deuterostome, Bourgeb, Gladissk, MMA rox, Miszatomic, Jimw338, Cyberbot II, ChrisGualtieri, Dantheman4297, InsaneInnerMembrane, Davidlwinkler, Kelvinsong, Dexbot, Randomizer3, CuriousMind01, Alcohkid, Royroydeb, FallingGravity, Iztwoz, AmericanLemming, Wikiuser13, Jwratner1, BruceBlaus, MBproj, SJ Defender, Ali Zifan, Ttlaz123, Mahusha, Monkbot, Rakeshyashroy, Trackteur, Arvind asia, TypingAway, KasparBot, DERPALERT, Sarthakniar and Anonymous: 1247

- **Cellular differentiation** *Source:* https://en.wikipedia.org/wiki/Cellular_differentiation?oldid=709897681 *Contributors:* Magnus Manske, Malcolm Farmer, AdamRetchless, Lexor, Wapcaplet, Ahoerstemeier, JWSchmidt, Mxn, Marshman, Robbot, Hadal, TPK, Giftlite, Sj, Everyking, Dullhunk, Neffk, Paniq, Rich Farmbrough, Dmr2, Bender235, El C, RoyBoy, Brim, Arcadian, Giraffedata, Srlasky, Alansohn, Bsadowski1, Miss Madeline, Graham87, BD2412, Rjwilmsi, DVdm, Wavelength, Reo On, Grafen, Nephron, Bota47, Theda, Josh3580, SmackBot, Hydrogen Iodide, EncycloPetey, Delldot, Gilliam, Betacommand, Chris the speller, RDBrown, MK8, MDChanderson−enwiki, Valich, TheLimbicOne, Drphilharmonic, Vina-iwbot−enwiki, Nehrams2020, GBuilder, Yoni bhonker, Christian75, Thijs!bot, Epbr123, Dr Aaron, Headbomb, Peter Znamenskiy, AntiVandalBot, Res2216firestar, Avjoska, JamesBWatson, Ciar, LookingGlass, Emw, DerHexer, MartinBot, R'n'B, FANSTARbot, Uncle Dick, 12dstring, Hodja Nasreddin, NewEnglandYankee, KylieTastic, Paulmch, A4bot, Lova Falk, Twoars, Dreamafter, Flyer22 Reborn, Harry−enwiki, Forluvoft, ClueBot, Franamax, NuclearWarfare, Ottawa4ever, AkashAD, Avoided, MystBot, Addbot, Some jerk on the Internet, Redheylin, Quercus solaris, DES04, Bratko4223, Tide rolls, OlEnglish, QuadrivialMind, Yobot, Bremenjock, Hadi1362−enwiki, AnomieBOT, Noq, Materialscientist, Citation bot, Tekks, LilHelpa, Xqbot, GrouchoBot, Jhbdel, SassoBot, Shadowjams, A.amitkumar, Citation bot 1, I dream of horses, AMAGOOCH, CANTFLAME, Rushbugled13, RedBot, SW3 5DL, Kalaiarasy, RjwilmsiBot, Garuh knight, EmausBot, Architeuthidae, Nyxhadanielle, TuHan-Bot, Wayne Slam, Cellmedsoc, Kleopatra, ClueBot NG, Gareth Griffith-Jones, Chester Markel, AaronAppelle, Mesoderm, Rezabot, Widr, WikiPuppies, Bibcode Bot, BG19bot, Opencircle, Austinprince, Snow Blizzard, Rob Hurt, Jakebarrington, The Illusive Man, ChrisGualtieri, Dexbot, Dmitry Dzhagarov, Fookedme, Nahrafsfa, U.SAIDM, Trustmeimdabest, Iztwoz, Cmekain14, Melbourne132, Revolution1221, DavidLeighEllis, Kharkiv07, Stemcellz, Wrightdi09, Monkbot, SongofSol, Zhuj wiki, Trogluddite, JMWSlack, Jfrabajante, Bashoh19988, Qaazwooj and Anonymous: 148

- **Epigenetics in stem-cell differentiation** *Source:* https://en.wikipedia.org/wiki/Epigenetics_in_stem-cell_differentiation?oldid=686625297 *Contributors:* Graeme Bartlett, Rjwilmsi, Tony1, Mild Bill Hiccup, AnomieBOT, RjwilmsiBot, Klbrain, Khazar2, Sms0610, Dmitry Dzhagarov and Anonymous: 3

- **Embryonic stem cell** *Source:* https://en.wikipedia.org/wiki/Embryonic_stem_cell?oldid=712832775 *Contributors:* Ed Poor, Ubiquity, Gabbe, Ixfd64, Andrewman327, Tomchiuke, Jfire, Aetheling, Giftlite, Orangemike, Alison, Btphelps, Adenosine, Chowbok, Beland, Israel Steinmetz, Rich Farmbrough, Kbh3rd, Sfahey, RoyBoy, Smalljim, OGoncho, Alansohn, Nicholas Cimini, Velella, ClockworkSoul, Tony Sidaway, Nightstallion, Carnw, Matthew Platts, GSlicer, Rjwilmsi, Koavf, Jivecat, Klortho, Titoxd, Ground Zero, Old Moonraker, AAMiller, Nivix, Smithbrenon, Ryddragyn, Wavelength, TexasAndroid, Saintjust, Gravecat, Cquan, Nick123, Closedmouth, Emc2, JLaTondre, DVD R W, Snalwibma, Crystallina, Royalguard11, McGeddon, C.Fred, Stepa, Delldot, J0lt C0la, Gilliam, Bluebot, QTCaptain, Deli nk, Kotra, Can't sleep, clown will eat me, Shalom Yechiel, Skidude9950, Snowmanradio, Hairouna, JR98664, Arbitrage54, Drphilharmonic, CN31808, Krashlandon, Scientizzle, Sir Nicholas de Mimsy-Porpington, Wickethewok, Slakr, Werdan7, Ryulong, Lunajurai, A Obeidat, Martious, Alecbings, Zachary Newton, Blehfu, Tawkerbot2, JForget, Tom872, Geremia, JohnCD, Mystylplx, FlyingToaster, Eubanks718, RelentlessRecusant, Billheller, TheCheeseManCan, Narayanese, Chachilongbow, Epbr123, Dr Aaron, Edupedro, Mellogirl, Mojo Hand, Marek69, Peter Znamenskiy, Mokkan88, CharlotteWebb, Tedford, Ju66l3r, AntiVandalBot, Schemish, TimVickers, JAnDbot, MER-C, Eurobas, Getaway, GGreeneVa, East718, VoABot II, Wikidudeman, Michael Goodyear, Polymerase−enwiki, Robert M. Hunt, Badreligion, Acdixon, MartinBot, Arjun01, Nashpaul, Spawn968, J.delanoy, Numbo3, Maurice Carbonaro, Colincbn, Rod57, Mustangblue, Pyrospirit, DadaNeem, Redblock, Id711, Mmoople, Qu3a, Darkfrog24, Ja 62, Useight, Conangle, Jeff G., Kevinkor2, Jomasecu, Jboh, Clarince63, Alehead, DoktorDec, Marleene, Cmcnicoll, TexasAggie2007, RaseaC, Agüeybaná, Brianga, Doc James, Quantpole, SylviaStanley, D. Recorder, Tomjhou, Whiskey in the Jar, Moonriddengirl, Illini91, Caltas, Dirty blonde, Keilana, Alexbrn, Harry−enwiki, Bagatelle, ResearchChannel, Sunrise, Hippie Metalhead, Americanprogress, Superbeecat, Harrybatty, Mikedsd, Regener, ImageRemovalBot, Forluvoft, ClueBot, GorillaWarfare, Healthwise, Negative and Positive, SuperHamster, Yadmat, Auntof6, Morozlm, SoxBot, ChrisHodgesUK, Light show, Yonskii, Thingg, Versus22, SoxBot III, DumZiBoT, Spitfire, PervyPirate, Rror, Yumito, Vojtěch Dostál, Avoided, Addbot, Jacopo Werther, Some jerk on the Internet, DOI bot, Jncraton, ContiAWB, Diptanshu.D, Bernstein0275, Aktsu, Tide rolls, Solid State, Wordday, LuK3, Luckas-bot, Yobot, Themfromspace, AnomieBOT, Noq, Jim1138, Kingpin13, Materialscientist, Limideen, The High Fin Sperm Whale, Citation bot, Gregmweir3, Frankenpuppy, Andrewmc123, Futureseer, Capricorn42, Penn Station, AVBOT, Wikireader41, Wiki emma johnson, Sirozha, Smallman12q, Shadowjams, DannyOreilly, Cliveglover, FrescoBot, Godfactauthor, Tobby72, Oldlaptop321, Danhomer, Cacryobank, Citation bot 1, Pinethicket, Jhbuk, Vrenator, Lcaiscool, DARTH SIDIOUS 2, RjwilmsiBot, Regancy42, John of Reading, BillyPreset, Mitonet, GoingBatty, Myoavi, Hardrockcrossing, Tommy2010, 100chris, Wikipelli,

K6ka, ZacBowlingAlt, Jojojlj, Joshp20, Brandmeister, Ipscellmate, ClueBot NG, Erik Lönnrot, Pokemonq1, Gareth Griffith-Jones, Serasuna, O.Koslowski, 1usMARINEcorps1, Michael Overland, 69imp, Katedoh, BG19bot, Mispy, Stoex, Aberdam, Srchen, Onlygother, Teammm, Geenest, Ducknish, Snydes21, Alazka87, Zt00358, Dexbot, ThCPS, Ruha9, TwoTwoHello, Lugia2453, Frosty, Dmitry Dzhagarov, Ginganinja34, Gsayah, BradYard, The Stringinator, CensoredScribe, Brotsv), The Herald, Ginsuloft, Manul, Hjzhou988, 7Sidz, Monkbot, Vieque, Happy Attack Dog, Haines6, WendyKally2015, 1115crocodileov, KasparBot, Alem1846 and Anonymous: 455

- **Embryo** *Source:* https://en.wikipedia.org/wiki/Embryo?oldid=708206170 *Contributors:* Magnus Manske, Youssefsan, William Avery, Montrealais, Frecklefoot, Patrick, Michael Hardy, Lexor, Gdarin, Ahoerstemeier, Darkwind, Salsa Shark, Rob Hooft, Fuzheado, Andrewman327, Time, Marshman, Phoebe, Samsara, Renato Caniatti~enwiki, Chuunen Baka, Robbot, Moink, Hadal, Wikibot, Diberri, Luis Dantas, Electric goat, Ceejayoz, Kpalion, Cjensen, Alexf, Knutux, OldZeb, Joeblakesley, MisfitToys, G3pro, PDH, Anythingyouwant, Icairns, Arcturus, Rich Farmbrough, KillerChihuahua, Lejean2000, Vsmith, Smyth, D-Notice, Paul August, Bender235, JoeSmack, RJHall, Bobo192, Nicke Lilltroll~enwiki, Arcadian, La goutte de pluie, John Fader, ADM, Storm Rider, Jigen III, Poweroid, Alansohn, Mlabar, Sarah the poet, Mac Davis, Snowolf, Wtmitchell, 2004-12-29T22:45Z, Bratsche, Sejessey, Grika, DocRuby, Palica, Paxsimius, RichardWeiss, Graham87, Alienus, Koavf, Eptalon, Boccobrock, Nandesuka, Falphin, Windchaser, Old Moonraker, RexNL, Maustrauser, Alphachimp, Chobot, Drvgaikwad, Sharkface217, DVdm, Gwernol, YurikBot, Wavelength, RobotE, Huw Powell, Tznkai, Severa, Chris Capoccia, Pseudomonas, DJ Bungi, Snek01, Grafen, Bjford, Muu-karhu, JHCaufield, Bota47, Lt-wiki-bot, Theda, Modify, GraemeL, Crystallina, SmackBot, Drwtsn32, Haza-w, Prodego, FloNight, Unyoyega, Kilo-Lima, EncycloPetey, Phaehe, Gilliam, Betacommand, Smeggysmeg, Hraefen, J.Steinbock, Kurykh, Audacity, Zacwee, Can't sleep, clown will eat me, Abyssal, Nakon, Alexandra lb, Andrew c, Nishkid64, DO11.10, Scientizzle, Mgiganteus1, JohnWittle, Smith609, Munita Prasad, Avedomni, Serephine, KJS77, Norm mit, Levineps, Iridescent, Mephij, Courcelles, Nibs208, George100, JForget, Hyphen5, Myasuda, Qrc2006, Cydebot, Metanoid, Was a bee, Anthonyhcole, Tawkerbot4, DumbBOT, Epbr123, Drumguy88, Peter Znamenskiy, Randomfrenchie, Abort73, Cooljuno411, Sean William, Escarbot, Mentifisto, AntiVandalBot, BokicaK, Prolog, Danny lost, Alphachimpbot, Wayiran, Erxnmedia, JAnDbot, Janejellyroll, Kipruss3, Euthman, PhilKnight, Hunny516, Acroterion, Dark Kubrick, Bongwarrior, VoABot II, Avjoska, Chevinki, Disconformist, Michael Goodyear, WhatamIdoing, Robotman1974, Hveziris, The cattr, DerHexer, NatureA16, MartinBot, Poeloq, Uriel8, Kateshortforbob, Joie de Vivre, Tgeairn, CFCF, Uncle Dick, SU Linguist, Shawn in Montreal, Mrjohns2, Mikael Häggström, Tmax64, Eleclair, 83d40m, Id711, Mr Willis, Lou500, Wikieditor06, VolkovBot, ABF, Macedonian, AlnoktaBOT, Vlmastra, Philip Trueman, TXiKiBoT, Rollo44, Tricky Victoria, ElinorD, JhsBot, Crempuff222, Anarchangel, CO, Hymiegladstone, Eubulides, Vector Potential, Bonb, Doc James, AlleborgoBot, Fanatix, SieBot, StAnselm, Classicstruggle2, Euryalus, Dawn Bard, Caltas, Andrewjlockley, Wageslave, Flyer22 Reborn, DivineBurner, Maelgwnbot, Florentino floro, Drgarden, Sasha Callahan, Bee Cliff River Slob, IRKAIN, Invertzoo, Twinsday, ClueBot, Framdamdidily, The Thing That Should Not Be, Rizessoc07, Rjd0060, Crookedpinky, Mr sean meers1, SuperHamster, CounterVandalismBot, Ottava Rima, Arunsingh16, Alexbot, Razorflame, The Red, Rds865, BalkanFever, Novjunulo, Dany4175, Pbizannes, Botpankonin, HappyJake, Callam23, Addbot, Jacopo Werther, Willking1979, Ironholds, CanadianLinuxUser, Fluffernutter, Jurj, Ccacsmss, Glane23, Tide rolls, ScAvenger, Jarble, Tartarus, Yobot, THEN WHO WAS PHONE?, KamikazeBot, AnakngAraw, AnomieBOT, Nony1502, Sonimom, Materialscientist, ImperatorExercitus, The High Fin Sperm Whale, OllieFury, Quebec99, Galata Kulesi, Xqbot, Jayarathina, Capricorn42, Almabot, Omnipaedista, Beansandveggies, Basharh, FrescoBot, 753951m, Pinethicket, Rushbugled13, Waldemahr, Jauhienij, WandaRMinstrel, Kalaiarasy, Fama Clamosa, Aoidh, Περίεργος, Tbhotch, TjBot, EmausBot, Super48paul, Smartdictionary, ETandWEIRDO, Winner 42, Wikipelli, Daonguyen95, Traxs7, Wikignome0530, PhantomPlugger, L Kensington, Donner60, Jewels301, Gongoozler123, Mcc1789, Uziel302, Jaggezi, TYelliot, George Makepeace, ClueBot NG, MRFazry, BarrelProof, LISChesBay, Frietjes, Delusion23, Shashankesh, WikiPuppies, Jorgenev, BG19bot, Hamster3~enwiki, Teivtaoht, Davidiad, Anatomist90, Docsufi, K2kush, BattyBot, LeniStudent, Khazar2, Chris247100, AliSartawi, 069952497a, Oink mR, Red-eyed demon, MartianCat, Iztwoz, DrLinguini, Zlelik2000, LT910001, Cyborg1981, Stamptrader, NJSfour, Vieque, Lewislumsden, DSCrowned, Oiyarbepsy, ToonLucas22, EliTap, Dr. Finglhimer, SocraticOath, Visembryo, Charlotte135, Meganemer4, MamadouuodamaM and Anonymous: 330

- **Germ layer** *Source:* https://en.wikipedia.org/wiki/Germ_layer?oldid=664575196 *Contributors:* Heron, Lexor, Jebba, Nikai, Zarius, RodC, Dave6, Curps, PDH, Fungus Guy, Nina Gerlach, Tinus, Bender235, CheekyMonkey, RoyBoy, Arcadian, Jag123, Alansohn, Wouterstomp, Tycho, Ceyockey, Woohookitty, Benbest, Rjwilmsi, Shao, Dj Capricorn, Nephron, Zwobot, Wknight94, RupertMillard, KnowledgeOfSelf, Kipmaster, Gilliam, J.Steinbock, Bluebot, Tamfang, Radagast83, TheLimbicOne, Bansp, Clicketyclack, Werlop, Epingchris, Jon186, Lottamiata, Vsoulremix, Jamoche, A876, Thijs!bot, Kilva, GAThrawn22, AntiVandalBot, Alphachimpbot, JAnDbot, RuthieK, STBot, Wlodzimierz, CFCF, Ginsengbomb, Jotunn, JBarno, VolkovBot, LeilaniLad, TXiKiBoT, BotKung, AlleborgoBot, SieBot, Kochipoik, Anchor Link Bot, ClueBot, Eric Van Bogaert, Addbot, 2enable, Yobot, AnomieBOT, Lapabc, RibotBOT, FrescoBot, D'ohBot, Fama Clamosa, TjBot, JaysonSunshine, EmausBot, GoingBatty, Donner60, ClueBot NG, Harps21, Mesoderm, Newyorkadam, BG19bot, Biolprof, Kenneth.jh.han, Mannintg, Biologize, Iztwoz, Sonicnation, Comp.arch, Mendoza.m420 and Anonymous: 76

- **Inner cell mass** *Source:* https://en.wikipedia.org/wiki/Inner_cell_mass?oldid=710571697 *Contributors:* Pabloes, Arcadian, Seans Potato Business, Woohookitty, BillC, Yamamoto Ichiro, Roboto de Ajvol, Mikalra, Wolfmankurd, Magn0lia, Bluebot, James McNally, Mgiganteus1, Novangelis, Dl2000, Lottamiata, George100, Amalas, Robotsintrouble, Phl3djo, Was a bee, Scarface., GAThrawn22, Beelaj., JAnDbot, Sepul^, Magioladitis, MartinBot, Petter Bøckman, J.delanoy, Mikael Häggström, Iru9k, Sfbergo, Adrienne, PixelBot, Addbot, Yobot, ArthurBot, FrescoBot, Rthistle, ZéroBot, MacDaid, Frietjes, Helpful Pixie Bot, KLBot2, Vokesk, Bjorklund21, Lugia2453, SteenthIWbot, Abdullah123456789012345678901234567890, Arielrinon and Anonymous: 15

- **Stem-cell line** *Source:* https://en.wikipedia.org/wiki/Stem-cell_line?oldid=711558015 *Contributors:* Naelphin, Discospinster, MPS, Perfecto, Bobo192, Alansohn, Keenan Pepper, Snowolf, Abanima, Liface, Rjwilmsi, Jake Wartenberg, Jw21, Ryddragyn, Wavelength, SLATE, Alynna Kasmira, Tony1, Mediatetheconflict, RenamedUser jaskldjslak904, SmackBot, Opiniastrous, Darth Panda, RyanEberhart, Dreadstar, TiCPU, Spook`, Kaarel, CoerciveUtopian, Codetiger, BCSWowbagger, Epbr123, Ueanlookitup, Peter Znamenskiy, Res2216firestar, Ph.eyes, Millstone, Delknee, NewEnglandYankee, Radon210, Sonicemi, Forluvoft, ClueBot, The Thing That Should Not Be, Razorflame, Some jerk on the Internet, Yobot, Themfromspace, AnomieBOT, Noq, JimVC3, Ginkomithu, FrescoBot, Mprasol, RjwilmsiBot, EmausBot, Manicjedi, Wiwach, Donner60, ClueBot NG, MerllwBot, Dmitry Dzhagarov, Ginsuloft, Monkbot, Daenarys and Anonymous: 73

- **Adult stem cell** *Source:* https://en.wikipedia.org/wiki/Adult_stem_cell?oldid=711615399 *Contributors:* SimonP, Timwi, Giftlite, MSGJ, Gamaliel, Utcursch, Beland, OverlordQ, Rich Farmbrough, Art LaPella, Eltomzo, Adambro, Bgeer, Shereth, Sam Korn, Alansohn, Arthena, Clarphimous, Nicholas Cimini, Bootstoots, Dalillama, Versageek, Bobrayner, Uncle G, Elmarco, Rjwilmsi, Nightscream, Nneonneo, Bhadani, Amelio Vázquez, AAMiller, RexNL, WriterHound, Briaboru, Chris Capoccia, Hydrargyrum, NawlinWiki, Aaron charles, Eth4n, Allens, Bill,

Luk, Crystallina, SmackBot, CelticJobber, Roope, Edgar181, Master of Puppets, Wykis, Kotra, Can't sleep, clown will eat me, RedHillian, Pascaweb, Sirgregmac, Lostart, Drphilharmonic, Ohconfucius, Attys, John, DaleEastman, Coldpaws, Joseph Solis in Australia, Thaddeusjwhoopie, Beno1000, Tawkerbot2, ChrisCork, Yicker, RelentlessRecusant, Was a bee, Nick2253, Wikipediarules2221, Narayanese, IComputerSaysNo, Hypnosadist, Dr Aaron, CopperKettle, Headbomb, Marek69, Peter Znamenskiy, James086, The Hybrid, Seaphoto, Kbthompson, Ellwyn, Hut 8.5, GoodDamon, Magioladitis, Bongwarrior, VoABot II, MastCell, Wookiepedian, Faizhaider, Robert M. Hunt, Adacus12, Squidonius, Yobol, Rob Lindsey, Sport woman, J.delanoy, James cudahy, 5Q5, DigitalCatalyst, Hodja Nasreddin, Jv9mmm, Enuja, NewEnglandYankee, Id711, Ronbo76, CardinalDan, Jeff G., Fences and windows, Vgranucci, Richwil, Sfmammamia, Dawn Bard, LeadSongDog, Ephenix, Hanaichi, Faradayplank, Steven Crossin, Fratrep, Speimer, Poopoo 73, Sean.hoyland, Forluvoft, Elassint, ClueBot, LAX, The Thing That Should Not Be, Getwood, Abbaroodle, Mheidara, Browni1992, Niceguyedc, Richerman, Yadmat, Excirial, JFseekingtruth, Estirabot, Dekisugi, Johnuniq, Roxy the dog, Addbot, DOI bot, Dpottier, Marinerblue, Diptanshu.D, Odemin, Formerly very active, now only occasional editor, Yobot, Themfromspace, Bunnyhop11, Senator Palpatine, Eric-Wester, Noq, Scienceandhonor, Citation bot, NinetyNineFennelSeeds, Hwahwahwa, LovesMacs, Spotfixer, Tad Lincoln, Machiavelli11, Appleonious, Hola Kotla, HG-evader, Se1ma1ntic1s, Se1ma1ntic1s233, Doulos Christos, Aaron10103, Shadowjams, Snapscan1236, Mudkip1989 2008, Citation bot 1, Edderso, Jonesey95, MaxDel, SkyMachine, Rebeccanichols09, Trappist the monk, Otc core, Shawtown, DARTH SIDIOUS 2, RjwilmsiBot, EmausBot, Immunize, Klbrain, K6ka, Sangeepp, Manicjedi, John Cline, 24Adrianus, Ajournaleditor, Cellmedsociety, Cellmedsoc, DASHBotAV, ClueBot NG, Visualone-, Watchedsuddenurn, Bibcode Bot, BG19bot, Roberticus, Stoex, Tayamoz, CitationCleanerBot, Eprofessa, Michaelswilliams84, The Pikachu Who Dared, Samwalton9, BattyBot, Cyberbot II, Dexbot, Tha1uw4nt, Makecat-bot, Jcor1, Maverick 1960, Dmitry Dzhagarov, Simmsjk20, Debouch, Peter13542, Stamptrader, Chaya5260, Monkbot, BethNaught, Ladams444, Lumican, Crystallizedcarbon, Cristo fourie, Biggmacc, Valdomri, Neuronguy1, Dlainee666 and Anonymous: 246

- **Progenitor cell** *Source:* https://en.wikipedia.org/wiki/Progenitor_cell?oldid=710145274 *Contributors:* Pgan002, Discospinster, Arcadian, Stansz, Rjwilmsi, DVdm, Lijealso, Tinlv7, Stepa, EOZyo, Drphilharmonic, StanfordProgrammer, Christian75, Alaibot, Dr Aaron, CopperKettle, Peter Znamenskiy, Robert M. Hunt, Lenticel, Yobol, Rod57, CardinalDan, Jeff G., Philip Trueman, Amaher, ترجمان05, Bagatelle, Kerrio, Mild Bill Hiccup, Addbot, VUBio Pieterjan, Poojans, Susan Carley, Luckas-bot, AnomieBOT, The High Fin Sperm Whale, Eumolpo, TheChymera, Obaid221, DrilBot, Canadian hockey1, Logical Gentleman, John of Reading, EWikist, ClueBot NG, Alaa al-Otaibi, Kerribergeron7197, Rob Hurt, Samwalton9, Bihawang, YFdyh-bot, Dexbot, Volodymyr Stetsyuk, Dmitry Dzhagarov, Me, Myself, and I are Here, Helixitta, Dough34, Monkbot, Scarlettail, EpicOrange, Godsy, Valdomri and Anonymous: 36

- **Cancer stem cell** *Source:* https://en.wikipedia.org/wiki/Cancer_stem_cell?oldid=709400951 *Contributors:* The Anome, Greenrd, Jeffq, Jae, Cyc~enwiki, Arcadian, Giraffedata, Pearle, Woohookitty, Hibaby, Rjwilmsi, Harro5, Bgwhite, Paul White, Dave the Explosive Newt, Banus, SmackBot, Chris the speller, Miguel Andrade, JonHarder, Kendrick7, CastorCanada, Stwalkerster, Skapur, Rhetth, CmdrObot, Amalve, Captainktainer, Imamathwiz, Narayanese, TillJE, Daniel, Headbomb, Peter Znamenskiy, Oreo Priest, AntiVandalBot, Kzrt, Ellwyn, Lfstevens, Richiez, Bifrost99, Esmhead, Sage2k6, JaGa, GermanX, STBot, R'n'B, Leyo, CFCF, Boghog, Rod57, Clerks, Doughd54, Loopback007, Jeff G., Kevinkor2, AlnoktaBOT, GDonato, Y tambe, Inductiveload, Rosseauwake, Adr1liano, WereSpielChequers, Serenity forest, Sunrise, Ryocharlesyang, Thoratmangesh, ImageRemovalBot, Dante Marx, DragonBot, SchreiberBike, Wdustbuster, XLinkBot, Crenim, Coastlinewalk, Snapperman2, Addbot, LaaknorBot, DubaiTerminator, Alasto Light, Yobot, AnomieBOT, JackieBot, Citation bot, ArthurBot, Cymothoa exigua, Wiki emma johnson, FrescoBot, BioMedV, Citation bot 1, Sab.spen, 蜡笔小, Jujutacular, Full-date unlinking bot, Trappist the monk, RjwilmsiBot, Ripchip Bot, Agent Smith (The Matrix), Visudoc, John of Reading, GoingBatty, Klbrain, Mz7, Jlscrub, Josve05a, Lalsingh, Daltonemma, Tobeprecise, Hazard-Bot, Cscbiolifebio, ClueBot NG, Paulasb, BG19bot, Mark Arsten, BattyBot, William.haskins, Dexbot, Hmainsbot1, Mogism, Dmitry Dzhagarov, Malymajo, Goodwinjinesh, Phupe1, AndreasKyttaro, Lyanesse, Anrnusna, Monkbot, Lcchong, Sjones008, RegenerativeMedicineFollower, Wiki CRUK John, Regrounding, Maytham85, Sparkerino, Mannan369, Mecllg and Anonymous: 83

- **Induced pluripotent stem cell** *Source:* https://en.wikipedia.org/wiki/Induced_pluripotent_stem_cell?oldid=704722564 *Contributors:* AxelBoldt, Kku, Gabbe, JWSchmidt, Wikiborg, Tomchiukc, Timrollpickering, Jondel, Giftlite, Spiffy sperry, Bender235, Spoon!, R. S. Shaw, Ceyockey, Rjwilmsi, Jivecat, DonSiano, Drsamir, Thecurran, Benlisquare, Wavelength, Mikalra, Nikkimaria, SmackBot, Phattonez, Tinz, Reedy, Zirconscot, Pwb, Zeteg, IronGargoyle, Ben Moore, CmdrObot, Agathman, Fnlayson, Rifleman 82, RelentlessRecusant, Anthonyhcole, Alaibot, Narayanese, Thijs!bot, Epbr123, Headbomb, Marek69, Batmo, Nick Number, Lfstevens, FCAlive, Wookiepedian, Yobol, R'n'B, LedgendGamer, Tonbo~enwiki, J.delanoy, Buhuzu, LPLT, Caspian blue, VolkovBot, Indubitably, Elkhouse, Oh Snap, Y tambe, Afireinside13t, Ahendric, Sunrise, PerryTachett, Forluvoft, Juandope, PipepBot, Niceguyedc, BenoFreedman, Excirial, Alexbot, Sun Creator, DatDoo, GotaForce, Wdustbuster, XLinkBot, Jytdog, Cvcc, Addbot, LaaknorBot, Musiclover23, سمرقندی, AndersBot, Keepcalmandcarryon, Bouncingball2, Susan Carley, Lightbot, Wordday, Luckas-bot, Yobot, Entoumo, Themfromspace, Amirobot, CHW100, AnomieBOT, Rubinbot, Rami.shinnawi, Materialscientist, KenLee318, Citation bot, Frederic Y Bois, Xqbot, Zad68, Futureseer, Strottiek, Penn Station, Addbc, J04n, Omnipaedista, SassoBot, Wiki emma johnson, FrescoBot, Slalger, Allowrocks2003040957, Waveguide2, RedBot, Phoenix7777, Trappist the monk, Prosody31, LilyKitty, Deebeedoubleyou, Onel5969, RjwilmsiBot, TjBot, Lotez, Dewritech, Uploadvirus, ZéroBot, Ὁ οἶστρος, Bamyers99, Ipscellmate, Monostitch, Bioupdate55, Audi tri harsono, ClueBot NG, Jupanula, Kasirbot, Korovin2, Katedoh, KLBot2, BG19bot, M0rphzone, Liz1988, Vokesk, MusikAnimal, Peterrune, Leekaiinthesky, Aberdam, Ginger Maine Coon, Minsbot, Michelino12, BattyBot, Dexbot, Дмитрий Дж., Lugia2453, Dmitry Dzhagarov, Nwkimberley, Cth2630, OskNe, Divergence5, Zaeema Zafar, Rabdolla, CellengEC, Giancarlobasile, Monkbot, NM02114, CellbioPhD, Jfriend2, Fyrpepr Red, Dtmdt, Tresvakana, WendyKally2015, Ands92, KasparBot, ClaraAA and Anonymous: 134

- **Induced stem cells** *Source:* https://en.wikipedia.org/wiki/Induced_stem_cells?oldid=710393326 *Contributors:* RHaworth, Rjwilmsi, Bgwhite, Wavelength, Vanisaac, Cydebot, Headbomb, Lfstevens, Magioladitis, CommonsDelinker, Skier Dude, Walor, Jytdog, Tassedethe, Yobot, EdwardLane, SwisterTwister, AnomieBOT, Citation bot, Mathonius, FrescoBot, Jonesey95, Phoenix7777, Trappist the monk, John of Reading, AManWithNoPlan, BG19bot, BorisVM, A1candidate, Dexbot, Дмитрий Дж., Eakopskvm, Dmitry Dzhagarov, Joeinwiki, Dave Braunschweig, Jodosma, Rebeccapearman, EtymAesthete, Muse2000, Mricheyp, Rodosaenz and Anonymous: 17

- **Cell potency** *Source:* https://en.wikipedia.org/wiki/Cell_potency?oldid=705833158 *Contributors:* Thpn, Atcack, Gatewaycat, BD2412, Rjwilmsi, Bhny, BOT-Superzerocool, SmackBot, Gilliam, Chris the speller, J. Spencer, Drphilharmonic, Valfontis, Magioladitis, Xris0, Notreallydavid, 1812ahill, EuTuga, Itemirus, Sunrise, ImageRemovalBot, CorenSearchBot, AdamFouracre, Jtle515, LiuMasters, Staticshakedown, Jytdog, Addbot, Yobot, KamikazeBot, Materialscientist, Citation bot, Shadowjams, 蜡笔小, Paine Ellsworth, Pinethicket, 10metreh, Trappist the monk, DARTH SIDIOUS 2, Hb2007, TjBot, The Stick Man, EmausBot, Orphan Wiki, ZéroBot, Fæ, Ajournaleditor, DASHBotAV, ClueBot NG, Jack Greenmaven, ผักไม้, Hiperfelix, TruPepitoM, Cntras, Helpful Pixie Bot, Pdyrjab, Gautehuus, CeraBot, BattyBot, Munzar lissemore,

Rdenu11, Dmitry Dzhagarov, Jurrehageman, Me, Myself, and I are Here, Joeinwiki, Melonkelon, Vdiaz3, Aaron.aude, Mrhomerjaysimpson, Monkbot, EpicOrange, AbhishekPal03012001 and Anonymous: 38

- **Zygote** *Source:* https://en.wikipedia.org/wiki/Zygote?oldid=711191923 *Contributors:* Kpjas, Mav, Tarquin, Css, PierreAbbat, LionKimbro, Icarus~enwiki, Lexor, Evercat, Ehn, Marshman, Samsara, Toreau, Robbot, Nurg, Diberri, Alan Liefting, Marc Venot, Robodoc.at, Perl, Jfdwolff, Zizonus, Alexf, Onco p53, OverlordQ, Anythingyouwant, Tail, Atemperman, Rich Farmbrough, Vsmith, Silence, Bender235, Violetriga, Nabla, Sharkford, Shanes, Bobo192, Smalljim, Robhu, Gingko, Nicke Lilltroll~enwiki, Arcadian, Giraffedata, Hooperbloob, Abstraktn, Alansohn, Anthony Appleyard, Ricky81682, Bantman, ClockworkSoul, RainbowOfLight, Woohookitty, Thorpe, Graham87, Rjwilmsi, The wub, Flavr-Savr, FlaBot, RexNL, Narvalo, Jaraalbe, DVdm, Bgwhite, YurikBot, Vagodin, Eraserhead1, Tznkai, Grafen, Arichnad, Lepidoptera, Jaufrec, Epipelagic, Bota47, Leptictidium, CWenger, SmackBot, Jfurr1981, Alksub, EncycloPetey, Bragador, BiT, MalafayaBot, Domthedude001, Михаил Јелисавчић, Dreadstar, Drphilharmonic, Ugur Basak Bot~enwiki, SashatoBot, ArglebargleIV, Rory096, Potosino, Sir Nicholas de Mimsy-Porpington, JoseREMY, Kleinburgerei, Mr. Lefty, Inoesomestuff, Manifestation, MTSbot~enwiki, Wilbiddle42, Courcelles, Woodshed, Tawkerbot2, Banedon, SEJohnston, WeggeBot, Treybien, MC10, Gogo Dodo, Chasingsol, Smeazel, Oliver202, West Brom 4ever, James086, Cyclonenim, Luna Santin, Seaphoto, TimVickers, Wayiran, JAnDbot, MER-C, OckRaz, LittleOldMe, VoABot II, JamesBWatson, WhatamI-doing, Allstarecho, Gwern, Jimthompson~enwiki, MartinBot, Anaxial, CommonsDelinker, Nono64, Tgeairn, Svetovid, Zezima8282, Eski-mospy, Katalaveno, Dr d12, Jasonasosa, StrayGoose, Bobianite, Leemyster, Bcnof, Lights, VolkovBot, SERSeanCrane, Jeff G., Philip True-man, Tameeria, Qxz, Blarvink, Madhero88, Burntsauce, Logan, NHRHS2010, Fcady2007, Midjungards, Fanatix, Bfpage, Earthelemental99, Xenophon777, Yerpo, Judicatus, Tombomp, WacoJacko, Onopearls, Hamiltondaniel, Pam519, TheCatalyst31, Atif.t2, ClueBot, Vladkornea, Psypherium, Moguls, Ktr101, Sean Steel, Leonard^Bloom, The Founders Intent, Peter.C, Thingg, 7, Versus22, Johnuniq, Novjunulo, Life of Riley, Lumenos, Jovianeye, Rror, Painking, Avoided, WikHead, NellieBly, Addbot, Brumski, Ronhjones, CactusWriter, Download, Glane23, Omnipedian, Favonian, 5 albert square, Dangles1989, Tide rolls, Bushyballz, Anxietycello, Zorrobot, WikiDreamer Bot, Jarble, Angrysockhop, Legobot, Luckas-bot, Vedran12, Yobot, Jacobs, Maxi, AnakngАraw, JackieBot, Materialscientist, Quebec99, Xqbot, Jayarathina, Saffle, Capri-corn42, Spotfixer, Almabot, RibotBOT, Kirin13, KenByers5, Doulos Christos, Thehelpfulbot, Stockprice1977, Prari, Eisengel, TruthIIPower, Doctorwhofan328, Pinethicket, CANTFLAME, Steve2011, Jauhienij, Kalaiarasy, Lotje, Pexego, Ranga e, Purple garden gnome, FoxLogick, Salvio giuliano, Tommy2010, K6ka, Wikmii, E-citizen, Monterey Bay, Mayur, Donner60, Benjaminalberto, ClueBot NG, This lousy T-shirt, Vacation9, Frietjes, Dictabeard, Zynwyx, Mesoderm, Rezabot, Widr, Meepdeedoo, Titodutta, MusikAnimal, Fontea, GoShow, Bear h, JYBot, Makecat-bot, Sidsandyy, Lugia2453, The Anonymouse, Clucaj, Tempuser00, Padraig Singal, Vuagunny2608, Dheer chudasama, Hubbard1231, Miakirsty123 and Anonymous: 246

- **Spore** *Source:* https://en.wikipedia.org/wiki/Spore?oldid=702970367 *Contributors:* AxelBoldt, 0, Zundark, Youssefsan, Ray Van De Walker, Jaknouse, Lexor, Menchi, Alfio, Stan Shebs, Hectorthebat, Marshman, Tlotoxl, David.Monniaux, Altenmann, Srtxg, Gracefool, Pne, Pgan002, Knutux, Sonjaaa, DanielCD, Random contributor, Discospinster, Rich Farmbrough, Bender235, CanisRufus, Tom, ZayZayEM, Kalyanvarma, Sasquatch, Larryv, Haham hanuka, Pearle, Jumbuck, Alansohn, Cdc, Velella, Jrleighton, Bsadowski1, Saxifrage, TigerShark, GregorB, Macad-dct1984, Cuchullain, Kbdank71, Drbogdan, Rjwilmsi, Sdornan, The wub, Fish and karate, FayssalF, Strobilomyces, FlaBot, RobertG, Margos-bot~enwiki, RexNL, Diza, Chobot, Dj Capricorn, Fleelloguy, YurikBot, Wavelength, NTBot~enwiki, Peter G Werner, Phantomsteve, Ytrottier, Chaser, Stephenb, GeeJo, Marcus Cyron, NawlinWiki, Curtis Clark, Keithonearth, Falcon9x5, CLW, Lt-wiki-bot, Ketsuekigata, DVD R W, SmackBot, Haza-w, Davewild, EncycloPetey, Aksi great, Ohnoitsjamie, Master of Puppets, Miquonranger03, Merlin Cox, Ctbolt, Gruzd, Can't sleep, clown will eat me, Hammer1980, Rodeosmurf, Scharks, Kazztawdal, Epingchris, John Cumbers, JHunterJ, Smith609, Serephine, Mets501, Mengsk, Sasata, Iridescent, Dansiman, Laurens-af, Kaarel, Marysunshine, 5 0 cent, Afghana~enwiki, SkyWalker, J Milburn, Ale jrb, Matty-boymr, Ilikefood, CWY2190, Jort227, Dgw, Kupirijo, Ramitmahajan, Gogo Dodo, DumbBOT, AndreasBlixt, Fazolli1, Epbr123, Wikid77, Opabinia regalis, Osborne, Hazmat2, Headbomb, John254, AgentPeppermint, Natalie Erin, AntiVandalBot, Mrmoocow, KP Botany, JAnDbot, Plantsurfer, Ygoloxelfer, GoodDamon, Propaniac, VoABot II, Georgian, JamesBWatson, Michael Goodyear, Fabrictramp, Animum, Adrian J. Hunter, Emw, MartinBot, CommonsDelinker, Nall256, Alfred Legrand, RockMFR, J.delanoy, 11010, Ian.thomson, Katalaveno, NewEng-landYankee, Elbowworm, Nadiatalent, STBotD, Signalhead, Wikieditor06, CWii, Pirex, Vlmastra, TXiKiBoT, Alan Rockefeller, Tameeria, JayC, Someguy1221, Una Smith, ^demonBot2, Madhero88, Brainmuncher, Isimbot, Insanity Incarnate, Trievil, Mario1952, Atubeileh, SieBot, Medmedmedmedmed, AS, WereSpielChequers, Oldag07, Calabraxthis, Gmathieu66, Oxymoron83, Not an anon anymore, Ceramic2metal, AMbot, Mygerardromance, Graminophile, Alwynj, Doyee5, ClueBot, Gits (Neo), Mriya, Blanchardb, Estirabot, DumZiBoT, Taas0, Vo-jtěch Dostál, Sergay, SilvonenBot, Johnelson, HexaChord, Cewvero, Addbot, Glane23, ChenzwBot, SamatBot, لقمان, Zorrobot, David0811, Slgcat, SSgt.Crapgame, Luckas-bot, Yobot, AnomieBOT, Andrewrp, JDavis680, IRP, Sdhsakjhdk, RandomAct, Materialscientist, Pitchitich, Johnny1995, Maulucioni, Lifecommand, PimRijkee, GrouchoBot, Riotrocket8676, Omnipaedista, Mettythepineapple, AlimanRuna, Žiedas, BoomerAB, Moofloowcazoo, Joeygram, KuroiShiroi, Jd56970, Citation bot 1, .xIGallardoIx., Pinethicket, John Elson, Tim1357, Catinator, Realn coralblue34, Seahorseruler, DARTH SIDIOUS 2, Whisky drinker, Onel5969, Mean as custard, EmausBot, John of Reading, Wikitanvir-Bot, TheHomiCide420, RenamedUser01302013, Winner 42, Wikipelli, MikeyMouse10, Maxviwe, ClueBot NG, Mysterio888884, HMSSolent, Bibcode Bot, Plantdrew, BattyBot, Sminthopsis84, Rorytheking, Lugia2453, Howicus, Helixitta, Vítězslav Maňák (SLU), Mykophile, Uptothe-sun89, Antrusna, Ritoxavi, UnicornerXX2, PotatoNinja, Spaetzle02, KasparBot, Дмитро Леонтьєв, Rakesh2994 and Anonymous: 240

- **Morula** *Source:* https://en.wikipedia.org/wiki/Morula?oldid=698483176 *Contributors:* Heron, Hephaestos, DennisDaniels, Lexor, Robbot, Roy-Boy, Bobo192, Arcadian, SpeedyGonsales, Wouterstomp, Metju~enwiki, Benbest, Eyu100, FlaBot, Margosbot~enwiki, YurikBot, Mfero, Lep-idoptera, Nolanus, Kubra, SmackBot, InvictaHOG, Jfurr1981, EncycloPetey, Audriusa, Ravi12346, Lottamiata, Robotsintrouble, Neelix, Peter morrell, Daniel, Escarbot, The prophet wizard of the crayon cake, W7347, JAnDbot, SilentWings, Ubiquita, CommonsDelinker, AlphaEta, Sollosonic, VolkovBot, Synthebot, Bfpage, Chhandama, Yerpo, ClueBot, Alexbot, Addbot, CarsracBot, SpBot, Anxietycello, Zorrobot, Fryed-peach, Luckas-bot, Yobot, AnomieBOT, Materialscientist, Maxis ftw, DynamoDegsy, Erud, GrouchoBot, Erik9bot, DrilBot, 3BBOOD, Rjwilm-siBot, Amerias, ZéroBot, ClueBot NG, Frietjes, Mesoderm, Helpful Pixie Bot, Snow Blizzard, Iztwoz, Monkbot, Shibbolethink, Tilifa Ocaufa and Anonymous: 41

- **Callus (cell biology)** *Source:* https://en.wikipedia.org/wiki/Callus_(cell_biology)?oldid=708098978 *Contributors:* Zfr, Raazer, BD2412, Rjwilmsi, Chobot, WriterHound, Jab843, Zephyris, TimBentley, Valenciano, Mgiganteus1, Igge, Myasuda, Cydebot, Thijs!bot, Plantsurfer, Nono64, Million Moments, TXiKiBoT, StAnselm, Jotterbot, Addbot, Jacopo Werther, Moosehadley, لقمان, Luckas-bot, Yobot, Obersachsebot, Sylwia Ufnalska, Shirik, BenzolBot, Trappist the monk, Siltloam, EmausBot, ClueBot NG, Lovingpatel, Mr Sheep Measham, Vishwanath31, Gangmembaz, Electriccatfish2, Blahnais, CitationCleanerBot, NotWith, BattyBot, Monkbot and Anonymous: 19

- **Endothelial stem cell** *Source:* https://en.wikipedia.org/wiki/Endothelial_stem_cell?oldid=707552548 *Contributors:* Skysmith, Dcoetzee, MBisanz, Arcadian, Uncle G, Rjwilmsi, Sarg, SmackBot, MDChanderson~enwiki, Dl2000, Loulougaga, Headbomb, Rod57, SieBot, Niceguyedc, SchreiberBike, Addbot, Citation bot, Wimpus~enwiki, Frietjes, UWOBio4920, DavidOSX and Anonymous: 5

- **Hematopoietic stem cell** *Source:* https://en.wikipedia.org/wiki/Hematopoietic_stem_cell?oldid=710477884 *Contributors:* Edward, Nina, Robbot, Jfdwolff, Thpn, Pascal666, Rich Farmbrough, Bender235, Jensbn, Arcadian, Jag123, Gigano, Melaen, TenOfAllTrades, Stemonitis, Benbest, Eleassar777, Rjwilmsi, Chobot, Bgwhite, YurikBot, Epolk, Cquan, Yahya Abdal-Aziz, Mysid, JSLR, Cyrus Grisham, Hiddekel, SmackBot, Victor M, Vicente Selvas, Triggtay, Gilliam, Chris the speller, Jjalexand, Deli nk, Miguel Andrade, Fuhghettaboutit, Drphilharmonic, A. Rad, DO11.10, Euchiasmus, Dr.saptarshi, Mgiganteus1, Mfourman, Martious, Schulj, CmdrObot, SpK, Thijs!bot, Anupam, Headbomb, Peter Znamenskiy, James086, Openlander, Calaka, Kauczuk, Osquar F, R'n'B, Nono64, Leyo, Felixcheung, Feonaway, Etparle, Akela1, Nbauman, Mikael Häggström, Harrisd5917, Kavanagh21, Ja 62, AlnoktaBOT, Cmcnicoll, Spitfire8520, Arcfrk, Antibody2000, SieBot, Linzhoo2u, Keilana, Fimbriata, Staylor71, ClueBot, The Thing That Should Not Be, Hsieburg, Excirial, Alexbot, Psinu, Muro Bot, GotaForce, Dthomsen8, Vojtěch Dostál, Addbot, DOI bot, Quercus solaris, Luckas-bot, Yobot, Pthbotgourou, Gronk, Anypodetos, AnomieBOT, Pyrrhus16, Justme89, Xqbot, Capricorn42, Some standardized rigour, FrescoBot, D'ohBot, Aardappelmesje, Citation bot 1, 吴国盛, Kristina Marton, 777sms, Amkilpatrick, EmausBot, ZéroBot, DASHBotAV, ClueBot NG, Amirmeiri, Wimpus~enwiki, Frietjes, Katedoh, Thomas.clapes, BG19bot, Puripakorn, Zoldyick, Lugia2453, SimonPerera, HeartOfClubs, Seppi333, YiFeiBot, Hiprofgrn, Chaya5260, Nyashinski, Monkbot, WyattAlex, CV9933, Ands92, Kavkins9, Shivang Bhawsar and Anonymous: 108

- **Mesenchymal stem cell** *Source:* https://en.wikipedia.org/wiki/Mesenchymal_stem_cell?oldid=706465177 *Contributors:* AxelBoldt, Jeffq, Fuelbottle, Thpn, Horatio, Gadfium, Imroy, Rich Farmbrough, Mwanner, Arcadian, Jag123, Ceyockey, Woohookitty, Benbest, Rjwilmsi, DonSiano, Chris Capoccia, Hydrargyrum, Gaius Cornelius, Meiquer, Cquan, Closedmouth, Andrew73, SmackBot, Chris the speller, Keantom, TechPurism, Drphilharmonic, Linnell, Novangelis, Sammyjo~enwiki, CapitalR, Yicker, Cydebot, Hebrides, Narayanese, Dr Aaron, Magioladitis, MastCell, Robert M. Hunt, Yobol, STBot, Boghog, Rod57, Maduskis, Mikael Häggström, DoktorDec, Zondi, Khartma1, Doc James, SieBot, Neggiem01, ClueBot, The Thing That Should Not Be, Abbaroodle, DumZiBoT, Shunju-kun, Addbot, DOI bot, 2enable, Luckas-bot, Yobot, Noq, Danandlollie, Piano non troppo, Citation bot, Hwahwahwa, Hullbay, 4twenty42o, J04n, Smallman12q, Citation bot 1, Natisto, Machn, Jonesey95, Trappist the monk, Fama Clamosa, Georgesgoossens, Amkilpatrick, Some Wiki Editor, Xnn, RjwilmsiBot, EmausBot, Costas Lyssiotis, Mhahnel, Bermanya, Ajournaleditor, Cellmedsociety, Sabvega, ChuispastonBot, Cellmedsoc, ClueBot NG, Wimpus~enwiki, Pascalrds2, Castncoot, Katedoh, BG19bot, Rdenu, MrBill3, Leo181, Chinmayamahapatra, Samwalton9, ChrisGualtieri, Frusak, Kelsey.ohaganwong, Umweb, VictorLucas, Mrhomerjaysimpson, Monkbot, Arielrinon, Mbadelj, EoRdE6, Kavkins9, Wikibiowriter, Valdomri and Anonymous: 74

- **Neural stem cell** *Source:* https://en.wikipedia.org/wiki/Neural_stem_cell?oldid=710412719 *Contributors:* Bearcat, Nagelfar, Rich Farmbrough, Arcadian, Benbest, Mandarax, Rjwilmsi, Wavelength, Chris Capoccia, Mais oui!, SmackBot, GerryShaw, DRPVM, Headbomb, Peter Znamenskiy, Yobol, Adavidb, Sudhir h, Arjayay, Addbot, Yobot, BlackRaspberry, Citation bot, 吴国盛, FrescoBot, Pinethicket, HRoestBot, Trappist the monk, Shtanto, Haljammy, RjwilmsiBot, KIbrain, Retstern, Wikitavanti, Wimpus~enwiki, Frietjes, U4000, Rob400024, Bibcode Bot, Stemcellhub, Wasbeer, Mct333, Aisteco, Nvirani3, Dmitry Dzhagarov, AndreasKyttaro, DarkComedy, Helixitta, Sriyash11, Rsjs1913, Monkbot, CreeperGamer, Neuronguy1 and Anonymous: 20

- **Precursor cell** *Source:* https://en.wikipedia.org/wiki/Precursor_cell?oldid=631341296 *Contributors:* Xezbeth, Arcadian, Dorie Loon, AKeen, SmackBot, Dicklyon, ShelfSkewed, IceHorse, Davehi1, Addbot, Wojder, Trurle, NifCurator1, ClueBot NG, Lysosome, KLBot2, Dinisoe, OHateYou and Anonymous: 2

- **Blastocyst** *Source:* https://en.wikipedia.org/wiki/Blastocyst?oldid=702822738 *Contributors:* AxelBoldt, Magnus Manske, Lexor, Darkwind, Greenrd, Jeffq, Bearcat, Robbot, Cyrius, Giftlite, Robodoc.at, Chowbok, Bender235, Nectarflowed, Arcadian, Giraffedata, Unused000701, Alansohn, Seans Potato Business, Aniketvartak, Tycho, Gimboid13, Graham87, Rjwilmsi, Joffan, FlaBot, Frappyjohn, Tznkai, Wolfmankurd, Chris Capoccia, SpuriousQ, Mfero, Tetsuo, Rmky87, BOT-Superzerocool, Elkman, Werdna, Eaefremov, Ashenai, Ksargent, Sct72, James McNally, SashatoBot, ArglebargleIV, Epingchris, Sir Nicholas de Mimsy-Porpington, Soulkeeper, Lottamiata, UncleDouggie, Xcentaur, Was a bee, Narayanese, Thijs!bot, Al Lemos, AgentPeppermint, AntiVandalBot, Opelio, JAnDbot, ZDrache, Wikipodium, JaGa, Timothy Titus, J.delanoy, Mikael Häggström, M-le-mot-dit, Mufka, Jonas094, Synthebot, OnlyWayne, SieBot, Milnivri, VVVBot, Heatring, Sunrise, Avenged Eightfold, The Thing That Should Not Be, Mild Bill Hiccup, -Midorihana-, Flatjosh, Novjunulo, Rungladwin, Vojtěch Dostál, Addbot, Diptanshu.D, Chzz, OlEnglish, Legobot, Luckas-bot, Yobot, Fraggle81, Takanjack, AnomieBOT, IRP, Citation bot, Capricorn42, GrouchoBot, Omnipaedista, Listerineman, FrescoBot, Strawbaby, Pknkly, Pinethicket, Kugalskaper, Trappist the monk, Some Wiki Editor, Bohemian89, EmausBot, Jcbsdhbc777, ClueBot NG, Frietjes, Save me, Barry!, Mesoderm, Mohamed CJ, AvocatoBot, Harimiao, Makecat-bot, JakobSteenberg, Manningt, I am One of Many, Iztwoz, Wildcator, NJSfour, Vicktory7, Singleembryotransfer, Amortias, Shibbolethink, KasparBot, Charleselionel and Anonymous: 116

- **Fibroblast** *Source:* https://en.wikipedia.org/wiki/Fibroblast?oldid=692519495 *Contributors:* AxelBoldt, -- April, Andre Engels, Tijnz, Lir, Robbot, Sbisolo, Altenmann, Nilmerg, Fuelbottle, Bensaccount, Thpn, Trevor MacInnis, Cacycle, El C, Arcadian, Jag123, Seans Potato Business, TenOfAllTrades, Darked~enwiki, Hendrik Fuß, The Wordsmith, Mlewan, FlaBot, Ryddragyn, YurikBot, Wavelength, Epolk, Splette, Hydrargyrum, Light current, Colin, KnightRider~enwiki, SmackBot, Tinz, Bomac, Wedian, TheLimbicOne, Jitterro, Dr.saptarshi, SubtleGuest, Noah Salzman, P199, Cerealkiller13, IvanLanin, Igoldste, Robotsintrouble, Harej bot, EnglishEfternamn, Mojo Hand, Dfrg.msc, JAnDbot, Roidroid, Professor marginalia, Appraiser, CTF83!, WhatamIdoing, Ciar, Adrian J. Hunter, Robert M. Hunt, Enquire, J.delanoy, Ibrmrn3000, Mbarden, Mikael Häggström, Chibi.akutenshi, Squids and Chips, CardinalDan, VolkovBot, Holme053, Leeearnest~enwiki, Corvus cornix, Temporaluser, SieBot, Jsc83, Oxymoron83, Dravecky, Mygerardromance, ClueBot, Franamax, CounterVandalismBot, Asdasder, XLinkBot, Laboratory, Addbot, Ka Faraq Gatri, Tide rolls, Luckas-bot, JohnnyCalifornia, Essam Sharaf, AnomieBOT, FBW, ArthurBot, Xqbot, Godering, MuffledThud, Shadowjams, Jhfortier, A little insignificant, Zfrenchee, Some Wiki Editor, Keegscee, Biologist2001, Wyang, MithrandirAgain, Koala0090, Minnsurfur2, ClueBot NG, Aqua112233, ArionVII, Jaybear, Dexbot, Melonkelon, DavidLeighEllis, Abbas daify, ZetiuxRBLX and Anonymous: 95

- **Trophoblast** *Source:* https://en.wikipedia.org/wiki/Trophoblast?oldid=659119768 *Contributors:* Glimz~enwiki, Iosif~enwiki, LastCaress, Violetriga, Brian0918, Arcadian, Seans Potato Business, SabineCretella, Tabletop, FlaBot, Jbarfield, NTBot~enwiki, Wolfmankurd, Draeco, Grafen, Mccready, Nephron, FloNight, Stepa, Zephyris, Drphilharmonic, Lottamiata, Alfirin, Was a bee, Chasingsol, Thijs!bot, Boghog, Theespuja, Mikael Häggström, CardinalDan, Ronsword, Winchelsea, Xenobiologista, Fimbriata, Vojtěch Dostál, Addbot, Anxietycello, Yobot, Wlady 2009,

AnomieBOT, Golgibody, Strawbaby, ZéroBot, Jackalackapot1, OrcLady, Wafaashohdy, ClueBot NG, Frietjes, Mesoderm, Iztwoz, LT910001 and Anonymous: 20

- **Stem cell controversy** *Source:* https://en.wikipedia.org/wiki/Stem_cell_controversy?oldid=710785614 *Contributors:* Malcolm Farmer, William Avery, Gabbe, Llull, Radiojon, Astronautics~enwiki, Goethean, Nurg, Dave6, Wolfkeeper, Orangemike, HangingCurve, Romanpoet, Gadfium, David Battle, DocSigma, Elembis, Anythingyouwant, Demeter~enwiki, Naus, Alperen, Flex, Diagonalfish, Discospinster, Rich Farmbrough, Xezbeth, El C, MPS, Bobo192, Circeus, Smalljim, ParticleMan, Slicky, MPerel, Alansohn, Anthony Appleyard, Arthena, AzaToth, Nicholas Cimini, Seans Potato Business, Bart133, Snowolf, Wtmitchell, Evil Monkey, Amorymeltzer, Bsadowski1, NPswimdude500, Josephf, Mazca, Chris Mason, Sejessey, WadeSimMiser, GregorB, Sin-man, BD2412, Edison, Rjwilmsi, Mystalic, KamasamaK, Tdowling, Yamamoto Ichiro, Cocheese805, The.valiant.paladin, SouthernNights, AJR, Gurch, Str1977, Atrix20, Alphachimp, Banaticus, Wavelength, Candy156sweet, AVM, KevinCuddeback, Rsrikanth05, Pseudomonas, Wimt, Bullzeye, Thane, Robertvan1, Cquan, Joel7687, Bubaloo, Chooserr, Irishguy, Ragesoss, Nescio, Hrvoje Simic, Wknight94, Herb West, Nikkimaria, Zr2d2, Modify, Generic Name, Matau, Ratagonia, Mais oui!, Djr xi, Nippoo, Veinor, SmackBot, Haymaker, Ksargent, KnowledgeOfSelf, Unyoyega, Wolf ODonnell, Gilliam, Portillo, Ohnoitsjamie, Skizzik, Chris the speller, Mattythewhite, Roseelese, Basalisk, Ctbolt, Darth Panda, Can't sleep, clown will eat me, Avb, JonHarder, Rrburke, BarryTheUnicorn, Addshore, RedHillian, Kyle sb, WhereAmI, JR98664, Escotff, Nakon, BDSIII, Dreadstar, Hammer1980, Drc79, WoodyWerm, Kukini, Rockpocket, Derekwriter, The undertow, Serein (renamed because of SUL), Krashlandon, Ser Amantio di Nicolao, Soap, Caim, Mike1901, Gobonobo, JoshuaZ, Deadflagblues, IronGargoyle, Jinny~enwiki, Kaos Klerik, BranStark, DouglasCalvert, Roland Deschain, Iridescent, Joseph Solis in Australia, Walton One, Johnthescavenger, Mazdapickup89, Tawkerbot2, George100, Deetdeet, Lahiru k, CmdrObot, Geremia, Zarex, Dycedarg, Agathman, Ilikefood, Jgwlaw, JohnCD, Enigma.Zealot, Mato, Billheller, Strom, Dr.enh, Tawkerbot4, Chrislk02, Dmbaty, CalculatinAvatar, Shrinkshooter, Lid, Epbr123, Dr Aaron, Wikid77, MHGinTN, Dubc0724, Pajz, Croove55, Anupam, Headbomb, Marek69, John254, Aquishix, Chrisdab, Steve crowder, 70114205215, Grayshi, CharlotteWebb, Mentifisto, AntiVandalBot, Luna Santin, Seaphoto, Opelio, Quintote, Prolog, BballJones, Gatemansgc, Leuko, Roving Wordslinger, Davewho2, MER-C, Instinct, SuperthEman, Gerash77, Getaway, Dpeacock, Hut 8.5, Kinzalow, Globalhealth, Lulurascal, Bongwarrior, VoABot II, JNW, JamesBWatson, Pugetbill, Jesse Reynolds, WODUP, Naustra, Pianoman123, GirlForLife, Ciaccona, Waxwon, Ai429, DerHexer, Pan Dan, Nisbetme, MartinBot, Ariel., Rettetast, Nashpaul, Runningamok, Iarescientists, J.delanoy, R. Baley, James cudahy, Tikiwont, Kemiv, Eliz81, Extransit, WarthogDemon, Gzkn, DanielEng, EsperantoStand, Arronax50, Antelope19, Dr d12, Maduskis, Jayden54, Blackhawk2588, M-le-mot-dit, InspectorTiger, NewEnglandYankee, ArchieCS, ThinkBlue, Rumpelstiltskin223, Ephexx, Id711, 2help, Challenged Reader, Vanished user 39948282, Gtg204y, Logiboy123, Useight, CardinalDan, Funandtrvl, Sweshroj, Black Kite, Lights, Hugo999, Jonwilliamsl, Alex.bikfalvi, ABF, Varun136, Yoloyo, Jeff G., Paxcoder, Philip Trueman, Oshwah, Cosmic Latte, Rightfully in First Place, Friar44, Pwnage8, Caster23, GDonato, Woodstock, Qxz, Someguy1221, Sam729, Cerebellum, Leafyplant, DonutGuy, Jackfork, LeaveSleaves, PMenchaca, Bearian, Wiae, Lawyerjusticeforall, Maxim, BigDunc, Epgui, Cmenicoll, Swordmonkey1, Dudewasup, PoeticX, Why Not A Duck, Brianga, Sein Majesty, HiDrNick, Onceonthisisland, Nagy, Logan, DivaNtrainin, Pdg223, Snobo52, Calliopejen1, Charliejackpot, Caltas, God Emperor, Grundle2600, QuarElf, Thesavagenorwegian, Happysailor, Flyer22 Reborn, Tiptoety, Alexbrn, Pharmregulations, Oxymoron83, Jadio, Bagatelle, Svick, SiameseTurtle, StaticGull, Americanprogress, Armorwind, Greenofroogle, Struway2, Denisarona, Regener, TheCatalyst31, Elassint, ClueBot, DrewClaire, Healthwise, The Thing That Should Not Be, Rjd0060, Metaprimer, MacroDaemon, Mx3, Arakunem, Saddhiyama, Cp111, CounterVandalismBot, Leodmacleod, Thegreedyturtle, Dutch courage uk, John J. Bulten, Llamalover33, Jusdafax, Winston365, Muhandes, KnowledgeBased, NuclearWarfare, Razorflame, Polly, Horselover Frost, Aadh, PCHS-NJROTC, Potatopwner, SoxBot III, Herunar, DumZiBoT, BigK HeX, Spitfire, PseudoOne, Rextalionis, WikHead, Beach drifter, Miagirljmw14, ZanDaMan429, HexaChord, Addbot, Proofreader77, C6541, DOI bot, Fyrael, Landon1980, Crazysane, Binary TSO, Reidlophile, Ronhjones, TutterMouse, CanadianLinuxUser, Dkived, Cst17, DFS454, Glane23, Kazenwil, Favonian, 5 albert square, Antnefungus, Tide rolls, OlEnglish, LuK3, Tohd8BohaithuGh1, Legobot II, THEN WHO WAS PHONE?, Untrue Believer, IW.HG, Synchronism, AnomieBOT, Noq, Marauder40, 1exec1, Jim1138, BlazerKnight, Piano non troppo, Kingpin13, Materialscientist, Sciencetruth, Citation bot, Bigbob262, Neurolysis, Capricorn42, Spotfixer, Prowler08, Jmundo, Jld6, Br anna, Jwaustin188, Abce2, Shelt1rt, Frankie0607, Azzayan, MerlLinkBot, Natural Cut, LincolnSt, Erik9, Dougofborg, Llfisher, FrescoBot, Grundynash, Basketballer196, Calibrador, Lagelspeil, Mr.lenk, Omniscientest, Joproch, 753951m, Mithrandir, Paladin20, Airborne84, Babyyoumyeverything, ClickRick, Citation bot 1, Skaminsky, Pinethicket, Nigtv, Trw sf, Full-date unlinking bot, Rebeccanichols09, Suffusion of Yellow, Andrea105, RjwilmsiBot, Bento00, WildBot, EmausBot, John of Reading, RA0808, Carsonyarb, Winner 42, Wikipelli, K6ka, Manicjedi, Nader ecl, Access Denied, H3llBot, Tolly4bolly, Wiwach, Donner60, Hpminimom2009, DASHBotAV, FeatherPluma, FruitSalad4225, Petrb, ClueBot NG, Bmh12311, D1GPS, Renderedaddison, Widr, Jolly8585, WikiPuppies, Theonethatgotaway32, Piltech, Drianmcniece, Helpful Pixie Bot, Thedabbydabby, Calabe1992, BG19bot, Murry1975, Kirsty2011, Nietvoordekat, MusikAnimal, Mark Arsten, Deke0852, Blaspie55, Snow Blizzard, Eldar123, Physicsch, Ellencavanaugh, Glacialfox, Klilidiplomus, Josh1016, MattFloyd, Iamseth101, Anbu121, BattyBot, Greg Nevers, KRAVISWH, Madeleined2, Babaoriley51, Cyberbot II, ChrisGualtieri, Ruha9, Lugia2453, Dmitry Dzhagarov, Wywin, Babitaarora, Magic1551, Manul, Therealpirateblue, Monkbot, KH-1, Eurodyne, Knowledgebattle, Ashabhargave, TeaLover1996, Gooddecisionbasedonknowledge, Ajp124819, Anarchyte, 420BlazinSkrubz, Brae1342, Catrichardson31, Ajdhevsisbbsjxjs and Anonymous: 910

29.7.2   Images

- **File:3d_tRNA.png** *Source:* https://upload.wikimedia.org/wikipedia/commons/f/f1/3d_tRNA.png *License:* CC BY-SA 3.0 *Contributors:* Own work *Original artist:* Vossman

- **File:Adhesion_diagram.jpg** *Source:* https://upload.wikimedia.org/wikipedia/commons/5/5d/Adhesion_diagram.jpg *License:* CC-BY-SA-3.0 *Contributors:* Transferred from en.wikipedia to Commons by Kelly using CommonsHelper. *Original artist:* The original uploader was JWSchmidt at English Wikipedia

- **File:Animal_cell_structure_en.svg** *Source:* https://upload.wikimedia.org/wikipedia/commons/4/48/Animal_cell_structure_en.svg *License:* Public domain *Contributors:* Own work using Adobe Illustrator. Image renamed from Image:Animal cell structure.svg *Original artist:* LadyofHats (Mariana Ruiz)

- **File:Average_prokaryote_cell-_en.svg** *Source:* https://upload.wikimedia.org/wikipedia/commons/5/5a/Average_prokaryote_cell-_en.svg *License:* Public domain *Contributors:* Own work (Source. Typical prokaryotic cell, Chapter 4: Mutagenicity of alkyl N-

- **File:Mesoderm.png** *Source:* https://upload.wikimedia.org/wikipedia/commons/e/e8/Mesoderm.png *License:* Public domain *Contributors:* ? *Original artist:* J.Steinbock

- **File:Microspore-formation.svg** *Source:* https://upload.wikimedia.org/wikipedia/commons/c/cd/Microspore-formation.svg *License:* CC-BY-SA-3.0 *Contributors:* Own work *Original artist:* Emmanuel Boutet

- **File:Morelasci.jpg** *Source:* https://upload.wikimedia.org/wikipedia/commons/6/61/Morelasci.jpg *License:* CC BY 3.0 *Contributors:* Own work *Original artist:* Peter G. Werner

- **File:Muse_Cell_Cluster.jpg** *Source:* https://upload.wikimedia.org/wikipedia/commons/0/07/Muse_Cell_Cluster.jpg *License:* CC BY-SA 3.0 *Contributors:* Own work *Original artist:* Upsilon0

- **File:NIEHScell.jpg** *Source:* https://upload.wikimedia.org/wikipedia/commons/a/ac/NIEHScell.jpg *License:* Public domain *Contributors:* U. S. government source *Original artist:* US Government

- **File:Neural_progenitors_in_olfactory_bulb.tif** *Source:* https://upload.wikimedia.org/wikipedia/commons/b/b4/Neural_progenitors_in_olfactory_bulb.tif *License:* CC BY-SA 4.0 *Contributors:* Own work *Original artist:* Oleg Tsupykov

- **File:Normal_stem_cells.png** *Source:* https://upload.wikimedia.org/wikipedia/commons/c/c9/Normal_stem_cells.png *License:* CC BY-SA 3.0 *Contributors:* Own work *Original artist:* Malymajo

- **File:Nuvola_kdict_glass.svg** *Source:* https://upload.wikimedia.org/wikipedia/commons/1/18/Nuvola_kdict_glass.svg *License:* LGPL *Contributors:*

- Nuvola_apps_kdict.svg *Original artist:* Nuvola_apps_kdict.svg: *Nuvola_apps_kdict.png: user:David_Vignoni

- **File:Overview_of_iPS_cells.png** *Source:* https://upload.wikimedia.org/wikipedia/commons/2/28/Overview_of_iPS_cells.png *License:* CC BY-SA 3.0 *Contributors:* Own work *Original artist:* Humanips

- **File:People_icon.svg** *Source:* https://upload.wikimedia.org/wikipedia/commons/3/37/People_icon.svg *License:* CC0 *Contributors:* OpenClipart *Original artist:* OpenClipart

- **File:Plant_cell_structure-en.svg** *Source:* https://upload.wikimedia.org/wikipedia/commons/d/d8/Plant_cell_structure-en.svg *License:* Public domain *Contributors:* Self-made using Adobe Illustrator. (The original edited was also made by me, LadyofHats) *Original artist:* LadyofHats

- **File:Portal-puzzle.svg** *Source:* https://upload.wikimedia.org/wikipedia/en/f/fd/Portal-puzzle.svg *License:* Public domain *Contributors:* ? *Original artist:* ?

- **File:Production_of_iPSC_Timeline.png** *Source:* https://upload.wikimedia.org/wikipedia/commons/4/41/Production_of_iPSC_Timeline.png *License:* CC BY-SA 3.0 *Contributors:* Own work *Original artist:* Liz1988

- **File:Proteinsynthesis.png** *Source:* https://upload.wikimedia.org/wikipedia/commons/0/09/Proteinsynthesis.png *License:* Public domain *Contributors:* ? *Original artist:* ?

- **File:Pv_callus_dark_3_3-11-2008.jpg** *Source:* https://upload.wikimedia.org/wikipedia/commons/e/e7/Pv_callus_dark_3_3-11-2008.jpg *License:* CC0 *Contributors:* Own work *Original artist:* Blahnais

- **File:Question_book-new.svg** *Source:* https://upload.wikimedia.org/wikipedia/en/9/99/Question_book-new.svg *License:* Cc-by-sa-3.0 *Contributors:*
  Created from scratch in Adobe Illustrator. Based on Image:Question book.png created by User:Equazcion *Original artist:*
  Tkgd2007

- **File:Regen2.svg** *Source:* https://upload.wikimedia.org/wikipedia/commons/e/e7/Regen2.svg *License:* Public domain *Contributors:* Own work (Original text: *self-made*) *Original artist:* Squidonius (talk)

- **File:Signal_transduction_pathways.svg** *Source:* https://upload.wikimedia.org/wikipedia/commons/b/b0/Signal_transduction_pathways.svg *License:* CC BY-SA 3.0 *Contributors:* http://en.wikipedia.org/wiki/File:Signal_transduction_v1.png *Original artist:* cybertory

- **File:Sporic_meiosis.svg** *Source:* https://upload.wikimedia.org/wikipedia/commons/7/78/Sporic_meiosis.svg *License:* CC BY-SA 3.0 *Contributors:* This file was derived from Sporic meiosis.png: <a href='//commons.wikimedia.org/wiki/File:Sporic_meiosis.png' class='image'><img alt='Sporic meiosis.png' src='https://upload.wikimedia.org/wikipedia/commons/thumb/8/86/Sporic_meiosis.png/100px-Sporic_meiosis.png' width='100' height='73' srcset='https://upload.wikimedia.org/wikipedia/commons/thumb/8/86/Sporic_meiosis.png/150px-Sporic_meiosis.png 1.5x, https://upload.wikimedia.org/wikipedia/commons/thumb/8/86/Sporic_meiosis.png/200px-Sporic_meiosis.png 2x' data-file-width='465' data-file-height='340' /></a>
  Original artist: Sporic meiosis.png: Original uploader was Menchi at en.wikipedia.

- **File:Stem_cell_division_and_differentiation.svg** *Source:* https://upload.wikimedia.org/wikipedia/commons/1/18/Stem_cell_division_and_differentiation.svg *License:* Public domain *Contributors:* Self built from scratch using File:Stem_cells2.png *Original artist:* User:Wykis

- **File:Stem_cell_treatments.svg** *Source:* https://upload.wikimedia.org/wikipedia/commons/b/bd/Stem_cell_treatments.svg *License:* Public domain *Contributors:* All used images are in public domain. *Original artist:* Mikael Häggström.

- **File:Stem_cells_diagram.png** *Source:* https://upload.wikimedia.org/wikipedia/commons/3/3c/Stem_cells_diagram.png *License:* CC BY-SA 2.5 *Contributors:* From English Wikipedia. Original description page is/was here.
  Comment: The source of pluripotent stems cells from developing embryos. Original work by Mike Jones for Wikipedia. *Original artist:* Mike Jones

- **File:Stemcellmodulators.tif** *Source:* https://upload.wikimedia.org/wikipedia/commons/c/c4/Stemcellmodulators.tif *License:* Public domain *Contributors:* Own work *Original artist:* Mawelsh

- **File:Stromatolites.jpg** *Source:* https://upload.wikimedia.org/wikipedia/commons/c/c0/Stromatolites.jpg *License:* Public domain *Contributors:* National Park Service - http://www.nature.nps.gov/geology/cfprojects/photodb/Photo_Detail.cfm?PhotoID=204 *Original artist:* P. Carrara, NPS

29.7.3 Content license

www.ingramcontent.com/pod-product-compliance
Lightning Source LLC
Chambersburg PA
CBHW080654190526
45169CB00006B/2108